Man-made Catastrophes and Risk Information Concealment

Dmitry Chernov
ETH Zurich
Zürich
Switzerland

Didier Sornette
ETH Zurich
Zürich
Switzerland

ISBN 978-3-319-37098-9 ISBN 978-3-319-24301-6 (eBook)
DOI 10.1007/978-3-319-24301-6

Springer Cham Heidelberg New York Dordrecht London

Softcover re-print of the Hardcover 1st edition 2016

Printed on acid-free paper

Springer International Publishing AG Switzerland is part of Springer Science+Business Media
(www.springer.com)

Dmitry Chernov · Didier Sornette

Man-made Catastrophes and Risk Information Concealment

Case Studies of Major Disasters and Human Fallibility

Springer

The original version of the book frontmatter was revised: For detailed information please see Erratum. The Erratum to this book frontmatter is available at DOI 10.1007/978-3-319-24301-6_6

Preface

The manuscript consists of five chapters. First, an introduction sets the landscape and motivations. The second chapter is a series of 25 detailed case studies (plus 20 other cases more superficially covered), where the concealment of risk played an important and obvious role in the causation of the reported catastrophes. In the third chapter, the relevant aspects, which contributed to the risk concealment, are systematically gathered and analyzed. In the fourth chapter, possible signs for imminent risks in current technologies—such as fracking, cyber risks, and genetic engineering—are investigated. In the fifth chapter, a few positive examples are collected that show how with an open, transparent culture of communication corresponding risks could be reduced.

Our main goal is to show by the force of characteristic examples how prevalent the problem of concealment of information is in its many incarnations. The reader, layman or risk management, should come out of the reading of our examples with an aha moment, realizing how ubiquitous and important this dimension of human behavior is with respect to critical objects. This paints quite a different picture of the risk of critical objects that we have to deal with than is often understood. We show that essentially any kind of organization (industry, finance, military, social, associated with natural disasters, and retail) is vulnerable to risk information failures, both decades ago and in the present, and very likely in the future, both in top-down centralized institutions and at the other extreme in firms espousing the free market ideology. We think that this is the main strength of the book to demonstrate the ubiquity and permanent nature of the problem of risk information concealment. Therefore, there is a strong need to expose the relevant agents to the cases that we have documented, showing them that any organization including theirs is vulnerable and that a proactive attention to the dangers is needed. Chapter 3 offers a checklist that the risk manager or director of an institution should turn to again and again to check whether his or her organization may be drifting astray to exhibit some of the listed vulnerabilities or favorable factors promoting or facilitating risk information concealment. By offering many examples, we provide the decision maker and responsible individual with many scenarios to compare with that can help put in perspective their own situation.

This book is different from the myriad of books at the popular science level that present all kinds of conspiracy theories on mismanagement as being responsible for every catastrophe. In contrast, we make serious efforts to verify such assumptions by well-researched case studies. However, it seems unavoidable that some experts will criticize our endeavor, pointing out that we already have assumed upfront that the seemingly obvious or at least most likely reason for such catastrophes was fraudulent risk concealment or mismanagement. Indeed, in hindsight, it seems always plausible to attribute mistakes to either fraudulent behavior or carelessness. Any a posteriori rendering of a catastrophe will invariably and inevitably contain its share of revealed concealment and mismanagement—even in an idealized reality one could not imagine a catastrophe without such factors. But this does not establish the causal links to the catastrophe itself (rather than, e.g., random or stochastic ones). In many cases where so far no catastrophe has happened, one cannot rule out that no concealment or mismanagement has happened at some stage. One could criticize that our interpretations are one-sided, with too much focus on the risks of information concealment, at the expense of potentially even more important other factors. We have tried to the best of our abilities to avoid these traps by presenting carefully balanced descriptions of the involved processes. But we cannot claim that we have been entirely successful, given the issue of false positives and false negatives discussed at the end of Chap. 3. In the end, only advanced forecasts can be free of data snooping and other statistical biases of ex post analyses. This book can thus be viewed as an inspiration for a full-fledged program to identify, classify, and make operational the symptoms that emerge from the case studies (or any data collection) as having a particularly strong incidence on the increase in the risk in leading up to catastrophes. This book provides the background for any manager, decision maker, and responsible person to develop an acute awareness of the nature and risks of risk information concealment in its various forms.

Our approach is more pragmatic than academic, in the sense that we are not engaging a specific model or theory and we are not positioning our study in the context of any specific academic debate or school of thought, in particular regarding any psychological or sociological modeling of human fallibility. Our approach starts with empirical facts to gather as much as possible evidence of both the universal patterns and the idiosyncratic structures, documented by the 25 studied cases (too many, some critics would say). This allows us to free ourselves from a priori (mis-)conceptions. By the evidence and sheer clarity of the cases, we delineate the general ubiquitous presence of risk concealment as well as the factors catalyzing it. This does not prevent us from referring to relevant concepts and theories when useful, providing a second layer of interpretation. We have always in mind the practical person and the risk manager as well as decision maker, focusing on the kind of information that could be useful to him or her and on how to present it, searching for the gaps that could be identified as dangerous. In short, in this work, we focus on evidence-based practical and useful factors and hope this will serve as useful inputs for future academic and scholarly investigations on the roots of the problem.

Other critics will point out that our coverage of cases is imperfect and incomplete. Indeed, we would have liked to add more cases but had to stop somewhere.

As Herbert A. Simon (the Nobel Memorial Prize winner in economics) would say, we stopped when we reached a level of "satisficing," feeling that we had sufficiently many cases to cover the broad universe of human activities linked with critical objects. Some of the events such as the Ufa train disaster have been chosen because we believe we had unique access to sources of information that would bring novel insights compared to what has been previously written on these cases. Indeed, in a number of cases especially occurring in the previous Soviet Union and then in Russia, we had the possibility to interview some of the main actors of the industry that were active before and during the disasters, providing novel understanding. Thus, in contrast to the criticisms sometimes directed to books such as those of Sagan[1] and Schlosser[2], which are 100 % US-centered and thus incomplete for a generalization to other countries and cultures, we have tried to offer a more balanced viewpoint, aiming at broader and general insights. While the USA is a central character in many of our reported cases, we also cover several important other cases in Russia, Japan, Europe, and elsewhere. We believe this makes the book and our arguments stronger.

It is likely that many of our conclusions might be considered controversial, but we have consciously refrained from any political agenda and biased preconception and have tried to dig into the facts and present them as fairly as possible. Consider the Toyota recall case as a vivid illustration of our process of trying to go to the core of the facts and present a balanced view, often different from the clear-cut case that many pundits and the media converge to, in our opinion, too hastily. In the Chernobyl case, we were careful not to blindly accept the main claim of Gorbachev himself in his memoirs that the Chernobyl disaster contributed significantly to the demise of the Soviet Union.[3] We explain in more detail in the text the politically motivated nature of such claim by Gorbachev and try to disentangle the complex nature of information and disinformation spread by the Politburo at the time.

How should one use this book? In addition to the intrinsic historical interest and often eye-opening insights revealed by our reconstructions, we believe that there is a lot of value in comparative history: While it is often claimed that history never repeats and that generals are always preparing for the last war, our own research and our compilation and detailed analyses of 25 case studies suggest broad universal factors that recur again and again. This is perhaps not surprising since managing a human organization involves people with their characteristic psychology and predictable behavioral traits. As Ariely emphasized in his book with its explicit self-explanatory title,[4] all of us are endowed with quick and dirty

[1]Scott D. Sagan (1995) The Limits of Safety, Princeton University Press, Philadelphia.

[2]Eric Schlosser, Command and Control (2014) Nuclear Weapons, the Damascus Accident, and the Illusion of Safety, Penguin Books.

[3]Mikhail Gorbachev (2006) Turning Point at Chernobyl, Project Syndicate, http://www.project-syndicate.org/commentary/turning-point-at-chernobyl.

[4]Dan Ariely (2010) Predictably Irrational: The Hidden Forces That Shape Our Decisions. Harper Perennial, New York.

moduli of calculations, the emotional parts of our brain, which systematically make us react in certain predictable ways that are not necessarily fully rational. By having constantly in mind these examples that provide as many scenarios for possible risk information concealment to occur, one can keep aware of these biases and potential mishaps, avoiding habituation and the illusion of control. In particular, the synthesis provided by the figure at the beginning of Chap. 3 is especially useful for anyone to check his or her own organization with respect to what could go wrong. Continuously checking how the processes unfolding in one's organization live up against this list is a particularly healthy, and simple, way of keeping alert and honest.

Zürich
September 2015

Dmitry Chernov
Didier Sornette

Acknowledgments

We are indebted to many colleagues for their encouragement and for informative and constructive feedback on earlier versions. We would like to mention especially Salavat Abdulin (Chelyabinsk, Russia), Marco Avellaneda (New York University, USA), David Basin (ETH Zurich), Peter Cauwels (ETH Zurich), Michael A.H. Dempster (University of Cambridge, UK), Ingrid Eckerman (Swedish Doctors for the Environment, Sweden), Antonio Foglia (Belgrave Capital Management Ltd. and Banca del Ceresio, Lugano), Stefan Frei (Swisscom), Leonid Grinin (Institute of Oriental Studies of the Russian Academy of Sciences, Russia), William H. Janeway (Warburg Pincus and the Institute for New Economic Thinking), Taisei Kaizoji (International Christian University, Tokyo, Japan), Victor Khramov (Chelyabinsk, Russia), Andrey Korotayev (Higher School of Economics, Russia), Wolfgang Kröger (ETH Zurich, Switzerland), Victor Kudryavy (Eurocement, Russia), Jean Laherrère (retired oil geologist, Total), Thomas Maillart (Swiss SNF fellow at UC, Berkeley), Lidia Mikhailova (Chelyabinsk, Russia), Sergey Nefedov (Ural Federal University, Russia), Nick Onley (United Kingdom), Shamil Rahmatullin (Ufa, Russia), Radalif Shamsutdinov (Ufa, Russia), Thierry Sornette (Black Arrow, Qatar), Bruno Sudret (ETH Zurich), Hideki Takayasu (Sony Computer Science Laboratories and Meiji University, Japan), Michael Teitz (InterMediaCom, Russia), Sergey Tsirel (Plekhanov Technical University, Russia), Tsutomu Watanabe (University of Tokyo, Japan), and Wei-Xing Zhou (East China University of Science and Technology, China). A special thanks to Christian Caron, our editor at Springer, who followed the development of this project and provided important constructive feedback. The cover figure and all the cartoons in this book are credited to Alexey Iorsh.

Contents

About the Authors

Dmitry Chernov is a researcher in the Chair of Entrepreneurial Risks at ETH Zurich. He has more than 15 years of experience as a corporate communication consultant in the former Soviet Union Republics, focusing on improving investor and government relations, as well as crisis communication within the following industries: oil and gas, chemical, electric power, metals and mining, telecommunication, transport, etc. He specializes on researching solutions for efficient risk information transmission that enables timely decision-making before and during industrial disasters.

Didier Sornette is Professor on the Chair of Entrepreneurial Risks at the ETH Zurich, Director of the Financial Crisis Observatory, Co-founder of the ETH Risk Center and member of the Swiss Finance Institute. He has received many prizes and honors, among them the 2000 Research McDonnell award and the Risques-Les Echos prize 2002. He is a Fellow of the World Innovation Foundation (2004) and, since 2013, of the American Association for the Advancement of Science. Prof. Sornette has edited and authored a number books, many of them published by Springer and is a member of the Editorial and Advisory Board of the Springer Complexity publishing program.

Chapter 1
Setting the Landscape

All men naturally strive for knowledge [πάντες ἄνθρωποι τοῦ εἰδέναι ὀρέγονται φύσει]
First sentence of Aristotle's Metaphysics (πρώτη φιλοσοφία)

As a general rule, the most successful man in life is the man who has the best information
Benjamin Disraeli (1804–1881)

Mr. Corleone insists on hearing bad news immediately
"The Godfather" screen version, 1972

A long-expected but unpredictable earthquake just struck the community. A few seconds before the shaking seismic waves hit the city, the early warning system has alerted sensitive infrastructures and people through the advanced communication network connected to the widely spread arrays of monitoring seismic stations. Schoolchildren and citizens, well trained in advance, have dropped and covered up, turned off stoves and stopped delicate operations. In businesses, automated systems have opened the elevator doors, shut down production lines and placed sensitive equipment in a safe mode. In no time, power stations and grid facilities have been put in safety position to protect from strong shaking. Emergency responders have started to prepare and prioritize response decisions. Through a decentralized sensor network systems coupled with crowd-based cell phone apps, decisions-makers are immediately informed and continuously updated. Emergency response centers are directing the layered responses to avert negative consequences. Health teams are rushing to cater to the physical and psychological needs of the victims. Damaging reverberations such as fires, landslides and pollution are factored in and the suitable counter-measures are implemented. All this unravels smoothly through a combination of well-informed decentralized units with autonomous decision responsibilities integrated into a centralized managing command system gathering information on the unfolding of the disaster, synthesizing

D. Chernov and D. Sornette, *Man-made Catastrophes and Risk Information Concealment*, DOI 10.1007/978-3-319-24301-6_1

understanding and prioritizing actions concerning the deployment of experts, teams and equipment.

This ideal scenario epitomizes one of the axioms of management theory, which states that managers oversee other people by means of information.[1,2] They receive information from different sources, process it, make a decision, and translate this decision to subordinates and other audiences. The quality of the information being received about real conditions of the external and internal environment influences the quality of decisions, and later on the adequacy of an organization's response.

Unfortunately, the reality is often far from this idealization of management. Indeed, a widely held misconception is that, right after disasters, executives and government officials have comprehensive information about the important facets of the catastrophe that allow for adequate decision-making to respond. Regrettably, the truth is different: the quality of information in the hands of managers is often very poor, which translates into inadequate decisions after the disaster. In fact, this sad diagnostic extends to the amount and quality of information in the possession of managers before disasters, which poses an even more pressing question, namely the responsibility of misinformed managers for facilitating, promoting or even creating the calamity. Moreover, as Lee Clarke documented extensively,[3] when organizations prepare for a disaster, it is often the case that the proposed policies are void of any substance: when tested, the plans turn out to fail because they have missed the essential weaknesses and have not identified the needed remedies. Yet, they are used by organizations and by the public as templates of control and stability. They inspire confidence in our ability to understand and control the complex critical objects that we have to deal with, encouraging the syndrome of the "illusion of control".[4]

This disparity between perception and reality is also manifested in most of the books on risk and crisis communication, which are generally concerned with what companies and organization have to do right after a disaster and how they should

[1]Hedberg, B., How organizations learn and unlearn, in: Nyström, P.C. & Starbuck, W.H., Handbook of Organizational Design, Oxford University Press, 1981.

[2]Mullins, L.J. and G. Christy, Management & Organisational Behavior, Financial Times Management, May 2010.

[3]Lee Clarke, Mission Improbable: Using Fantasy Documents to Tame Disaster, University Of Chicago Press; 1 edition (June 1, 2001).

[4]Ellen J. Langer, The Illusion of Control, Journal of Personality and Social Psychology 32 (2), 311–328 (1975).

react to a crisis.[5,6,7,8,9,10,11,12,13,14,15,16] These books explain different approaches on how companies should focus on the preparatory actions that organizations should develop before potential accidents regarding risk communication towards external audiences in general and also on how companies should communicate with concerned audiences after the disaster. Only a small number of books mention internal audiences as an important part of adequate risk transmission. Overall, the focus of this literature is to explain how to communicate information to external audiences for fast and adequate perception. Most presentations, handbooks and monographs provide clear and simple recommendations regarding the needed actions of executives in crises, but all of them assume that these executives and their companies have a clear and quite complete understanding of the risks facing their organization and the nature of their occurrence. However, reality is different as we document extensively: right after the disaster, executives and managers receive distorted information from their staff or have developed year-long practice of risk concealment within their organization. In such cases, recommendations about proper risk communication to external audiences are pointless because the quality of risk-related information available to executives is initially poor. We found no publications that explain comprehensively the causes for the existence of such distorted information accessed by company staff and regulators, which lead to or amplify the magnitude of the disaster. This is disturbing given that concealment is a main cause for the inadequate actions of the organization and its staff during the normal practice of their duties. The magnitude of an accident is

[5]M. Granger Morgan, Baruch Fischhoff, Ann Bostrom and Cynthia J. Atman, Risk Communication: A Mental Models Approach, Cambridge University Press, 2001.

[6]Regina E. Lundgren and Andrea H. McMakin, Risk Communication: A Handbook for Communicating Environmental, Safety, and Health Risks, Battelle Press, 2004.

[7]Regina E. Lundgren and Andrea H. McMakin, Risk Communication: A Handbook for Communicating Environmental, Safety, and Health, Wiley-IEEE Press, 2013.

[8]Timothy L. Sellnow, Robert R. Ulmer, Matthew W. Seeger and Robert Littlefield, Effective Risk Communication: A Message-Centered Approach, Springer, 2009.

[9]Peter Bennett, Kenneth Calman, Sarah Curtis and Denis Smith, Risk Communication and Public Health, Oxford University Press, 2010.

[10]Robert L. Heath and H. Dan O'Hair, Handbook of Risk and Crisis Communication, Routledge, 2010.

[11]Pamela (Ferrante) Walaski, Risk and Crisis Communications: Methods and Messages, Wiley, 2011.

[12]James E. Lukaszewski, Lukaszewski on Crisis Communication: What Your CEO Needs to Know About Reputation Risk and Crisis Management, Rothstein Associates Inc., 2013.

[13]Joseph Arvai and Louie Rivers III, editors, Effective Risk Communication, Routledge, 2013.

[14]Robert R. Ulmer, Timothy L. Sellnow and Matthew W. Seeger, Effective Crisis Communication: Moving From Crisis to Opportunity, SAGE Publications, 2014.

[15]Hyunyi Cho, Torsten Reimer and Katherine A. McComas, Editors, The SAGE Handbook of Risk Communication, SAGE Publications, 2014.

[16]Valerie- November and Yvan Leanza, Risk, Disaster and Crisis Reduction (Mobilizing, Collecting and Sharing Information), Springer, 2015.

often influenced by inadequate decision-making just after the disaster caused by incomplete understanding of the severity of the crisis by involved parties, regulators, and victims. Other studies emphasize the phenomenon of social amplification of risks,[17] in which seemingly minor risk events often produce extraordinary public concern and social and economic impacts, cascading across time, geography, and social institutions. Our interest is at the other end, when risks are underestimated and hidden.

In contrast to the emphasis on disaster communication developed by other works, the present book concentrates on the importance of a proper understanding and transmission of the related information concerning the risks within a company, an industry or a society before a disaster strikes and the problems associated with internal risk transmission right after accidents. Severe reputation, material as well as human losses may result from the communication to external audiences of an incorrect understanding of the disaster in the first hours and days. Based on the analysis of past and on-going accidents, our aim is therefore to complement existing materials about proper risk communication processes, focusing on (i) the causes and consequences and (ii) the nature of the mistakes, which result from information gaps and concealments.

Professor Nancy G. Leveson (MIT) summarized masterfully the critical need of a proper information flow, whose many deficiency types are dissected in the present book: "*Flawed human decision making can result from incorrect information and inaccurate process models... Proper decision making often requires knowledge about the timing and sequencing of events. Because of system complexity and built-in time delays due to sampling intervals, however, information about conditions or events is not always timely or even presented in the sequence in which the events actually occurred... Enforcing safety constraints on system behavior requires that the information needed for decision making is available to the right people at the right time, whether during system development, operations, maintenance, or reengineering... Safety-related decision making must be based on correct, complete, and up-to-date information... Communication is critical... Communication channels, resolution processes, adjudication procedures must be created to handle expressions of technical conscience... Risk perception is directly related to communication and feedback. The more and better the information we have about the potential causes of accidents in our system and the state of the controls implemented to prevent them, the more accurate will be our perception of risk*".[18]

Before joining ETH Zurich as a researcher in March 2013, the first author, Dmitry, had been consulting in crisis communication after disasters for over seven years to some of the largest Russian companies (Gazprom, Gazprom-Neft, Russian

[17]Roger E. Kasperson, Ortwin Renn, Paul Slovic, Halina S. Brown, Jacque Emel, Robert Goble, Jeanne X. Kasperson, and Samuel Ratick, The Social Amplification of Risk: A Conceptual Framework, Risk Analysis 8 (2), 177–187 (1988).

[18]Nancy G. Leveson, Engineering a Safer World: Systems Thinking Applied to Safety, MIT Press, 2011, pp. 100, 198, 205, 301, 307, 379, 424.

Railways, Winter Olympic Games in Sochi 2014, RusHydro, EuroChem, Aeroflot, Russian Post, MegaFon, etc.). The second author, Didier, has been dismayed many times during his academic career by his observations of the divide between the standard post-mortem stories told about disasters, in particular in spaceflight accidents and financial crises, and the understanding that he has come to develop through his work on the failure of engineering structures and on financial bubbles and crashes. Moreover, the shock created by the 2011 Fukushima Daiichi disaster led him to conceive of a "civil super-Apollo project in nuclear R&D"[19] on how to manage civil nuclear risks over the required time scales of tens of years to thousands and even perhaps up to millions of years, given the short span and unstable nature of human societies. When Dmitry contacted Didier to come join him to hone his practical expertise learning from the quantitative engineering approach of ETH Zurich, it soon became clear to us that the roots of the Fukushima Daiichi disaster could be found in the (unlearned lessons of the) 1986 Chernobyl catastrophe, which itself had strong connections to (the unlearned lessons of) the 1979 Three Mile Island nuclear accident. During our investigations, we discovered important lessons that could be useful for the world industrial community and for policy makers on the management mistakes of such severe accidents.

One of the most important causes of the Chernobyl disaster was the tremendous information distortion of the real severity of the accident at different levels of the Soviet hierarchy during the first days following the explosion of one of the nuclear power plant cores. This led to an inadequate crisis response that magnified the severity and adverse consequences of the accident. The event became arguably one of the triggers of the collapse of the USSR[20], as a result of the destruction of common people's faith in the ability of the Politburo to run the country adequately and fairly, since its behavior seemed to contradict the Glasnost initiative (literally "publicity"), a policy that called for increased openness and transparency in government institutions and activities in the Soviet Union. Introduced by Mikhail Gorbachev in the second half of the 1980s, Glasnost referred to the specific period in the history of the USSR when there was less censorship and greater freedom of information. Our detailed investigation of the Chernobyl case and of the causes of the disaster revealed in addition a deeply ingrained practice of concealment of the design mistakes made over previous decades on the class of RBMK reactors operated at Chernobyl.[21] The apparently careless actions of the plant staff, which were the proximate causes of the disaster, should be put in the broader context of personnel operating a highly dangerous object without actually understanding the whole picture of risks. Indeed, different actors withheld information on the design problems as well as on the existence of previous incidents, accidents and

[19]D. Sornette, A civil super-Apollo project in nuclear R&D for a safer and prosperous world, Energy Research & Social Science 8, 60–65 (2015).

[20]Mikhail Gorbachev, Turning Point at Chernobyl, Project Syndicate, April 14, 2006, http://www.project-syndicate.org/commentary/turning-point-at-chernobyl

[21]RBMK Reactors, Appendix to Nuclear Power Reactors, Updated June 2010, http://www.world-nuclear.org/info/Nuclear-Fuel-Cycle/Power-Reactors/Appendices/RBMK-Reactors/.

near-misses plaguing these RBMK reactors. Our reconstruction leads us to conclude that the Chernobyl disaster was literally programmed to occur as a result of the Soviet civil nuclear energy organization planting the seeds of an inevitable disaster. As a result, the inadequate actions of the Politburo during the Chernobyl disaster were deeply connected with the poor transmission of reliable information to subordinates about the condition of the system, not only after the disaster but also long before it happened.

Given the evidence that we accumulated on the disinformation of the Soviet public by the Politburo and in turn of the Politburo by the Soviet nuclear establishment, we asked ourselves: was this just an isolated case, perhaps resulting from the failed model of top-down centralized planning and management extolled by the Soviets (see however[22]). This was the start of our journey to the wilderness of risk information concealment related in this book. With a focus towards practical implementation, we endeavored to research and document the causes of information concealment and managerial errors based on the detailed elaboration of 25 past disasters, augmented by more concise analyses of 20 additional notable accidents. Our analyses of these events, such as the Three Miles Island nuclear accident, the Bhopal disaster, the Deepwater Horizon oil spill, the Fukushima Daiichi nuclear disaster, and many others, revealed the presence of similar practices of withholding risk-related information. This impelled us to study the detailed sequences of information flow and of decisions by managers before, during and after disasters in these 45 cases covering most types of human activities. We uncovered a quite generic proclivity for risk information concealment, cover-up, distortion, gaps and deficits resulting in inadequate decisions regarding the exploitation of the system and consequently promoting the subsequent disaster. The inadequate crisis communication stemming from the absence of reliable information available to officials was generally found to increase the severity of the disaster.

This book is thus mainly about the risks of information concealment. We use the term 'concealment' to precisely represent the two distinct meanings of the behaviors uncovered in our investigations: (i) the condition that facts and knowledge about an organization and its functioning are hidden from those that should use them; the concealment can be due to many causes, including complexity, miscommunication, and so on; (ii) the conscious and deliberate action of keeping important information secret or misrepresenting it; this second meaning is a surprisingly important part of the pieces of evidence that we present. We will dissect the motives and origins of these obfuscations.

The book is organized by presenting first the in-depth analysis of 25 historical disasters augmented by an additional set of 20 others covered more superficially, developing a systemic dissection of the causes and consequences of concealment practice. For this, we have made use of official investigation reports, which

[22]Mark R. Beissinger, Scientific Management, Socialist Discipline and Soviet Power, I.B.Tauris, 1988.

constitute our main sources of information (when they exist), complemented by mainstream media publications and by interviews with decision makers and experts who were involved in the events. Finally, we have sought the feedbacks of world-class specialists in each field to double-check our reported facts and to obtain a more diverse international perspective. This last action turned out to be extremely useful to balance conclusions sometimes biased by a lack of broad perspective existing even in the most authoritative sources, as in the case of the 1941 Nazi invasion of the Soviet Union or the Toyota problems.

We then draw on the pile of evidence on these tens of historical disasters to develop a general taxonomy of the major origins of risk information concealment. A systematic analysis of the causes of managerial errors and their repetition is constructed with the goal of providing current managers with lessons from history that can be usefully transferred to improve current risk management practice. Our identification of the causes of risk concealment in these past events provides irreplaceable lessons for current managers to avoid repeating past mistakes, as we have observed to occur again and again. Following Otto von Bismarck, "*Only a fool learns from his own mistakes. The wise man learns from the mistakes of others*".

But the book would fall short of its ambitions if the lessons learned would not be used to examine on-going cases. We thus present our analysis, from the vantage offered by the tens of historical cases, of four outstanding on-going occurrences of suspicious risk concealment practice: (i) shale energy development in the USA; (ii) genetically modified organisms; (iii) real debt and liabilities of U.S, government and real GDP of China, and (iv) concealment of vulnerabilities in software industry. We selected these four cases for their critical nature in the future development of the World. We hope that our study will contribute to triggering timely reactions of the public and of current decision makers and managers.

Finally, we end by presenting a few successful examples of preventive anti-concealment practice. These examples obviously leave out many cases and are just put forward to suggest that efficient and transparent information transmission exists as an endless objective, and that forgetting it produces crises sooner or later, as illustrated by the success story of the Toyota Production System followed by its recall problems in 2009–2011.

A final word of caution is in order before diving into the subject. Notwithstanding our attempt to sample a large and representative population of tens of catastrophes in all possible sectors of human practice, it is still open whether the problems diagnosed here can be generalized to every organization in the world or could serve as a platform to establish a "universal theory of risk obfuscation", because of the high complexity of modern technical and organizational systems and multi-cultural differences. With this caveat in mind, this book represents a significant effort towards the goal of developing an operational understanding and best practice for the management of sensitive organizations with full awareness of human fallibilities to avoid the cascade effects and entanglement of system complexities leading to catastrophes.

Lastly, dear reader, if you possess important information that we missed in our reports, please contact us. We will be more than happy to analyse your pieces of evidence and include your perspective in a corrigendum or a new edition (contact emails: chernovd@ethz.ch or info@dmitrychernov.ru and dsornette@ethz.ch).

Chapter 2
Examples of Risk Information Concealment Practice

2.1 Industrial Sector

2.1.1 Vajont Dam Disaster (Italy, 1963)

The Vajont hydropower station dam and reservoir was located at the foot of Mt. Toc in the Dolomite region of the Italian Alps. In October 1963, a large landslide of 260 million m^3 of rock (equivalent of cube with a 650 m side) filled the reservoir of the dam, initiating a 150–250 m high wave, which overtopped the dam and wiped out several villages in the nearby Piave valley, resulting in the death of at least 1921 people.[1]

[1]F. Guzzetti, G. Lollino, Book Review of "The Story of Vaiont Told by the Geologist Who Discovered the Landslide", Natural Hazards and Earth System Sciences, 11, 2011, pp. 485–486.

D. Chernov and D. Sornette, *Man-made Catastrophes and Risk Information Concealment*, DOI 10.1007/978-3-319-24301-6_2

2.1.1.1 Risk Concealment Before the Disaster

Vajont dam was an important industrial project for post-war Italy. A general plan was proposed for the erection of seven dams in Piave valley, but from the beginning of the Vajont dam development, the project met with fierce resistance from local communities protesting against the forced sale of land to Società Adriatica di Elettricità (SADE) to construct the highest arch dam in the world. SADE, a private electricity company in North-Eastern Italy, had support from Democrazia Cristiana (the Italian Christian democratic political party, which promoted pro-American and pro-capitalist ideology and was in power at the time). Opposition from locals in villages around the Vajont reservoir was suppressed by the police. After this, the Italian Communist Party, the main opponent of Democrazia Cristiana, stayed on the side of local residents and supported their struggle against SADE and the government during the construction of the dam.[2]

Engineers and geologists focused on the permeability of the Vajont dam foundation, but did not study carefully the geology and the stability of the slopes surrounding the upstream reservoir of the dam, which consisted of soft materials like sand, limestone and clay.[3] The geological instability of the surrounding mountains was well known among local residents[4]: Mt. Toc had many nicknames like "crazy", "rotten", "loose" or "walking mountain" among local people due to its propensity for huge unexpected landslides.[5,6] The construction of the dam was launched in January 1957 with the goal of completing it by 1959 (construction started without government approval, nor serious geological research of the surrounding mountains). Moreover, to meet the need of Italian industry for electricity and maximize the profitability of the project, SADE proposed to increase the height of the dam up to 722.5 m, and triple the volume of the reservoir. In June 1957, government approval was obtained for construction without a geological study of the consequences of expanding the reservoir.

In March 1959 in the nearby Pontesei dam reservoir (owned by SADE), a landslide of 10 million m^3 of rock occurred (equivalent of cube with a 220 m side), resulting in the death of one worker, killed by a 20-m wave which overtopped the dam and destroyed a nearby bridge. Moreover, clefts were discovered during mountain road constructions around the Vajont reservoir. Protests from local residents against SADE broke out again. In May 1959, l'Unità (the official newspaper of the Italian Communist Party) conjectured that the Pontesei dam accident could

[2]Rose Marco Delle, Decision-making errors and socio-political disputes over the Vajont dam disaster, Disaster Advances, Vol. 5(3), 2012, pp. 144–152.

[3]Mountain Tsunami, documentary of "Seconds from Disaster" serious, National Geographic Channel, 2012.

[4]Ibid.

[5]Sara Pavan, The Vajont Dam, 1998, http://www.vajont.info/eNGLISH/saraPavan.html.

[6]Mountain Tsunami, documentary of "Seconds from Disaster" serious, National Geographic Channel, 2012.

recur in the Vajont region: "*when there is water in the reservoir, the mountain will fall down and cause a tragedy*".[7] (Later SADE filed a lawsuit against the journalist for "*disclosure of false, exaggerated and biased information aimed at disturbing public order*"[8] and "*defamation and spreading false information*"[9]). After the Pontesei dam accident, SADE ordered German and Italian geologists to investigate the geology of mountains around the dam. After several months, they confirmed a potential instability in the southern slope of the reservoir: the possible volume of a landslide could exceed 200 million m^3 if the reservoir was filled completely due to undercutting of the foundation by an ancient landslide. They passed on information about a possible rockslide to the architect and the chief engineer of the Vajont system, who asked them to moderate some of the report conclusions and suggested testing these hypotheses with another round of studies.[10] These more detailed studies stated that the evidence of an ancient landslide was absent, the slope was potentially immovable and only a small landslide could occur.[11] Apparently, any geological survey demonstrating the dangers of further exploitation of the Vajont system was unacceptable for the engineering team, SADE and the government, which was promoting the Vajont arch dam as the highest in the world and as a historic masterpiece of Italian engineering. Frankly admitting that there were errors in the design could lead to question how safely the government was expanding Italian industry. It could also attract attention to its control over private companies. This could change the political landscape, with the Communists using any blunder for political capital.[12] It would also cause losses for SADE and bring their shares down in the market. Nobody among the managerial team wanted to take responsibility for this honest but painful recognition of the dangers.

In February 1960, SADE started filling the reservoir. In the process, small landslides were noticed. On November 4, 1960, when the water level reached 636 m above sea level after weeks of heavy rains, a 0.7 million m^3 landslide occurred creating a 2-m wave. SADE geologists revealed a direct correlation between the water level in the reservoir and movement in the southern slope of the lake. They proposed to bring down the level of the reservoir to reduce the observed increasing shift of the southern slope. When the level dropped to 600 m, the movement of land mass went down from 3 cm to 1 mm per day. During 1961, the construction

[7]Rose Marco Delle, Decision-making errors and socio-political disputes over the Vajont dam disaster, Disaster Advances, Vol. 5(3), 2012, pp. 144–152.

[8]Marco Paolini, Vajont timeline (from 1928 to 1960), 1998, http://www.vajont.info/engTimeline1.html.

[9]French Ministry for Sustainable Development, Release of 50 million m3 of water at the Vajont Dam October 9, 1963. Erto e Casso (PN) Italy, No. 23607, 2010.

[10]Ibid.

[11]Mountain Tsunami, documentary of "Seconds from Disaster" serious, National Geographic Channel, 2012.

[12]Rose Marco Delle, Decision-making errors and socio-political disputes over the Vajont dam disaster, Disaster Advances, Vol. 5(3), 2012, pp. 144–152.

of a bypass tunnel kept the level of the reservoir down to around 600 m, and there were no serious landslides (even during the cold winter of 1961–1962). In 1961, SADE sponsored a hydraulic study of worst-case scenarios using a simulation model of the reservoir and the dam (1/200 of real size) at Padua University. In July 1962, the results of this research showed that the maximum likely wave from a landslide up to a volume of 40 million m^3 would not exceed 25 m, if the minimum sliding duration was 1.5 min[13]; in reality, the volume of the final rockslide was 260 million m^3, the slide lasted only 45 s and the height of the wave generated was 150–250 m. The geologists assumed that keeping the maximum water level of the reservoir below 700 m would prevent a possible landslide wave from overtopping the dam crest. There is no documented evidence that the results of the hydraulic study and the possibility of a 25-m wave were transmitted to the government, local authorities, residents or onsite staff at the dam and hydropower station. Supporting such interpretation is the fact that, right before the disaster, SADE personnel and their families had not left the nearby city of Longarone, which was perceived by local residents as a sign that there was no serious threat. After the disaster, it was revealed that the inspectors of the dam and the commissions responsible for regulation of hydropower industry never received any final reports—in particular they never received the studies by geologists who had identified the fault. Nor did they see any of the results from the model tests or their ensuing recommendations, which emphasized the importance of the water level of the retention dam.[14] Moreover, some sources claim that the vice-president of SADE decided not to communicate about the seismic activity registered by the seismographic station at the dam, and even deleted some records about serious tremors in his reports to government officials.[15]

In 1960, political debates began about the possible advantages for Italy of the nationalization of 1270 electricity companies and the creation of uniform standards for the use of electrical infrastructure. Nationalization could reduce the selling price of electricity for industrial and retail customers. Intensive discussions about nationalization occurred during 1960–1962. To increase their profits before nationalization, senior managers of SADE decided to fill the reservoir up to 700 m by the end of 1962; the velocity of ground movement increased as a response from 1 mm to 1.5 cm per day.[16] Finally in December 1962, ENEL (the Italian National Agency for Electric Energy) was established and united all private players,

[13]Ibid.

[14]French Ministry for Sustainable Development, Release of 50 million m^3 of water at the Vajont Dam October 9, 1963. Erto e Casso (PN) Italy, No. 23607, 2010.

[15]Marco Paolini's monologue performance, https://www.youtube.com/watch?v=ULm8T8ySs1A, https://www.youtube.com/watch?v=MQjBgS2_5yU, http://www.vajont.info/eNGLISH/athensMlover.html, 1998.

[16]French Ministry for Sustainable Development, Release of 50 million m^3 of water at the Vajont Dam October 9, 1963. Erto e Casso (PN) Italy, No. 23607, 2010.

including SADE and all its assets, by July 1963.[17] Managers of the private SADE could become managers of the state ENEL after this acquisition. Immediately after nationalization was declared, the level of the reservoir began to go down, and by the spring of 1963, it reached a low of 650 m; ground movement returned to 1–2 mm per day and seismic activity ceased.[18] Nobody within the managerial team of ENEL/SADE wanted to reveal the shortcomings of the Vajont system during the process of transfer of assets. Therefore, in order to demonstrate the quality of the dam to government officials and to present the Vajont system as a fully-functioning project, the reservoir was filled to the limit of 715 m by autumn 1963; by this time, the total cumulative ground movement surpassed 3 m.[19] ENEL/SADE top managers were still confident that they could manage the ground movement by reducing the reservoir level. But from September 1963, in spite of permanent water drainage, the ground movement velocity reached 20 cm per day[20,21] and residents of villages located above the reservoir were registering cracks in their houses. Within the ENEL/SADE managerial team, the opinion prevailed that, if the reservoir had a water level of 700 m, the possible wave from any size of landslide would not be dangerous for the dam and nearby villages. Because of this, SADE management did not discuss the results of the hydraulic study of 1961–1962 with external and independent geologists, and ignored the necessity to continue investigations of the dynamics of the southern slope of Mt. Toc. They misjudged the possible volume and speed of potential landsides, and the resulting wave height. This underestimation led to a situation where only a few small villages above the reservoir were evacuated. Neither the residents of villages located below the dam nor staff at the Vajont system were informed by ENEL/SADE executives about the possible threats.

On October 9, 1963 at 10:29 pm, dozens of ENEL/SADE workers were on the crest of the dam. They—and thousands of residents of nearby villages—were completely unprepared for the large landslide or for the wall of water, hundreds of meters high, which killed them. Meanwhile, the decision makers in the ENEL/SADE managerial team were at a distance from what they knew to be a dangerous area.

After the accident, an investigation commission stated that the main cause of the disaster was "*bureaucratic inefficiency, muddled withholding of alarming*

[17]ENEL, History of establishment and growth of ENEL (1962–1977), http://www.enel.com/en-GB/group/about_us/history/1962_1977.

[18]French Ministry for Sustainable Development, Release of 50 million m^3 of water at the Vajont Dam October 9, 1963. Erto e Casso (PN) Italy, No. 23607, 2010.

[19]Mountain Tsunami, documentary of "Seconds from Disaster" serious, National Geographic Channel, 2012.

[20]Rinaldo Genevois, Monica Ghirotti, The 1963 Vaiont Landslide, Giornale di Geologia Applicata 1, 2005, pp. 41–52.

[21]D. Sornette, A. Helmstetter, J.V. Andersen, S. Gluzman, J.-R. Grasso and V.F. Pisarenko, Towards Landslide Predictions: Two Case Studies. Physica A 338, 2004, pp. 605–632.

information, and buck-passing among top-officials"[22]—not an unforeseen natural event, an act of God, as Democrazia Cristiana and ENEL tried to present it. Four years later, the court found 11 executives of ENEL/SADE and government officials guilty.

Unfortunately, thirty years after the Vajont dam disaster, a quite similar cover up of local geological instability and faulty design at Val Di Stava led to the collapse of another Italian dam in 1985, which resulted in 268 deaths.[23]

Vajont Dam Disaster: Why Risks Were Concealed

- **Cozy relations between SADE executives and Italian government officials**, which allowed the operator of the dam to construct and exploit it in blatant violations of the existing legislation.
- **The political struggle** between Democrazia Cristiana and the Italian Communist Party: if SADE and Democrazia Cristiana had disclosed defects in the design of the dam and the reservoir, or had revealed the illegal practice used in obtaining the construction permits, a serious political crisis would have erupted in Italy.
- **The short-term profitability** of a private enterprise took priority over the long-term resilience of the Italian electric power industry.
- Geologists and managers at SADE were **unwilling to admit mistakes** in the inadequate preliminary study of the geology and of the stability of slopes surrounding the upstream reservoir of the Vajont dam. They were reluctant to incur the massive losses that would follow from the release of information that would lead to much higher construction costs. The goal was to **save the dam project and avoid the collapse of SADE's shares** in the market.
- False **reassurance/self-suggestion/self-deception** among decision makers about the maximum possible volume and speed of the landslide.
- SADE geologists and managers were **afraid of being accused of incompetence**. They were also keen not to lose public confidence in the ability of Italian private business to implement complex industrial projects.

[22]Norbert J. Delatte, Beyond Failure: Forensic Case Studies for Civil Engineers, ASCE Press, 2009, pp. 234–248.

[23]Flood at Stava Dam, documentary of "Seconds from Disaster" series, National Geographic Channel, 2004.

2.1.2 *Three Mile Island Nuclear Accident (USA, 1979)*

The Three Mile Island Nuclear Power Plant (NPP) is located 15 km from Harrisburg, Pennsylvania, 140 km from Washington, DC and 240 km from New York. The plant has two pressurized water reactors (PWRs) with a generating capacity of 1700 megawatts (MW). When the largest civil nuclear accident the world had ever seen occurred there at the end of March 1979, Unit 2 (TMI-2) had only been in commercial service for about three months and was operating at 97 % capacity. Unit 1 was shut down for refueling. The reactor core of TMI-2 contained around 100 tons of uranium fuel.[24]

2.1.2.1 Brief Technical Summary of the Accident

At 5:00 p.m. on March 7, 1979, during regular servicing of the feedwater system on Unit 2, the polisher machines—which remove dissolved minerals from the system—were being repaired when a leakage of water whose causes are not fully understood occurred into the air-controlled system that opens and closes the polisher valves. Eleven hours later, at 4:00 a.m. on March 28, 1979, this problem triggered a stoppage of the secondary feedwater pumps, which were responsible for sending re-heated water to the steam generators to remove heat from the reactor core of Unit 2.[25] This in turn provoked the automatic shut down of steam generators, and thus of the entire TMI-2 reactor. "Scramming" (emergency shutdown) of the reactor stopped nuclear fission completely; nevertheless, decaying radioactive materials left from the fission process continued to heat the reactor's coolant water. Immediately after shutdown, the decay heat power generation was about 160 MW of the thermal power of 2575 MW, corresponding to the 850 MW generating capacity of TMI-2. One hour after the reactor shutdown, decay heat power generation was approximately 33 MW. Ten hours after shutdown, it was about 15 MW. Over time, the decay heat power generation decreased more slowly.[26] In spite of the fact that this post-shutdown decay released far less energy than that released during fission, operators of the plant had to continue cooling the reactor for several days to reach a total cold shutdown.[27]

Because the secondary feedwater pumps had tripped, heat was not anymore being removed from the reactor. An activation of three auxiliary pumps occurred but the valves have been closed earlier for routine maintenance; the system was unable to pump any water, in violation of a key US Nuclear Regulatory Commission rule.

[24]Report of the President's Commission on the Accident at Three Mile Island: The Need for Change: The Legacy of TMI, Oct. 1979, p. 83.

[25]Ibid, p. 90.

[26]Three Mile Island: Report to the Commissioners and to the Public, M. Rogovin and G. Frampton, U.S. Nuclear Regulatory Commission, Jan. 1980, Vol. II-Part 2, p. 311.

[27]Report of the President's Commission on the Accident at Three Mile Island: The Need for Change: The Legacy of TMI, Oct. 1979, p. 83.

This led to rising pressure within the system, so a relief valve at the top of the pressurizer tank—the so-called pilot-operated relief valve (PORV)—was automatically opened in order to reduce pressure by draining the steam and water from the reactor core into a tank on the floor of Unit 2. The valve should have closed when the pressure fell to proper levels, but it remained stuck open. Instruments in the control room of TMI-2, however, indicated to the plant operators that the valve was closed as the light did not indicate the position of the valve, only the status of the solenoid being powered or not.[28] The design of the reactor and the control room design included instruments that could not show how much water was covering the core.[29] As a result, the plant staff was unaware that cooling water was pouring out of the stuck-open valve and assumed that, as long as the pressurized water level was high, the core was properly covered with water.[30,31] The PORV was open for 2 h and 19 min until operators of the next shift found the leakage of coolant from the reactor and closed the valve. Furthermore, during the first few minutes after the accident, the automatic emergency cooling system was turned off, reducing the emergency cooling water flow into the reactor to 10 times less than the designed level due to fears that the system was overfilled. The combination of these factors led to overheating and severe damage of the nuclear fuel due to the shortage of coolant within the reactor.[32] Later investigations found that about half of the core melted during the early stages of the accident.[33] The US President's Commission on the Accident at Three Mile Island stated: "*We estimate that there were failures in the cladding around 90 percent of the fuel rods. Fuel temperatures may have exceeded 4000 °F in the upper 30 to 40 percent of the core (approximately 30 to 40 tons of fuel). Temperatures in parts of the damaged fuel that were not effectively cooled by steam may have reached the melting point of the uranium oxide fuel, about 5,200 °F*".[34] This deterioration of the nuclear fuel induced a powerful upsurge of radioactivity within the containment building of TMI-2, and caused a dangerous hydrogen gas bubble to form within the reactor vessel likely produced by the reaction between the zirconium alloy of the melting fuel rod cladding and the steam. If this hydrogen gas had reacted with oxygen, it could have ignited a blow out, damaging the reactor vessel and leading to severe radioactive contamination. Fortunately, the hydrogen bubble was eliminated in the first few days after the accident.

---

[28]Backgrounder on the Three Mile Island Accident, United States Nuclear Regulatory Commission, Feb. 11, 2013.

[29]Ibid.

[30]Ibid.

[31]Ibid.

[32]Three Mile Island: Report to the Commissioners and to the Public, M. Rogovin and G. Frampton, U.S. Nuclear Regulatory Commission, Jan. 1980, Vol. I, pp. 3–4.

[33]Backgrounder on the Three Mile Island Accident, United States Nuclear Regulatory Commission, Feb. 11, 2013.

[34]Report of the President's Commission on the Accident at Three Mile Island: The Need for Change: The Legacy of TMI, Oct. 1979, pp. 30–31.

The accident happened because of a combination of factors. Firstly, plant operators were ignorant of the risks of water leakage into the polisher valve control system and of the PORVs getting stuck open—even though these incidents had occurred many times before the TMI accident at other American NPPs—because information about both problems had been concealed by suppliers of the nuclear steam system and by other NPP operators. Secondly, nobody was really facing up to the challenge posed by the interaction between human and machine in running a nuclear power plant: appeals from the plant's staff about poor control room design, and the imperfection of instrumentation, were being ignored; training was inadequate and operating procedures poor; and neither operators nor management had sufficient specialist knowledge about pressurized water reactors, or skill in diagnosing problems.[35] A tremendous amount of nationwide public outrage—and panic within the local community—was induced by unconscious misleading statements by the operators and management of TMI-2, and by executives of Metropolitan Edison (Met Ed), the involved utility company. Incorrect statements by representatives of the federal Nuclear Regulatory Commission (NRC) were influenced by a mistaken evaluation of conditions at the reactor during the first days after the accident, when operators did not realize that coolant was being lost and the plant was experiencing a meltdown.

Fortunately, there was a containment building sited directly above the reactor, the steam generators and the pressurizer of TMI-2. Thus, in spite of the severe core meltdown, the major part of the radioactive material remained within the unit's containment vessel, with minimal threat to the environment.[36] The total release of radioactivity to the environment has been established as just 13 to 17 curies of iodine, while 10.6 million curies of iodine were retained in water tanks in

[35]Three Mile Island: Report to the Commissioners and to the Public, M. Rogovin and G. Frampton, U.S. Nuclear Regulatory Commission, Jan. 1980, Vol. I, p. 102.

[36]Report of the President's Commission on the Accident at Three Mile Island: The Need for Change: The Legacy of TMI, Oct. 1979, p. 12.

the containment building and 4 million curies were in the auxiliary building tanks.[37] The total cost of the 14-year cleanup operation on the TMI-2 site was evaluated at US$1 billion in 1993 US$.[38] In spite of the contamination of the TMI-2 site, the nearby TMI-1 has worked properly for decades since the accident—in fact in 2009, the NRC approved an extension of the TMI-1 operating license for a further 20 years.[39]

The event, which was rated as a Level 5 accident out of a maximum Level 7 according to the International Nuclear and Radiological Event Scale, led to wider consequences. Nearly 150,000 people were evacuated from their homes during the accident,[40] which turn out to be unnecessary. The Federal Emergency Management Agency (FEMA) was established on April 1, 1979 to coordinate evacuation efforts during any such accident that could occur in the future. After the accident, strong public resistance to civil nuclear energy, which was manifested very clearly on May 6, 1979 when 65,000 antinuclear demonstrators gathered in Washington, led to the suspension of the construction of new nuclear power stations within the United States.

At the beginning of 1980, the US President's Commission on the Accident at Three Mile Island and the Nuclear Regulatory Commission both published detailed reports, open for public evaluation, about the TMI-2 accident. However, executives of the Soviet and Japanese civil nuclear industries obviously did not pay serious attention to the findings of these commissions regarding organizational imperfections before and after the accident, as would be revealed later in their corresponding disasters from the fact that they did not implement many of the commissions' recommendations in their own industries. Unfortunately, many of the organizational mistakes, and the pervasive climate of poor communication about risks that occurred during TMI-2, were repeated before and during the 1986 Chernobyl and 2011 Fukushima disasters.

2.1.2.2 Risk Concealment Before the Disaster

Rapid Growth of the American Civil Nuclear Industry at the Expense of Safety Considerations

The American civil nuclear industry originated from the Manhattan Project, a US military nuclear program launched in 1942. In 1946, the first civil reactor was

---

[37]Report of the President's Commission on the Accident at Three Mile Island: The Need for Change: The Legacy of TMI, October 1979, p. 31.

[38]14-Year Cleanup at Three Mile Island Concludes The New York Times, August 15, 1993.

[39]NRC Issues Final Safety Evaluation Report For Three Mile Island Nuclear Plant License Renewal Application, U.S. Nuclear Regulatory Commission Press Release-09-119, June 30, 2009.

[40]Robert A. Stallings, Evacuation behavior at Three Mile Island, International Journal of Mass Emergencies and Disasters, № 2, 1984, p. 12.

constructed at Oak Ridge National Laboratory. By 1955, the first nuclear submarine, based on a pressurized-water reactor design, had taken to the water.[41] In parallel with military nuclear development, the Eisenhower administration tested five different types of reactors in order to choose the most effective designs for a national civil nuclear program based on tenders from private companies to "*design, construct and operate … atomic power plants with [their] own capital*".[42] Following these tests, the Navy's pressurized-water reactor (PWR) design and the boiling-water reactor (BWR) design were selected—nowadays, 69 of the 104 reactors operating in the United States are PWR and 35 are BWR.[43] The 1960s and 1970s were boom years for the industry: 91 reactors were ordered in 1969 and 160 by the end of 1972.[44] While reactor projects were small in terms of capacity, safety concerns were not adequately emphasized. But when high-capacity reactors were projected in densely populated areas in order to minimize transmission costs and power losses, the U.S. Atomic Energy Commission (AEC) had to pay attention to quality assurance programs, to the redundancy of certain critical equipment, to the addition of an emergency core cooling system and to improvements in containment design.[45] At the same time, the industry was reluctant to implement additional safety measures because of the desire to reduce production costs in comparison with other fuels.[46] According to TMI-2's NRC report "*[t]he industry wanted a 'streamlined' licensing process to reduce the lengthening lag time between application for permits and licensing, and actual issuance. … In the promotional atmosphere of the AEC, such arguments had appeal*".[47] After the 1973 oil crisis, the energy independence of the United States and the development of domestic energy sources assumed greater importance, and the government gave additional support to the civil nuclear industry by promoting "Project Independence", an ambitious plan to build 1000 nuclear reactors by 2000.[48,49] In 1974, the AEC was split into the Energy Research and Development Administration (the promotional side) and the Nuclear Regulatory Commission

---

[41]Three Mile Island: Report to the Commissioners and to the Public, M. Rogovin and G. Frampton, U.S. Nuclear Regulatory Commission, January 1980, Volume I, p. 179.

[42]Ibid, p. 180.

[43]US Nuclear Operating Plant Basic Information, Nuclear Energy Institute, March 2013.

[44]Three Mile Island: Report to the Commissioners and to the Public, M. Rogovin and G. Frampton, U.S. Nuclear Regulatory Commission, January 1980, Volume I, p. 182.

[45]Ibid, p. 182.

[46]Ibid, p. 183.

[47]Three Mile Island: Report to the Commissioners and to the Public, M. Rogovin and G. Frampton, U.S. Nuclear Regulatory Commission, January 1980, Volume I, p. 182.

[48]Richard Nixon, Address to the Nation About Policies To Deal With the Energy Shortages, November 7, 1973.

[49]Leslie Kaufman, Japan Crisis Could Rekindle U.S. Antinuclear Movement, The New York Times, March 18, 2011.

(the regulatory role); but in spite of this separation of interests, a NRC commissioner stated right after the TMI-2 accident that "*I still think it [the NRC] is fundamentally geared to trying to nurture a growing industry*".[50] *The* US President's Commission report concluded that, because of the need to ensure national energy independence, "*the NRC is so preoccupied with the licensing of plants that it has not given primary consideration to overall safety issues. ... NRC has a history of leaving generic safety problems unresolved for periods of many years*".[51] The commission primarily focused on nuclear reactor designs, licensing of new plants and equipment malfunction on existing plants, but paid less attention to systematic safety concerns—the day to day running of plants, serious operator errors, critical areas of operator training, engineering with concern for human factors, utility management, the technical qualifications of staff and the protection of public health and safety.[52,53] The President's Commission report stated: "*Two of the most important activities of NRC are its licensing function and its inspection and enforcement activities. We found serious inadequacies in both. In the licensing process, applications are only required to analyze 'single-failure' accidents. They are not required to analyze what happens when two systems fail independently of each other, such as the event that took place at TMI. ... The accident at TMI-2 was a multiple-failure accident. ...insufficient attention has been paid [by the NRC] to the ongoing process of assuring nuclear safety. ... NRC is vulnerable to the charge that it is heavily equipment-oriented, rather than people-oriented. ... [I]nspectors who investigate accidents concentrate on what went wrong with the equipment and not on what operators may have done incorrectly, in the lack of attention to the quality of procedures provided for operators, and in an almost total lack of attention to the interaction between human beings and machines*".[54]

Lack of Communication About Minor Incidents Within the American Civil Nuclear Industry

The industry had the very serious problem that decision makers had a fragmented perception of the risks, because information about operating experience, including dangerous incidents, was not routinely and reliably exchanged between the NRC, utility companies that operated plants, NPP designers, manufacturers of reactor systems, and contractors and suppliers of critical components. The President's

[50]Report of the President's Commission on the Accident at Three Mile Island: The Need for Change: The Legacy of TMI, October 1979, p. 51.

[51]Ibid.

[52]Three Mile Island: Report to the Commissioners and to the Public, M. Rogovin and G. Frampton, U.S. Nuclear Regulatory Commission, January 1980, Volume I, p. 89.

[53]Report of the President's Commission on the Accident at Three Mile Island: The Need for Change: The Legacy of TMI, October 1979, p. 20.

[54]Ibid, pp. 20, 21, 52.

Commission outlined this problem: "*The NRC accumulates vast amounts of information on the operating experience of plants. However, prior to the accident, there was no systematic method of evaluating these experiences, and no systematic attempt to look for patterns that could serve as a warning of a basic problem... The major offices within the NRC operate independently with little evidence of exchange of information or experience. For example, the fact that operators could be confused due to reliance on pressurizer level had been raised at various levels within the NRC organization. Yet, the matter 'fell between the cracks' and never worked its way out of the system prior to the TMI-2 accident*".[55]

Moreover, the President's Commission found out that the mistaken shutdown of the emergency cooling system was not unique to this incident, but a problem well known to representatives of the nuclear steam system suppliers. It had occurred at PWR plants on several occasions, but nobody had transmitted this information to other plants: "*The same problem of water leaking into the polisher valve control system had occurred at least twice before at TMI-2... During the 18-month period before the accident, no effective steps were taken to correct these problems... Had Met Ed [the operator of the TMI] corrected the earlier polisher problem, the March 28 sequence of events may never have begun. ... A senior engineer of the Babcock & Wilcox Company noted in an earlier accident [on Davis-Besse NPP in 1977], bearing strong similarities to the one at Three Mile Island, that operators had mistakenly turned off the emergency cooling system. He pointed out that we were lucky that the circumstances under which this error was committed did not lead to a serious accident and warned that under other circumstances (like those that would later exist at Three Mile Island), a very serious accident could result. He urged, in the strongest terms, that clear instructions be passed on to the operators. This memorandum was written 13 months before the accident at Three Mile Island, but no new instructions resulted from it... Nine times before the TMI accident, open pressurizer relief valves (PORVs) stuck open at B&W plants. B&W did not inform its customers of these failures, nor did it highlight them in its own training program so that operators would be aware that such a failure causes a small-break LOCA [loss of coolant accident]*".[56]

In addition, the excessive complexity of control room design, which made it difficult for operators to quickly grasp the condition of a nuclear plant and so make decisions adequately, had been recognized at the design phase but ignored until the TMI case: "*Burns and Roe, the TMI-2 architect-engineer, had never systematically evaluated the control room design in the context of a serious accident to see how well it would serve in emergency conditions. Over 100 alarms went off in the early stages of the accident with no way of suppressing the unimportant ones and identifying the important ones. The danger of having too many alarms was recognized by Burns and Roe during the design stage, but the problem was never resolved... The TMI-2 control room operator complained to his superiors*

[55]Ibid, pp. 21, 52.

[56]Ibid, pp. 10, 43, 93.

about problems with the control room. No corrective action was taken by the utility…".[57]

There was a huge problem of risk information transmission between the different players during the development of the nuclear industry, and an inadequate response even to identified risks: "*In a number of important cases, [the companies] failed to acquire enough information about safety problems, failed to analyze adequately what information they did acquire, or failed to act on that information. Thus, there was a serious lack of communication about several critical safety matters within and among the companies involved in the building and operation of the TMI-2 plant. … [C]ompanies … have little communication with those responsible for operator training and, therefore, the content of the instructional program does not lead to sufficient understanding of reactor systems… A similar problem existed in the NRC… The information and direction issued by NRC to licensees based on operating experience was, at times, fragmented and misleading… [I]mportant safety issues are frequently raised and may be studied to some degree of depth, but are not carried through to resolution; and the lessons learned from these studies do not reach those individuals and agencies that most need to know about them*".[58]

Once the plants were operational, it was common practice to focus on eliminating any potential large incidents, whereas fixing minor errors and flaws was generally seen by nuclear executives as less important: "*It was natural for the regulators and the industry to ask: 'What is the worst kind of equipment failure that can occur?' A preoccupation developed with such large-break accidents as did the attitude that, if they could be controlled, we need not worry about the analysis of 'less important' accidents… This was true in the B&W incident described above, it was true about various warnings within NRC that inappropriate operator actions could result in the case of certain small-break accidents… TMI illustrated a situation where NRC emphasis on large breaks did not cover the effects observed in a smaller accident*",[59] which can be attributed to the concept of "deterministic design", which does not incorporate the complexity of the possible cascades that can develop along the multiple branches of the tree of scenarios.

The NRC's post-accident investigation confirmed the findings of the President's Commission: "*[Similar incidents to the TMI-2 accident] occurred in 1974 at a Westinghouse reactor in Beznau, Switzerland, and in 1977 at Toledo Edison's Davis Besse plant in Ohio, a Babcock & Wilcox reactor similar in design to the one at Three Mile Island. Both involved the same failed open pressurizer relief valve (PORV), and the same misleading indications to operators that the reactor coolant system was full of water. In both cases, operators diagnosed and solved the problem in a matter of minutes before serious damage could be done. The NRC never learned about the incident at the Beznau reactor until after the TMI-2 accident, because Westinghouse was not required to report to the NRC such*

[57] Ibid, pp. 29–30.

[58] Ibid, pp. 11, 23, 43, 55.

[59] Ibid, pp. 9, 11, 30.

occurrences at foreign reactors. Westinghouse concluded that the actions by the Swiss operators proved the validity of an earlier Westinghouse study showing that, in this kind of incident, operators would have enough time to react to a stuck-open valve and correct the situation. A brief account of this earlier study had, in fact, previously been submitted to the NRC. But neither the Beznau incident nor the earlier study had prompted Westinghouse to notify its customers or the NRC that operators might well be misled by their instruments if a valve stuck open. The Davis Besse accident was intensively analyzed by Toledo Edison, by Babcock & Wilcox, and by the NRC. Each of these studies identified what should have been perceived to be a significant safety issue. But because no effective system for evaluating operating experience was in effect, none of the results of these studies were ever communicated to [Met Ed] or its operators at the TMI-2 plant. ... Toledo Edison, at the insistence of an NRC inspector and his supervisor from the agency's regional office, which is a part of NRC's Office of Inspection and Enforcement (IE), eventually adopted new operator precautions. But they were not communicated to B&W or to other utilities, and IE's regional office did not flag the issue to NRC headquarters".[60]

The American nuclear regulator admitted that its inspectors had not given TMI plant managers their conclusions from the experience of erroneous shutdowns of the cooling system at several other NPPs, over a number of years before the TMI accident. In later case studies, we will see exactly the same behavior by regulators of The Soviet Ministries of Medium Machine Building, and Energy and Electrification, during the 1970s and 1980s. The staff at both Chernobyl NPP and the Sayano-Shushenskaya hydropower station operated complex and dangerous technology without understanding the technical shortcomings of the equipment, or the need to implement special safe operation regimes—even though these had been revealed years if not decades earlier at other plants. In all these cases, the regulators concerned knew the risks, but for different reasons did not pass this knowledge on to operators—and disasters occurred as a result. The TMI investigation report included this example: "*In January 1978, a NRC reviewer in NRR prepared a memo based on ... the Davis Besse incident, which noted that, in certain circumstances, operators could be misled by their instruments to turn off the emergency core cooling system. But the reviewer's memo was not circulated outside NRR and the issue was not identified as a possible generic safety problem for operating plants; it was simply filed away... In sum, the agency's fragmented bureaucracy, its preoccupation with hardware and design questions, and the lack of any clear-cut responsibility for identifying significant operating problems and warning operators about them combined to prevent the real message of Davis Besse from getting to Three Mile Island... The structure of the nuclear industry has not been conducive to the effective sharing and integration of operating data. The utilities that operate the plants have never mobilized an industry-wide effort*

[60]Three Mile Island: Report to the Commissioners and to the Public, M. Rogovin and G. Frampton, U.S. Nuclear Regulatory Commission, January 1980, Volume I, pp. 94–95.

to concentrate on safety-related operational problems. As for the four principal U.S. manufacturers of reactors ("vendors") – General Electric, Westinghouse, Combustion Engineering, and B&W – we found a great deal of variation in the extent to which they monitor at their own expense operating problems in the plants they have built, after those plants are tested and turned over to the utilities that have purchased them. And the relationship between the vendor and its utility customer after operation of a plant begins is largely determined by the individual utility's choice of how much technical assistance it is willing to buy from the vendor on an ongoing, contract basis. Moreover, there is no requirement that utilities report failure data to the vendors... [A]lthough NRC requirements result in a great deal of material on reactor operations being generated and sent to the NRC by the utilities, this information has not been systematically reviewed to extract potentially important safety problems or trends... The situation is made more complex because the reporting requirements differ from plant to plant: incidents reportable at some plants do not have to be reported at others. As a result, the NRC is flooded with a mass of undifferentiated data on reactor operations... NRC publishes a computerized listing of [Licensee Event Reports], each described in a few sentences at most, and a periodical called 'Current Events-Power Reactors' containing more detailed descriptions of major problems... The lessons learned from malfunctions and mistakes at nuclear plants both here and abroad were never effectively shared within the industry... Coordination among these parties and between them and the NRC, as well as within the NRC, is inadequate".[61] This report was openly published in 1980—but remarkably, Soviet energy industry regulators did not learn the lessons of TMI: no system was established to continuously transmit detailed information about incidents occurring at Soviet nuclear and hydropower plants, so operators remained unaware of the risks. Decades later this lack of communication led to disasters.

In short, all the key organizations accountable for the safe operation and regulation of TMI-2 played their part in the accident, but none of them understood the whole picture of the risks involved in running a pressurized-water reactor: no one fully grasped what could develop during a multi-failure hardware malfunction, under the control of staff who had not been trained for such failures.

2.1.2.3 Challenges of Adequate Risk Transmission After the Disaster

Misreading of Instruments Led to Mistakes by the Operators at TMI-2

There were more than 750 alarms in the control room at TMI-2, and when multi-factor malfunction occurred in the early morning of March 28, 1979, more

[61]Three Mile Island: Report to the Commissioners and to the Public, M. Rogovin and G. Frampton, U.S. Nuclear Regulatory Commission, January 1980, Volume I, pp. 3, 33, 89, 94, 95, 96, 97.

than 100 of these alarms immediately went off.[62] Operators there recalled that the console was "*lit up like a Christmas tree*".[63] The control room's alarm printer was overloaded: it could type one line every 4 s, but several alarms per second were occurring during the first few minutes.[64] Moreover, there was no system to prioritize alarms, so operators could not trace back the sequence of emergency events in time to make decisions adequately. In addition, the control room and instrumentation were designed for normal, not conditions when an accident occurs.[65] The control room was far too large and there was no orderly grouping of instruments by function—in particular, emergency controls and instruments were not sited in a common location.[66] For the four TMI-2 operators, it was difficult to run the plant based on information from instruments that were not designed to show, for example, how much water covered the core, or which quickly went off scale, as was the case with the radiation-monitoring equipment.[67] The infamous PORV alarms were on a panel remote from the central console and facing away from the operators, and the indicator light on the control panel for the PORV was wired to show only what the valve had been "instructed" by the electrical system to do, not the valve's actual position.[68,69] This combination of circumstances misled the operators, who did not realize that the PORV had been open for 2 h and 19 min and that the plant had lost a critical amount of coolant. A year before the accident, a TMI-2 operator had informed Met Ed management about the problem: "*The alarm system in the control room is so poorly designed that it contributes little in the analysis of a casualty. The other operators and myself have several suggestions on how to improve our alarm system-perhaps we can discuss them sometime, preferably before the system as it is causes severe problems*".[70] The company made several improvements on TMI-1, but nothing had been done on TMI-2 at the time of the accident. Also the operators of the plant—despite having a rigorous background of working on nuclear submarines for the US Navy—had never been trained to understand all the plant parameters, and lacked theoretical knowledge of the operating principles

[62]Three Mile Island: Report to the Commissioners and to the Public, M. Rogovin and G. Frampton, U.S. Nuclear Regulatory Commission, January 1980, Volume II-Part 2, p. 577.

[63]Gregory Rolina, Human and Organizational Factors in Nuclear Safety: The French Approach to Safety Assessments, CRC Press, 2013, p. 42 and Three Mile Island: Report to the Commissioners and to the Public, M. Rogovin and G. Frampton, U.S. Nuclear Regulatory Commission, January 1980, Volume II-Part 2, p. 593.

[64]Three Mile Island: Report to the Commissioners and to the Public, M. Rogovin and G. Frampton, U.S. Nuclear Regulatory Commission, January 1980, Volume II-Part 2, p. 318.

[65]Three Mile Island: Report to the Commissioners and to the Public, M. Rogovin and G. Frampton, U.S. Nuclear Regulatory Commission, January 1980, Volume I, p. 128.

[66]Ibid, p. 123.

[67]Ibid, p. 128.

[68]Ibid p. 123.

[69]Ibid, p. 126.

[70]Ibid, p. 124.

of a pressurized water reactor.[71] After the accident, the NRC concluded: "*Not all utilities [in the United States had] either as large an engineering staff or executives with appropriate backgrounds to enable them to direct actual plant operations during emergencies... [Nevertheless,] we have concluded that the utility [Met Ed], in terms of technical capability, is as good as the median nuclear utility*".[72] Investigators ruled out any deliberate withholding of information: "*However, based on the evidence, we could not conclude that the causes of this breakdown in information flow went beyond confusion, poor communications, and a failure by those in the control room, including NRC and B&W employees, to comprehend or interpret the available information, a failing shared to some extent by offsite organizations as well. A number of factors other than deliberate attempts to downgrade the seriousness of the situation could have accounted for the failure of the control room crew to communicate critical information. The failure to recognize and act on significant data in our view demonstrates a lack of technical competency by site employees to diagnose and cope with an accident. Moreover, the inability of the utility's management to comprehend the severity of the accident and communicate it to the NRC and the public was a serious failure of the company's management. [Nevertheless,] there is no evidence to show willful withholding of information by Met Ed from NRC*".[73]

The NRC investigation also outlined the atmosphere among TMI-2 staff at the time of the accident: "*[N]o one appears to be theorizing about the cause of the increased radiation levels in the plant. No one postulates an uncovered core. If anyone is thinking such thoughts, he is keeping them to himself... Intellect tells them they don't really know what is going on; ego tells them none of the rest of these guys do either; on the evidence, both are right... Understanding what is happening to the core itself will not come until much, much later*".[74] Another important factor influencing the misreading of the situation by operators and management at TMI-2 was a prevailing mindset about the impossibility of a meltdown on the plant: "*[The] inability to recognize and comprehend the full significance of the information, and certain psychological factors: the difficulty of accepting a completely unexpected situation, the fear of believing that the situation was as bad as the instruments suggested, and a strong desire to focus on getting the reactor stable again rather than dwelling on the severity of the accident*".[75] For example, TMI station manager Gary Miller testified that "*I don't believe in my mind I really believed the core had been totally uncovered, or*

---

[71]Ibid, pp. 102–103.

[72]Three Mile Island: Report to the Commissioners and to the Public, M. Rogovin and G. Frampton, U.S. Nuclear Regulatory Commission, January 1980, Volume II-Part 3, p. 811.

[73]Three Mile Island: Report to the Commissioners and to the Public, M. Rogovin and G. Frampton, U.S. Nuclear Regulatory Commission, January 1980, Volume I, p. 70.

[74]Ibid, pp. 22–23.

[75]Ibid, p. 70.

uncovered to a substantial degree at that time".[76] According to the NRC investigation, during the first hours after the accident, Miller sent "*Lead Instrumentation Control Engineer Ivan Porter down below the control room to take more instrument readings directly off the wires that lead to the incore thermocouples. Porter has his technicians take four or five initial readings. Several are too low to be believable, but at least two are above 2000 °F. The technicians express concern that the core is uncovered ... [T]he technicians are taking dozens of additional readings. Many of them are far too high for comfort... Porter shrugs them off and returns upstairs to brief Miller. He tells Miller of the readings, but says he does not believe the high ones are accurate – after all, the low ones cannot be right... Apparently there was a lack of skepticism or a lack of willingness to believe the worst*".[77,78]

Misjudgments of the Status of TMI-2 Resulted in Misleading Information Being Given to External Audiences

These misjudgments of the plant status led the operators and management of TMI-2 to send misleading information to their supervisors at Met Ed and its parent company General Public Utility—who in their turn informed the NRC, the designers of the plant, federal, state and local government representatives and the general public about the unimportance of the accident. For instance, 5 h after the accident, around 20 engineers and managers from Babcock & Wilcox assembled in Lynchburg, Virginia, for a speaker-phone conference with the B&W representative at TMI, but the meeting was "*...under the circumstances, a surprisingly placid gathering, marked by a dearth of information from the plant site. 'B&W's most prevalent feeling,' according to one of the people present, 'was we're just in the dark*'".[79] Their reaction—and that of other external audiences—would have been very different if TMI-2 staff had been able, during the first hours after the accident, to recognize the possible consequences of the PORV being open for several hours and the reactor core being uncovered (which raised the temperature within the reactor and damaged the fuel rods), evaluate the real cause of the radioactivity and hydrogen bubble, correctly deduce the possibility of a core meltdown and immediately inform their supervisors. This would have enabled prompt federal response measures to be taken—instead of which, about 4 h after the accident,

[76]Report of the President's Commission on the Accident at Three Mile Island: The Need for Change: The Legacy of TMI, October 1979, p. 103.

[77]Three Mile Island: Report to the Commissioners and to the Public, M. Rogovin and G. Frampton, U.S. Nuclear Regulatory Commission, January 1980, Volume I, pp. 30–31.

[78]Three Mile Island: Report to the Commissioners and to the Public, M. Rogovin and G. Frampton, U.S. Nuclear Regulatory Commission, January 1980, Volume II-Part 3, p. 827.

[79]Three Mile Island: Report to the Commissioners and to the Public, M. Rogovin and G. Frampton, U.S. Nuclear Regulatory Commission, January 1980, Volume I, p. 21.

Met Ed manager of communications services was telling the media that "*[t]here was a problem with a feedwater pump. The plant is shut down. We're working on it. There's no danger off-site. No danger to the general public*".[80]

Misinformation About Real Condition of TMI-2 Led to Inadequate Crisis Response

These unwittingly inaccurate reports from TMI just postponed necessary action by external decision makers. It was only on the third day after the start of the accident that the possible meltdown of the reactor and the existence of a hydrogen bubble were officially confessed, in spite of the fact, shown in later investigations, that only 2 h into the accident at least a few of the reactor's fuel rod claddings had ruptured and zirconium alloy had already reacted with the steam to generate hydrogen. Moreover, there was already severe damage to the reactor core 3-4 h from the start of the accident.[81,82,83] The NRC concluded: "*In sum, ...the evidence failed to establish that Met Ed management or other personnel willfully withheld information from the NRC. There is no question that plant information conveyed from the control room to offsite organizations throughout the day was incomplete, in some instances delayed, and often colored by individual interpretations of plant status... Lack of understanding also affected the public's perception of the accident because early reports indicated things were well in hand, but later reports indicated they were not. [Only on the third day after the accident started], when the continuing problems were generally recognized, the utility management and staff began effective action to obtain assistance, plan for contingencies, and direct daily plant operations to eliminate the hazards. The recovery effort was massive, involving hundreds of people and many organizations*".[84,85] Thus it was only on the third day that General Public Utility executives began to ask for scientific and operational assistance from other utilities, reactor manufacturers, firms of architects and engineers, and national nuclear laboratories. And it was only on the afternoon of the fourth day that 30 people from 10 organizations of the Industry Advisory

---

[80]Report of the President's Commission on the Accident at Three Mile Island: The Need for Change: The Legacy of TMI, October 1979, p. 104.

[81]Ibid, p. 99.

[82]Three Mile Island: Report to the Commissioners and to the Public, M. Rogovin and G. Frampton, U.S. Nuclear Regulatory Commission, January 1980, Volume I, p. 21.

[83]Three Mile Island: Report to the Commissioners and to the Public, M. Rogovin and G. Frampton, U.S. Nuclear Regulatory Commission, January 1980, Volume II, pp. 535–537.

[84]Three Mile Island: Report to the Commissioners and to the Public, M. Rogovin and G. Frampton, U.S. Nuclear Regulatory Commission, January 1980, Volume I, p. 69.

[85]Three Mile Island: Report to the Commissioners and to the Public, M. Rogovin and G. Frampton, U.S. Nuclear Regulatory Commission, January 1980, Volume II-Part 3, p. 811.

Group arrived at TMI-2 and started to blueprint solutions to the core-cooling problem.[86]

In its turn, the President's commission declared that the NRC was not ready to conduct adequate response measures in such a situation: "*[W]e are extremely critical of the role [the NRC] played in the response to the accident... During the most critical phase of the accident, the NRC was working under extreme pressure in an atmosphere of uncertainty. The NRC staff was confronted with problems it had never analyzed before and for which it had no immediate solutions*".[87] According to the NRC investigation: "*They have no time to assess the situation themselves*".[88] The inadequate NRC assessment of the plant's status—based on information from the utility and mistakes in estimates of the hydrogen bubble size—led to correspondingly inadequate response measures: "*On the first day of the accident, there was an attempt by the utility to minimize its significance, in spite of substantial evidence that it was serious. Later that week, NRC was the source of exaggerated stories. Due to misinformation, and in one case (the hydrogen bubble) through the commission of scientific errors, official sources would make statements about radiation already released... The response to the emergency was dominated by an atmosphere of almost total confusion. There was lack of communication at all levels... The fact that too many individuals and organizations were not aware of the dimensions of serious accidents at nuclear power plants accounts for a great deal of the lack of preparedness and the poor quality of the response... Communications were so poor [more than 48 hours from the accident] that the senior management could not and did not develop a clear understanding of conditions at the site. As a result, an evacuation was recommended to the state by the NRC senior staff on the basis of fragmentary and partially erroneous information. ... The President asked us to investigate whether the public's right to information during the emergency was well served. Our conclusion is again in the negative*".[89]

The situation was aggravated by the fact that many decision makers were informed about the accident not by Met Ed managers or emergency agencies, but by media news representatives. An example of this is Paul Doutrich, at the time the mayor of Harrisburg—the state capital of Pennsylvania situated at about 15 km from TMI. He only found out about the accident when a radio station in Boston called him 5 h and 15 min after the beginning of the accident—despite the fact that a general emergency had been declared after 3 h and 24 min because of high

[86]Three Mile Island: Report to the Commissioners and to the Public, M. Rogovin and G. Frampton, U.S. Nuclear Regulatory Commission, January 1980, Volume I, pp. 68, 83.

[87]Report of the President's Commission on the Accident at Three Mile Island: The Need for Change: The Legacy of TMI, October 1979, pp. 21, 30.

[88]Three Mile Island: Report to the Commissioners and to the Public, M. Rogovin and G. Frampton, U.S. Nuclear Regulatory Commission, January 1980, Volume I, p. 36.

[89]Report of the President's Commission on the Accident at Three Mile Island: The Need for Change: The Legacy of TMI, October 1979, pp. 17, 18, 21, 39, 40.

radiation levels within the containment building: "*They asked me what we were doing about the nuclear emergency. My response was, 'What nuclear emergency?' They said, 'Well, at Three Mile Island.' I said, 'I know nothing about it*'".[90] Around the same time, the NRC notified the White House about the event at Three Mile Island. Seven hours after the accident, Robert Reid—mayor of Middletown, a small city located near TMI—called the Met Ed headquarters in Reading, who assured him that there was no escape of radioactive particles; but 20 s later, when he turned on the radio, he heard that radioactive particles had been released. Around the same time, William Scranton, Pennsylvania's Lieutenant Governor, said in a briefing with press representatives that "*The Metropolitan Edison Company has informed us that there has been an incident at Three Mile Island, Unit-2. Everything is under control. There is and was no danger to public health and safety... There was a small release of radiation to the environment. All safety equipment functioned properly*".[91] The President's Commission report described this incident: "*While some company executives were acknowledging radiation readings off the Island, low-level public relations officials at Met Ed's headquarters continued ... to deny any off-site releases [8 hours after the accident]. It was an error in communications within Met Ed, one of several that would reduce the utility's credibility with public officials and the press. 'This was the first contradictory bit of information that we received and it caused some disturbance'... 'I think they were defensive,' Scranton told the Commission in his testimony*".[92] Another such discrepancy on the third day after the accident showed that top management at Met Ed were still not coordinating the measures they were taking within their own organization: journalists were aware that the radioactivity released during the dumping of wastewater from TMI-2 into the Susquehanna River had been reported at 1200 millirems per hour, but Met Ed's vice president for power generation was not. During the regular press briefing, the vice president revealed data referring to a radiation level of 300 to 350 millirems per hour. This provoked suspicion that Met Ed was trying to conceal the real radiation reading, but the vice president declared that he had not heard the number 1200 and let drop: "*I don't know why we need to tell you each and every thing that we do specifically*".[93] Consequently the NRC concluded that: "*The TMI accident was a first of a kind for the nuclear power industry. Neither the utility nor the NRC was prepared to cope with the public's need for information. As a result, the residents around TMI were unduly confused and alarmed, and the level of anxiety nationwide about the safety of nuclear plants was unnecessarily raised. The information Met Ed and NRC provided to the news media during the course of the TMI accident was often inaccurate, incomplete, overly optimistic, or ultraconservative. Errors in judgment by*

---

[90]Ibid, p. 104.

[91]Ibid, p. 104.

[92]Ibid, pp. 106–107, 109.

[93]Report of the President's Commission on the Accident at Three Mile Island: The Need for Change: The Legacy of TMI, October 1979, p. 120.

Met Ed and NRC officials were major contributors to the inadequate public information effort at TMI... At the same time, the NRC failed to coordinate its internal flow of public information, resulting in speculative reports from Washington which conflicted with statements made by NRC to officials in Harrisburg. The NRC made the problem of conflicting reports even worse by refusing to participate in joint press conferences with the utility. The State's public information effort, which relied almost entirely on information from Met Ed and later the NRC, suffered accordingly. While both the public information performance of Met Ed and the NRC can be faulted in many instances, we found no evidence that officials from either the utility or the regulatory agency willfully provided false information to the press or public".[94] According to a White House representative "*many conflicting statements about TMI-2 reported by the news media were increasing public anxiety*".[95]

The invisibility of radiation also aggravated the perception of the accident by local residents: "*Never before have people been asked to live with such ambiguity. The TMI accident – an accident we cannot see or taste or smell ... – is an accident that is invisible. I think the fact that it is invisible creates a sense of uncertainty and fright on the part of people that may well go beyond the reality of the accident itself*".[96] In addition, nationwide public nervousness during the accident was likely intensified by the Hollywood blockbuster "The China Syndrome". The movie was introduced at cinemas all over the country 12 days before Three Mile Island. The plot was about an accident at a fictitious nuclear power plant near Los Angeles. In the film, the investigation that followed revealed massive cover-ups during construction of the plant, and deliberate attempts by the plant's management to conceal facts from the public. Remarkably, in one scene of the movie, a physicist is trying to evaluate the possible consequences of a total reactor core meltdown, and says that "*an area the size of Pennsylvania*" would be permanently uninhabitable.[97] Many of the 400 reporters who had arrived at TMI were under the influence of this movie, and assessed unintentionally misleading statements by Met Ed and the NRC as deliberate risk concealment. The President's Commission also mentioned that "*[a]nother severe problem was that even personnel representing the major national news media often did not have sufficient scientific and engineering background to understand thoroughly what they heard, and did not have available to them people to explain the information. This problem was most serious in the reporting of the various releases of radiation and the explanation of the severity (or lack of severity) of these releases... We therefore conclude that, while the extent of the coverage was justified, a combination of confusion and weakness in*

[94]Three Mile Island: Report to the Commissioners and to the Public, M. Rogovin and G. Frampton, U.S. Nuclear Regulatory Commission, January 1980, Volume I, p. 120.

[95]Report of the President's Commission on the Accident at Three Mile Island: The Need for Change: The Legacy of TMI, October 1979, p. 120.

[96]Three Mile Island: Report to the Commissioners and to the Public, M. Rogovin and G. Frampton, U.S. Nuclear Regulatory Commission, January 1980, Volume I, p. 48.

[97]Ibid, p. 29.

the sources of information and lack of understanding on the part of the media resulted in the public being poorly served... [N]either the utility nor the NRC nor the media were sufficiently prepared to serve the public well".[98]

On the fifth day after the accident, when it became clear that the risk of a hydrogen explosion within the reactor vessel had been mitigated, US President Jimmy Carter—formerly a senior officer on a nuclear submarine—visited TMI-2. He tried to convince the public that the reactor was stable, but stated that certain actions may yet have to be taken to bring it to cold shutdown.[99] And on the seventh day Dick Thornburgh, Governor of Pennsylvania, announced: "*The threat of any immediate catastrophe is over*".[100]

Three Mile Island Nuclear Accident: Why Risks Were Concealed

- The US government and the NRC shared an interest in developing the domestic civil nuclear industry, as part of a larger program to ensure the energy independence of the country after the severe oil crises of 1973 and 1979. This led to a perception among industry executives that **increasing the production of electricity took priority over safety matters**.
- **Wishful thinking/self-deception among decision makers**, who persuaded themselves that minor accidents did not merit close scrutiny; that the probability of a multi-factor malfunction of hardware was marginal; that the influence of human factors on the operation of a reactor during an emergency was minimal; and that the worst-case scenario—meltdown or decapsulation of a reactor vessel—could never happen.
- Government and the nuclear industry had weak control over the complex systems involved, and had only a **fragmentary perception of the whole picture of risks**. Key decision makers were ignorant of other accidents or near-miss cases within the organization or the wider industry, nationally or abroad.
- There was **no system for managing knowledge about risks within the industry** (exchange, accumulation, systematization and transmission).
- There was **no industry-wide risk assessment system for timely evaluation of the condition of nuclear power plants**. Both operators and management at TMI-2 misjudged the status of the plant, causing them to give misleading information to other audiences and delaying the measures that needed to be taken to cool the reactor.

[98]Report of the President's Commission on the Accident at Three Mile Island: The Need for Change: The Legacy of TMI, October 1979, p. 19.

[99]Three Mile Island: Report to the Commissioners and to the Public, M. Rogovin and G. Frampton, U.S. Nuclear Regulatory Commission, January 1980, Volume I, p. 85.

[100]Ibid, p. 76.

2.1.3 Bhopal Pesticide Plant Gas Leak (India, 1984)

During the night between December 2 and 3, 1984, at the pesticide plant in Bhopal, India, more than 40 tons of methyl isocyanide (MIC) and other gases leaked into the atmosphere. MIC is an intermediate in pesticide production processes and has an extremely toxic impact on human health. Over the days following the accident, from 3,000 to 10,000 citizens of Bhopal died, 100,000 were injured with irreversible changes in their health and more than 500,000 were exposed to toxic gases,[101] out of a total population of around 850,000 residents. After the disaster, no measures were taken to clean up the site of the plant. Since 1984, contaminated soil and water sources around the plant have continued to affect the environment of Bhopal.[102] Thirty years after the disaster, the death toll amounts to more than 15,000 victims of the lingering effects of MIC poisoning.[103] In terms of casualty numbers, this makes Bhopal the second largest industrial accident in world history; the largest was the breach of the Banqiao and Shimantan Dams in Central China in 1975 due to Typhoon Nina, when 26,000 people according to official estimates—or 83,000 according to unofficial data—were killed by the destruction of the dams and ensuing floods; 145,000 perished in the following months from disease and famine.[104,105]

[101]Ingrid Eckerman, The Bhopal Saga: Causes and Consequences of the World's Largest Industrial Disaster, Universities Press, 2005, p. 9.

[102]Clouds of injustice. Bhopal disaster 20 years on. Amnesty International Publications. London, 2004.

[103]Rishi Lekhi, 30 years later, disaster haunts Bhopal survivors, Associated Press, Dec 3, 2014.

[104]Typhoon Nina–Banqiao dam failure, Encyclopaedia Britannica.

[105]David Longshore, Encyclopedia of Hurricanes, Typhoons, and Cyclones, Infobase Publishing, 2009, p. 124.

2.1.3.1 Risk Concealment Before the Disaster

In 1969, Union Carbide Corporation (UCC), an American company, opened a pesticide plant in Bhopal (Madhya Pradesh state, India). In the early years, the plant produced pesticides extracted from US-imported concentrate. However, the Indian government pushed UCC to organize a full-cycle chemical output at the Bhopal plant. They motivated UCC to hire, train and develop local staff for the management of the plant, and allowed it to own 50.9 % of its Indian branch (Union Carbide India Ltd (UCIL)) while reserving less than 50 % for Indian businesses and local investors. This was an exception granted to UCC to the norm imposed at that time by the Indian government that it should own more than 50 % of the shares of any foreign investment.[106] Such exception made Union Carbide Corporation with headquarters in the United States clearly responsible for all matters concerning the Indian plant. In 1979, UCIL launched production of an insecticide called carbaryl pesticide under the trademark SEVIN, using locally produced methyl isocyanide (MIC). UCC invested reluctantly in MIC production in India because the cost of local production exceeded US costs by a factor of 3 to 4. Due to severe droughts in India in 1977, 1982 and 1984, and the resulting decline in demand for pesticide from local farmers, the Bhopal plant became unprofitable (up to $4 M losses from 1980 to 1984). By 1982, the plant was working at just 50 % of its capacity, and by 1984 just 20 %.[107]

In spite of the fact that more than 20 Indian engineers were flown to the USA in 1978–1979 at the UCC's West Virginia MIC plant to learn how to run the MIC process safely, there were at least five major chemical leaks—in which one worker died and close to 50 were injured—at the Bhopal plant between 1981 and 1984.[108] Moreover, due to the policy of the Indian government to have 100 % replacement of all positions in industry by Indians, the last American engineer left the Bhopal plant by the end of 1982.[109] From that time, the plant was operated only by Indian citizens and employed around 650 people. In 1982, UCC began to put pressure on the management of UCIL to reduce production costs, that resulted in decrease of morale at the plant: "*There was widespread belief among employees that the management had taken drastic and imprudent measures to cut costs and that attention to details that ensure safe operation was absent*".[110,111] UCIL started to extend the time between full safety checks from every 6 months to every 12 months and,

---

[106]Personal communication with Dr. Ingrid Eckerman (February 28, 2015).

[107]M.J. Peterson, Case Study: Bhopal Plant Disaster (with appendixes), University of Massachusetts—Amherst, 2009.

[108]M.J. Peterson, Case Study: Bhopal Plant Disaster, Appendix A: Chronology, University of Massachusetts—Amherst, 2009.

[109]M.J. Peterson, Case Study: Bhopal Plant Disaster (with appendixes), University of Massachusetts—Amherst, 2009.

[110]Ibid.

[111]Nancy G. Leveson, Engineering a Safer World: Systems Thinking Applied to Safety, MIT Press, 2011, pp. 25–27.

instead of replacing rusted pipes by stainless pipes every 6 months, they replaced them by common steel pipes every 2 years.[112] UCIL fired the best trained and most experienced (and therefore most highly paid) engineers and hired lower paid staff with little experience of working with dangerous chemicals and equipment (for instance, the chemical engineer, who was responsible for managing the MIC unit "*resigned because he disapproved of falling safety standards*" one year before the disaster and an electrical engineer replaced him[113]); and the length of training courses declined from 6 to 2 months.[114] Thirty percent of the staff at the Bhopal plant were fired and, for several years, turnover of staff at the plant exceeded 80 %; by December 1984, only a negligible number of employees remained who had been trained in the United States on the original MIC unit.[115,116] In addition, the managers of UCIL decided to reduce the number of workers in every shift on the MIC unit: only one manager and six workers were required on a given shift in spite of UCC stating that they should keep three supervisors and twelve workers on each shift on the MIC unit.[117] Despite these desperate attempts to economize, the SEVIN production plant remained unprofitable, and UCC had plans to sell the plant or disassemble it and ship it to Brazil and Indonesia.[118] By the autumn of 1984, plant operators were ordered to produce SEVIN from the remaining stocks of chemicals in anticipation of the possible shutdown of the plant in the near future; the MIC production unit had been halted six weeks prior to the incident. As a result, the plant accumulated a large amount of MIC in its tankers: stocks of this lethal chemical reached 62 tons, of which only 3–4 tons were required daily for production of SEVIN. It was in contravention of common practice in the chemical industry, which is to "*always keep only a strict minimum of dangerous materials on site*".[119] Due to cost cuts on refrigeration, the MIC mixture began to be stored at the plant at temperatures of nearly 20 °C, while technical requirements for the mixture required it to be stored below 5 °C in order to avoid uncontrolled reactions. At the same time, plant managers had updated the settings of temperature

---

[112] Ingrid Eckerman, The Bhopal Saga: Causes and Consequences of the World's Largest Industrial Disaster, Universities Press, 2005, p. 32.

[113] Nancy G. Leveson, Engineering a Safer World: Systems Thinking Applied to Safety, MIT Press, 2011, pp. 25–27.

[114] Ingrid Eckerman, The Bhopal Saga: Causes and Consequences of the World's Largest Industrial Disaster, Universities Press, 2005, p. 30.

[115] M.J. Peterson, Case Study: Bhopal Plant Disaster (with appendixes), University of Massachusetts—Amherst, 2009.

[116] Bridget Hanna, Ward Morehouse, Satinath Sarangi, The Bhopal reader: remembering twenty years of the world's worst industrial disaster, The Apex Press, New York, The Other India Press, Mapusa, Goa, 2005, p. 41.

[117] Ingrid Eckerman, The Bhopal Saga: Causes and Consequences of the World's Largest Industrial Disaster, Universities Press, 2005, p. 31.

[118] Ibid, p. 25.

[119] Ibid, p. 25.

alarm activation, so that operators did not receive early warnings of the temperature rise in the MIC tanks.[120] To make matters worse, on 31 October 1984, Indira Gandhi, the 3rd Prime Minister of India, was assassinated by two of her Sikh bodyguards—and during the massive social riots that followed, the plant could not safely produce SEVIN from its tremendous stocks of MIC. In November 1984, the Indian government announced nearly two weeks of national mourning and brought in a curfew to stop communal and religious violence in the country. Consequently, workers on the second and third shifts at the Bhopal plant had trouble fulfilling their duties and production of SEVIN from existing MIC stocks was slow.[121]

For many years, Bhopal was considered an attractive place to get a job, and many poor people from the countryside moved to the city and seized empty land to build slums. The population of Bhopal increased from 300,000 at the end of the 1960s to 900,000 in the mid 1980s.[122] Local authorities were reluctant to fight illegal land grabbing and construction and, as a result, shantytowns built up around the plant. During an inspection of the plant in 1979, UCC engineers emphasized that their Indian colleagues should build a complex contingency plan to respond to a possible leak, however small, of hazardous MIC. UCIL managers said that there was a contingency plan but, as the investigation following the 1984 disaster revealed, the city and state governments were not aware of any such plans.[123] In spite of the possible threat of MIC to human health, and a series of MIC-related accidents, management at the Bhopal plant never informed the authorities about these risks to the city. However, plant workers complained several times to the government of the Madhya Pradesh state about poor safety conditions on the plant, but the resulting inspections did not lead to a halt in production at the plant, because the involved state government representatives had insufficient technical experience of the chemical industry.[124]

In 1982, UCC again sent American engineers to inspect the plant at Bhopal. They found many shortcomings in the safety system and recommended UCIL to fix them. During the following years, UCIL was sending reassuring reports to UCC about safety measures, but some of these were either temporary or were never fully implemented across the plant.[125] Meanwhile, risk concealment was also shown to have happened at UCC's operations back in the U.S. Indeed, 67 leakage events occurred in the West Virginia MIC plant between 1980 and

[120]Nancy G. Leveson, Engineering a Safer World: Systems Thinking Applied to Safety, MIT Press, 2011, pp. 25–27.

[121]M.J. Peterson, Case Study: Bhopal Plant Disaster (with appendixes), University of Massachusetts—Amherst, 2009.

[122]Ibid.

[123]Ibid.

[124]M.J. Peterson, Case Study: Bhopal Plant Disaster (with appendixes), University of Massachusetts—Amherst, 2009.

[125]M.J. Peterson, Case Study: Bhopal Plant Disaster, Appendix A: Chronology, University of Massachusetts—Amherst, 2009.

1984.[126] In September 1984, UCC engineers reported the following to UCC management in a survey on operational safety and health: "*There is a concern that a runaway reaction could occur in one of the MIC unit storage tanks and that response to such a situation would not be timely or effective enough to prevent catastrophic failure of the tank*"—but investigators could not find any proof that this information was transmitted to the Bhopal plant, where the design of the MIC unit was similar to that at the West Virginia plant.[127,128]

On December 2, 1984, the plant and MIC unit were manned by incompetent staff. The safety systems installed were inadequate in the face of the existing amount of dangerous materials (more than 60 tons of MIC). Poor and low-cost maintenance over years had led to the progressive destruction of the integrity of the plant production system. There are two theories[129] about the origin of the disaster: the first holds that water poured into MIC tanks due to the operation of washing pipes. The second states that there was sabotage among Indian staff for unknown reasons. Either way, a significant amount of water poured into a tank of MIC and ignited a powerful chemical reaction. After hours of unskilled attempts by the operators to control the reaction, approximately 40 tons of MIC were released into the atmosphere of Bhopal.

2.1.3.2 Risk Concealment After the Disaster

The tremendous number of casualties at Bhopal was caused by a lack of transmission by the plant management of the information concerning the MIC leakage to local authorities and UCC headquarters.

Water was introduced in the E610 MIC tank at 10 p.m. on December 2, 1984 and the reaction started. By 0:50 a.m. on December 3, the operators understood that they could not control this leakage, and fled the plant. Earlier, at 0:30 a.m., they switched on a large siren which was heard outside the plant, but soon turned it off. When operational staff left the plant, a low-power siren was turned on, which could only be heard within the plant area.[130] Before the disaster, "*alarms at the plant sounded so often (the siren went off twenty to thirty times a week for various purposes) that*

[126] Ingrid Eckerman, The Bhopal Saga: Causes and Consequences of the World's Largest Industrial Disaster, Universities Press, 2005, p. 22.

[127] Clouds of injustice. Bhopal disaster 20 years on. Amnesty International Publications. London, 2004.

[128] Internal correspondence of UCC: operations safety/health survey MIC II Unit at Institute, West Virginia MIC plant, September 1984, http://bhopal.net/source_documents/institute.sep84.pdf.

[129] Ashok S. Kalelkar, Investigation of large-magnitude incidents: Bhopal as a case study, Arthur D. Little, Inc., Cambridge, Massachusetts, USA, presented at the Institution of Chemical Engineers Conference on Preventing Major Chemical Accidents, London, England, May 1988.

[130] M.J. Peterson, Case Study: Bhopal Plant Disaster, Appendix A: Chronology, University of Massachusetts—Amherst, 2009.

an actual alert could not be distinguished from routine events or practice alerts".[131] None of the plant executives informed local authorities about the accident. Moreover, when the police began to receive hundreds of phone calls from Bhopal residents about a strange gas, bouts of coughing and people lying unconscious on the streets, the plant operators continued to reassure local police that "*Everything is OK*" and "*We do not know what [is happening]*". The plant manager flatly denied any leakage there: "*The gas leak just can't be from my plant*".[132] This all resulted in a situation in which local officials, police and the military stalled the evacuation of local residents. Unbelievably, the authorities had never expected to evacuate people because nobody thought that the plant was a hazardous operation: the management of the plant never revealed the risks of MIC. Moreover, according to the Madhya Pradesh Town and Country Planning Board, the plant was classified in the "*general industry*" rather than in the "*hazardous industry*" category.[133]

The second cause of mass mortality at Bhopal was the fact that medical institutes were unprepared for a large influx of patients. This city of almost a million people had only 5 hospitals, with 1800 hospital beds and 300 doctors, who were never trained to serve patients with chemical poisoning. Before the accident, UCIL did not provide any recommendations about medical treatment appropriate to the list of chemicals in use at the plant, and afterwards they declined to reveal the composition of the leaked gas.[134] This led to a complete absence of adequate treatment for the thousands of victims. Moreover, according to a statement by UCC executives in the United States, the first—still contradictory—information about the disaster reached headquarters more than 12 h after the accident.[135] By 8 a.m. on December 3, 1984, managers at the plant had been arrested by the local police, and UCC did not have top-level contacts within Bhopal to get reliable information about the accident. In addition, communication between UCC and the Bhopal plant and authorities became complicated by the insufficient capacity of the existing telephone network (a city of nearly a million people had only 10,000 numbers in the telephone network). All together, this led to the absence of detailed information about the accident reaching UCC executives, and consequently to a shortage of clear and strict recommendations about antidotes from UCC's advanced medical service to hospitals in Bhopal in the first few critical hours after the leak.

In addition, Bhopal also lacked the infrastructure and ground level preparation for an adequate and fast evacuation: there were no channels for the transmission of public information (e.g. networks of loudspeakers); 80 % of the people affected by

---

[131]Nancy G. Leveson, Engineering a Safer World: Systems Thinking Applied to Safety, MIT Press, 2011, pp. 25–27.

[132]Ingrid Eckerman, The Bhopal Saga: Causes and Consequences of the World's Largest Industrial Disaster, Universities Press, 2005, p. 81–82.

[133]M.J. Peterson, Case Study: Bhopal Plant Disaster (with appendixes), University of Massachusetts—Amherst, 2009.

[134]Ingrid Eckerman, The Bhopal Saga: Causes and Consequences of the World's Largest Industrial Disaster, Universities Press, 2005, pp. 19, 64.

[135]Browning, Jackson B. Union Carbide: Disaster At Bhopal. Union Carbide Corporation, 1993.

the accident had an income of less than US$6 per month (half of the Indian subsistence level)[136] and could not even afford to have a radio receiver; and only a small number of residents had access to electricity, so most had no stable reception of radio or TV. In October 1982, after major MIC leaks, the labor union published a leaflet about possible threats to people from the plant to Bhopal, but the majority of slum residents could not understand the message and ignored it due to illiteracy. Nevertheless, due to the existence of a series of minor incidents on the plant over the preceding years, word-of-mouth communication has informed the residents of the slums of the potential danger posed by the plant and they understood the necessity to run away from the plant in case of emergency. But nobody had anticipated the magnitude of the potential disaster,[137] which occurred in December 1984, when the leaked chemicals covered almost the whole city of Bhopal.

After the accident the Indian government paid only US$800 to relatives of the deceased, and just $100 to the 20,000 victims who developed chronic diseases because of the disaster.[138] Moreover, in 1989, out-of-court agreements on compensation between the Indian government and the mother company Union Carbide Corporation stipulated a payment of just US$470 million[139]; the majority of this amount did not even reach the victims of the disaster. It took until 2010, that is 26 years after the disaster, for seven UCIL employees to be convicted by Indian courts, each receiving a two-year prison sentence and a fine of about US$2000.[140]

Bhopal Pesticide Plant Gas Leak: Why Risks Were Concealed

- The Indian government's desire to reach national industrial independence, and its negligence to reveal details of deliberate violations of safety rules at the plant. The **lack of experience or qualifications of government representatives**, which did not allow them to recognize the disastrous state of the plant years before the accident. In addition, without sufficient control by the parent corporation over Union Carbide India Limited, **management at the plant could manipulate data about real conditions at the plant** without fear to be punished by representatives of Union Carbide Corporation and Indian authorities.
- The **desire of Indian managers to appear in a good light** in the eyes of Union Carbide Corporation executives motivated them to **play down the existence of massive safety imperfections at the plant**.

---

[136] Ingrid Eckerman, The Bhopal Saga: Causes and Consequences of the World's Largest Industrial Disaster, Universities Press, 2005, p. 13.

[137] Personal communication with Dr. Ingrid Eckerman (Feb. 28, 2015).

[138] Paul Shrivastava, Long-term recovery from the Bhopal crisis, UN University Press, 1994.

[139] Ingrid Eckerman, The Bhopal Saga: Causes and Consequences of the World's Largest Industrial Disaster, Universities Press, 2005, p. 132.

[140] Bhopal trial: Eight convicted over India gas disaster, BBC News, 7 June 2010.

- The chronic unprofitability of the Bhopal plant, and **reluctance of plant managers to reveal the risks involved to local authorities that would likely oblige them to incur additional expense on safety measures**, or to suffer from increased wages to reward employees for hazardous work that would be known as such, or to support the costs for relocating the shantytowns, and so on.
- The **reluctance of Union Carbide Corporation executives to reveal statistics of accidents** at the West Virginia MIC plant, and the extreme danger posed by MIC, to their international subdivisions.
- **False reassurance/self-suggestion/self-deception among American and Indian executives** about the maximum possible scale of any chemical accident at the plant.
- The **absence of a prompt risk assessment system**: nobody among the managers of the plant could evaluate the possible volume of the chemical release or the direction of its movement, nor could they provide recommendations regarding antidotes, how the residents of Bhopal should respond, and so on.

2.1.4 Challenger Space Shuttle Disaster (USA, 1986)

You don't concentrate on risks. You concentrate on results.

No risk is too great to prevent the necessary job from getting done

Chuck Yeager

On January 28, 1986 at 11:39 a.m., the Space Shuttle Challenger exploded in the second minute after lift-off from the Kennedy Space Center. This resulted in the deaths of all seven astronauts.

2.1.4.1 Risk Concealment Before the Disaster

Constant Struggle Within the US Space Shuttle Program to Increase the Launch Frequency and Face US Government Financing Shortages

The US Space Shuttle program began in 1972. It was based on the idea that a reusable space shuttle system could make regular civil space launches possible, with the goal to achieve 24 flights per year.[141] Expenditure would be reduced by

[141] Investigation of the Challenger accident. Report of the Committee on Science and Technology House of Representatives, Oct. 29, 1986, p. 22.

reusing the Shuttles and by having more frequent launches through economies of scale. Over the following 30 years, 135 Shuttle were launched at a total cost of US$192 billion[142]—or around US$1.5 billion per launch in 2010 prices—and the annual launch rate did not exceed 4.5 flights per year. The Space Shuttle Program has been NASA's single most expensive activity.[143] Compared with unmanned space cargo programs, the cost of one kilogram of the Shuttle's payload exceeded the payload of existing programs by a factor between 2 and 10.[144,145,146] High usage rates were critical to the Shuttle's economy because its huge development costs needed to be recouped within a reasonable amount of time.[147] For example, in 1976, NASA anticipated 49 flights in 1984 and 58 in 1985.[148] In contrast, in 1981 there were two launches, in 1982 three, in 1983 four, in 1984 five, and in 1985 nine (which is the record in the history of the Shuttle program). In 1985, NASA published a projection of about 24 flights per year by 1990. There were 14 flights scheduled for 1986.[149] The Space Shuttle flights were manned, imposing on the engineers of the Shuttle to improve its reliability, at the cost of expansive additional safety systems. As a result, launches were permanently delayed ("*Manpower limitations due to high workload created scheduling difficulties and contributed to operational problems*"[150]). From the beginning of the program, underestimation of the cost of launches and the irregularity of flights became major managerial problems for NASA executives.

---

[142]Roger Pielke Jr, Radford Byerly, Shuttle programme lifetime cost, Nature, 472(7341), 07 April 2011.

[143]William Starbuck, Moshe Farjoun, Organization at the Limit: Lessons from the Columbia Disaster. Oxford: Blackwell, 2005, p. 31.

[144]The final cost of one kilogram of payload amounted to US$60,000; the Shuttle's payload, which was taken up to low earth orbit (LEO), was 24,400 kg. For example, the payload cost of the American expendable Titan IV space rocket, used by the U.S. Air Force, was calculated as US$27,000/kg—US$588 million per launch at 2010 prices with 21,680 kg of payload to low earth orbit; the Russian Proton rocket had a payload cost of US$5,300/kg—US$110 million per launch with 20,700 kg of payload to low earth orbit.

[145]According to a report of the United States Air Force, the average unit cost of Titan IV rocket was US$450 mln in 1999. The cumulative rate of US inflation between 1999 and 2000 is 30.9 %. This implies that expenditures on every launch was US$588 in 2010 price. "Titan IV" report, United States Air Force, 1999, http://www.dote.osd.mil/pub/reports/FY1999/pdf/99titaniv.pdf.

[146]Peter B. de Selding, ILS May Pitch Proton as Cost-saver Over Soyuz for Galileo Satellites, Space News, January 15, 2010.

[147]John R. London III, LEO on the Cheap Methods for Achieving Drastic Reductions In Space Launch Costs, Air University Press, Maxwell Air Force Base, Alabama, Oct. 1994, p. 45.

[148]Investigation of the Challenger accident. Report of the Committee on Science and Technology House of Representatives, Oct. 29, 1986, p. 120.

[149]Report of the Presidential Commission on the Space Shuttle Challenger Accident, Chap. VIII: Pressures on the system, Washington, D.C., June 6th, 1986.

[150]Report of the Presidential Commission on the Space Shuttle Challenger Accident, Vol. 2: Appendix I—NASA Pre-Launch Activities Team Report, Washington, D.C., June 6th, 1986.

The initial plan implied developing towards self-sufficiency, but during program development, it became clear that NASA would always rely on Congress and government spending. The design of the Shuttle's solid rocket boosters was primarily based on the U.S. Air Force's Titan III solid rocket.[151] In 1983, Ronald Reagan proposed the Strategic Defense Initiative—ground-based and space-based systems to protect the United States from attack by Soviet strategic nuclear ballistic missiles. On August 28, 1985, a Titan 34D rocket laden with military equipment exploded after take-off from the Vandenberg Air Force Base. This gave NASA additional leverage to convince Congress that the shuttle transportation system could deliver military staff and equipment, including components of the Strategic Defense Initiative program, to orbit in any conditions. Accordingly, NASA requested that military funding for developing the Titan IV program—US$17.6 billion was to be spent for this purpose up to 1999[152]—should be transferred to NASA. Ultimately, "*the nation's reliance on the Shuttle as its principal space launch capability created a relentless pressure on NASA to increase the flight rate*".[153]

Because of this, NASA executives could not accept cancellations or serious delays of Shuttle flights due to weather conditions or minor technical problems. Such confidence was based on the statistics of previous flights and the false perception that the probability of Shuttle failure was extremely low. By January 1986, NASA management interpreted the previous 24 successful Shuttle launches as a transition of the space shuttle program from the experimental phase to the operational phase, which meant that the Shuttle's design was now proven to be adequate for serial launches.

Masterful Encapsulation of the Problem

Richard Feynman, American theoretical physicist, participant in the Manhattan Project to develop an American atomic bomb, Nobel Prize laureate in Physics and member of The Rogers Commission Report, which was created to investigate the Space Shuttle Challenger disaster, wrote after the disaster what can be considered as an authoritative last word: "*There are enormous differences of opinion as to the probability of a failure with loss of vehicle and of human life. The estimates range from roughly 1 in 100 to 1 in 100,000. The higher figures come from the working engineers, and the very low figures from management. What are the causes and consequences of this lack of agreement? Since 1 part in 100,000 would imply that*

[151] Investigation of the Challenger accident. Report of the Committee on Science and Technology House of Representatives, Oct. 29, 1986, p. 106.

[152] "Titan IV" report, United States Air Force, 1999, http://www.dote.osd.mil/pub/reports/FY1999/pdf/99titaniv.pdf.

[153] Report of the Presidential Commission on the Space Shuttle Challenger Accident, Chapter XV: Return to Flight: Richard H. Truly and the Recovery from the Challenger Accident by John A. Logsdon, Washington, D.C., June 6th, 1986.

one could put a Shuttle up each day for 300 years expecting to lose only one, we could properly ask 'What is the cause of management's fantastic faith in the machinery?" An estimate of the reliability of solid rockets was made by the range safety officer, based on the study of all previous rocket flights. Out of a total of nearly 2,900 flights, 121 failed (1 in 25). This includes, however, what may be called early errors, rockets flown for the first few times in which design errors are discovered and fixed. A more reasonable figure for the mature rockets might be 1 in 50. With special care in the selection of parts and in the inspection process, a figure of below 1 in 100 might be achieved but 1 in 1000 is probably not attainable with today's technology. Since there are two rockets on the Shuttle, these rocket failure rates must be doubled to derive the Shuttle failure rates from the Solid Rocket Booster failure rate... Engineers at Rocketdyne, the manufacturer, estimate the total probability as 1/10,000. Engineers at Marshal estimate it as 1/300, while NASA management, to whom these engineers report, claims it is 1/100,000. An independent engineer consulting for NASA thought 1 or 2 per 100 to be a reasonable estimate. NASA officials argued that the figure is much lower. They point out that these figures are for unmanned rockets but since the Shuttle is a manned vehicle 'the probability of mission success is necessarily very close to 1.0.' It is not very clear what this phrase means. Does it mean it is close to 1 or that it ought to be close to 1? They go on to explain 'Historically, this extremely high degree of mission success has given rise to a difference in philosophy between manned space flight programs and unmanned programs; i.e., numerical probability usage versus engineering judgment'. It is true that, if the probability of failure was as low as 1 in 100,000, it would take an inordinate number of tests to determine it. Official management ... claims to believe the probability of failure is a thousand times less [the engineers' estimation of 1 in 100]. One reason for this may be an attempt to assure the government of NASA perfection and success in order to ensure the supply of funds. The other may be that they sincerely believed it to be true, demonstrating an almost incredible lack of communication between themselves and their working engineers... The astronauts, like test pilots, should know their risks".[154]

Problems of Timely Launching of Space Shuttles in 1985–1986

The 25th launch of the Shuttle (STS-51-L mission, Challenger) had been planned for July 1985. It was postponed until late November to accommodate changes in payloads. The launch was subsequently delayed again and finally was rescheduled

[154]Report of the Presidential Commission on the Space Shuttle Challenger Accident, Appendix F. R. P. Feynman, Personal observations on the reliability of the Shuttle, Washington, D.C., June 6th, 1986.

to January 22, 1986.[155] However, due to problems with weather conditions (bad weather at transoceanic abort landing sites and unacceptable weather at the Kennedy Space Center itself[156]), the launch was rescheduled for the morning of January 28, 1986. However, a major cause of delay to the STS-51-L mission was the delay of the previous mission (STS-61-C, Columbia), which launched only on January 12, 1986, after one month and 7 delays caused by a series of technical problems. During this month, journalists based at the Kennedy Space Center published critical and negative articles about NASA's ability to manage technical aspects of the Shuttle and launch schedule (the STS-61-C mission was sarcastically called "*Mission Impossible*"[157]). For the next mission (STS-51-L, Challenger), more than 500 journalists were accredited to cover the launch from the Kennedy Space Center and Johnson Space Center in Houston. After several delays to the new mission, the media continued to ridicule NASA capabilities to adhere strictly to the schedule. Moreover, the delays meant that the STS-51-L mission could miss opportunities to reach the target orbit for the correct deployment of satellites.[158]

Outdoor Temperature and O-Rings Problems

Weather forecasts for the night before the Challenger launch and early morning of January 28, 1986 predicted favorable conditions. However, engineers at Morton Thiokol Inc. (MTI), the main supplier of solid rocket boosters to NASA's Shuttle

[155]Report of the Presidential Commission on the Space Shuttle Challenger Accident, Chapter II—Events Leading up to the Challenger Mission, Washington, D.C., June 6th, 1986.

[156]NASA—STS-51L history, NASA's John F. Kennedy Space Center, http://www.nasa.gov/mission_pages/shuttle/shuttlemissions/archives/sts-51L.html.

[157]David Shayler, Disasters and Accidents in Manned Spaceflight, Springer, 2000, p. xxviii.

[158]Documentary "Challenger: The Untold Story", National Geographic Channel, 2006.

program and USAF's Titan rocket family, were alarmed by one detail: extremely cold weather was predicted for Florida, dropping to minus 6.6 °C (20 °F) in the last 11 h before the launch.[159] Indeed, previous successful lift-offs all occurred at temperatures above +11.5 °C (53 °F). This low temperature of +11.5 °C (53 °F) was registered during the launch of mission STS-51-C (Discovery) on January 24, 1985. After this mission, the solid rocket boosters were salvaged from the Atlantic Ocean and Roger Boisjoly, one of the senior engineers on the MTI team, found out that the O-rings, which sealed the hot gases inside the combustion chambers of the solid rocket boosters while firing, were damaged. The O-rings were among 700 parts included on the "Critical 1" list. Of the 2 million components comprising the Shuttle, a failure of any one of the parts on this list would result in the loss of the spacecraft and/or crew. Boisjoly concluded that the main cause of damage to the O-rings was the low temperature on the day of the launch: +11.5 °C (53 °F). The rings had excellent resistance to high temperatures up to 327 °C (621 °F),[160] but lose their flexibility in cold conditions. The manufacturer of the O-rings expected that the product would retain resilience below −3.8 °C (25 °F), but there were no practical tests of O-rings on MTI solid rocket boosters in cold temperature conditions. In view of this and other factors, the recommended temperature range for the entire Shuttle launch was between +0.5 °C (31 °F) and +37.2 °C (99 °F).[161] Some erosion of the O-rings was recorded during the 51-B mission (29 April, 1985, Challenger).

After the launches, MTI formally mentioned some problems with the rings in a report to NASA, but no action was taken. In July 1985, Boisjoly sent an internal report to MTI executives about his concerns about the need for an immediate redesign of the solid rocket boosters, but received an informal reply from a top MTI manager that "*this material is too sensitive to release to anybody. We will keep it a secret*".[162] Professor Leveson from MIT stated "*schedule and launch pressures in the Shuttle program created a mindset that dismissed all concerns, leading to overconfidence and complacency. This type of culture can be described as a culture of denial where risk assessment is unrealistic and credible risks and warnings are dismissed without appropriate investigation. Managers begin to listen only to those who provide confirming evidence that supports what they want to hear. Neither Thiokol nor NASA expected the rubber O-rings sealing the joints to be touched by hot gases during motor ignition, much less to be partially burned. However, as tests and then flights confirmed damage to the sealing rings, the*

---

[159]Report of the Presidential Commission on the Space Shuttle Challenger Accident, Chapter V: The Contributing Cause of the Accident, Washington, D.C., June 6th, 1986.

[160]Information about fluoroelastomer (FKM), http://en.wikipedia.org/wiki/FKM.

[161]Diane Vaughan, The Challenger launch decision: risky technology, culture and deviance at NASA. Chicago: The University of. Chicago Press; 1996, p. 155.

[162]Documentary "Challenger: The Untold Story", National Geographic Channel, 2006.

reaction by both NASA and Thiokol was to increase the amount of damage considered 'acceptable'".[163]

In fact, the problem of the O-rings was known from 1977,[164,165] but "*NASA and contractor management first failed to recognize it as a problem, then failed to fix it and finally treated it as an acceptable flight risk. [MTI] did not accept the implication of tests early in the program that the design had a serious and unanticipated flaw. [NASA executives] did not accept the judgment of its engineers that the design was unacceptable and, as the joint problems grew in number and severity, NASA minimized them in management briefings and reports. [MTI also stated that] the condition is not desirable but is acceptable*".[166] NASA executives were not informed in detail about the seriousness of the problem with solid rocket boosters during cold weather launches before January 27, 1986.[167] On the day before the launch of Challenger, during conference calls with NASA, Boisjoly tried to convince MTI executives and NASA managers to cancel the flight until the temperature on the launch pad at the Kennedy Space Center reached at least +11.5 °C (53 °F) when all of the solid rocket boosters would be defrosted. He demonstrated that damage had occurred to the material of the O-rings during flights in January and April 1985. Although the STS-61C mission (January 12, 1986, Columbia) was launched at temperatures lower than +5 °C (41 °F), MTI engineers did not provide any information concerning a possible erosion of the O-rings during the conference calls.[168] NASA executives exclaimed that they were "*appalled at the given recommendations*", that "*we can't launch, we won't be able to launch until April*"[169] and argued that the evidence was "*incomplete*".[170] Nevertheless, NASA officials emphasized that they would "*not agree to launch*

[163]Nancy G. Leveson, MIT, Technical and Managerial Factors in the NASA Challenger and Columbia Losses: Looking Forward to the Future published within Kleinman, Cloud-Hansen, Matta, and Handelsman Controveries in Science and Technology Volume 2, Mary Ann Liebert Press, 2008.

[164]Investigation of the Challenger accident: hearings before the Committee on Science and Technology, U.S. House of Representatives, Ninety-ninth Congress, June 15, 16, 23, 24, 1986, Vol.1, p. 69.

[165]Diane Vaughan, The Challenger launch decision: risky technology, culture and deviance at NASA. Chicago: The University of. Chicago Press; 1996, p. 267.

[166]Report of the Presidential Commission on the Space Shuttle Challenger Accident, Chapter VI: An Accident Rooted In History, Washington, D.C., June 6th, 1986.

[167]Diane Vaughan, The Challenger launch decision: risky technology, culture and deviance at NASA. Chicago: The University of. Chicago Press; 1996, p. 295.

[168]Diane Vaughan, The Challenger launch decision: risky technology, culture and deviance at NASA. Chicago: University of Chicago Press; 1996.

[169]Report of the Presidential Commission on the Space Shuttle Challenger Accident, Chap. V: The Contributing Cause of the Accident, Washington, D.C., June 6th, 1986.

[170]Richard P. Feynman, What Do You Care What Other People Think? 1988, W W Norton, p. 141.

against the contractor's recommendation".[171] After NASA's comments, MTI managers organized a caucus for intensive discussion about the final decision regarding the launch, during which one executive proposed to one of the skeptical managers: "*It is time to take off your engineering hat and put on your management hat*".[172] Ultimately, MTI executives approved the launch with the following comments: "*(1) there is a substantial margin to erode the primary O-ring by a factor of three times the previous worst case, and (2) even if the primary O-ring does not seal, the secondary is in position and will*".[173] NASA managers were satisfied with this decision from the contractor. In their turn, they informed their superiors (Levels I and II program officials and the Launch Director for 51-L) that the issue had been resolved and MTI did not have objections for the launch.

The next day, the Shuttle Challenger exploded in the 72nd second after lift-off. Hot gases from the combustion chambers had leaked through a breach, created by tremendous pressure—from 900 to 1200 psi (pounds per square inch) or 62 to 87 bars—on the frosted and stiff O-rings. The temperature at the launch pad during the launch was +2.2 °C (36 °F).[174,175]

Disclosure of Continual Flawed Decision-Making Processes Within the Program

After the disaster, the Roger Commission stated…"*that testimony reveals failures in communication that resulted in a decision to launch 51-L based on incomplete and sometimes misleading information, a conflict between engineering data and management judgments, and a NASA management structure that permitted internal flight safety problems to bypass key Shuttle managers… Organizational response to the technical problem was characterized by poor communication, inadequate information handling, faulty technical decision making, and failure to comply with regulations instituted to assure safety*".[176] The Commission found that NASA's safety system had many faults, including "*a lack of problem reporting requirements, inadequate trend analysis, misrepresentation of criticality, and lack of involvement in critical discussions… Problem reporting requirements are not concise and fail to get critical information to the proper levels of*

[171]Diane Vaughan, The Challenger launch decision: risky technology, culture and deviance at NASA. Chicago: University of Chicago Press; 1996, p. 312.

[172]Report of the Presidential Commission on the Space Shuttle Challenger Accident, Chap. V: The Contributing Cause of the Accident, Washington, D.C., June 6th, 1986.

[173]Ibid.

[174]Report of the Presidential Commission on the Space Shuttle Challenger Accident, Chap. V: The Contributing Cause of the Accident, Washington, D.C., June 6th, 1986.

[175]Report of the Presidential Commission on the Space Shuttle Challenger Accident, Chap. IV: The Cause of the Accident, Washington, D.C., June 6th, 1986.

[176]Report of the Presidential Commission on the Space Shuttle Challenger Accident, Chap. V: The Contributing Cause of the Accident, Washington, D.C., June 6th, 1986.

management".[177] "*The decision to launch the Challenger was flawed. Those who made that decision were unaware of the recent history of problems concerning the O-rings and the joint and were unaware of the initial written recommendation of the contractor advising against the launch at temperatures below 53 degrees Fahrenheit [+ 11.5 °C] and the continuing opposition of the engineers at Thiokol after the management reversed its position... If the decision makers [Levels I and II program officials, or Launch Director for 51-L] had known all of the facts, it is highly unlikely that they would have decided to launch 51-L on January 28, 1986*".[178]

Boisjoly declared during the commission testimony: "*I felt personally that management was under a lot of pressure to launch and that they made a very tough decision, but I didn't agree with it*".[179] Twenty years later, Boisjoly stated that "*I must emphasize that MTI Management fully supported the original decision to not launch below 53 °F [+ 11.5 °C] prior to the caucus. The caucus constituted the unethical decision-making forum resulting from intense customer intimidation. NASA placed MTI in the position of proving that it was not safe to fly instead of proving that it was safe to fly. Also, note that NASA immediately accepted the new decision to launch because it was consistent with their desires and, please note, that no probing questions were asked*".[180] The Commission concluded that "*the Thiokol Management reversed its position and recommended the launch of 51-L, at the urging of Marshall and contrary to the views of its engineers in order to accommodate a major customer*".[181]

Richard Feynman finally found out about the O-ring problem, about which NASA managers had known since 1977: "*If all the seals had leaked, it would have been obvious even to NASA that the problem was serious. But only a few of the seals leaked on only some of the flights. So NASA had developed a peculiar kind of attitude: if one of the seals leaks a little and the flight is successful, the problem isn't so serious... Mr. Weeks said there was a rumor that the history of the seals' problem was being leaked to the newspapers. That bothered him a little bit, because it made NASA look like it was trying to keep things secret... We had our emergency closed meeting to hear from the guy whose story was in the New York Times. His name was Mr. Cook. He was in the budget department of NASA when he was asked to look into a possible seals problem and to estimate the costs needed to rectify it. By talking to the engineers, he found out that the seals had*

---

[177]Report of the Presidential Commission on the Space Shuttle Challenger Accident, Chap. VII: The Silent Safety Program, Washington, D.C., June 6th, 1986.

[178]Report of the Presidential Commission on the Space Shuttle Challenger Accident, Chap. V: The Contributing Cause of the Accident, Washington, D.C., June 6th, 1986.

[179]Ibid.

[180]Roger M. Boisjoly, Telecom Meeting (Ethical Decisions—Morton Thiokol and the Challenger Disaster), 2006, http://www.onlineethics.org/cms/7061.aspx.

[181]Report of the Presidential Commission on the Space Shuttle Challenger Accident, Chapter V: The Contributing Cause of the Accident, Washington, D.C., June 6th, 1986.

been a big problem for a long time. So he reported that it would cost so-and-so much to fix it – a lot of money. From the point of view of the press and some of the commissioners, Mr. Cook's story sounded like a big expose, as if NASA was hiding the seals problem from us".[182]

The commission summarized major organizational problems related to the accelerated launch schedule. We here select a few that are relevant to our discussion[183]:

"*1. The capabilities of the system were stretched to the limit to support the flight rate in winter 1985/1986. Projections into the spring and summer of 1986 showed a clear trend; the system, as it existed, would have been unable to deliver crew training software for scheduled flights by the designated dates. The result would have been an unacceptable compression of the time available for the crews to accomplish their required training.*
2. Spare parts are in critically short supply. The Shuttle program made a conscious decision to postpone spare parts procurements in favor of budget items of perceived higher priority. Lack of spare parts would likely have limited flight operations in 1986...
3. The scheduled flight rate did not accurately reflect the capabilities and resources
4. Training simulators may be the limiting factor on the flight rate: the two current simulators cannot train crews for more than 12–15 flights per year.
5. When flights come in rapid succession, current requirements do not ensure that critical anomalies occurring during one flight are identified and addressed appropriately before the next flight".

During hearings before the US Congress Committee on Science and Technology, one committee member made the following assessment of NASA's organizational culture prior to the disaster: "*[O]ne difficult question is this whole attitude, this whole new culture that grew up in NASA and perhaps in the Marshall Center, this culture that has been called arrogance, conceit that they knew it all; they didn't need to include in the information circle outside experts. They didn't need to listen to the Rockwell fears, expressed fears of the subzero temperatures. They pressured Morton Thiokol not to bother with a lot of chintzy concerns about safety. They excluded the astronauts themselves from the information circle. They had the feeling that they knew it all and didn't need any outside information. They didn't want anything to interfere with the schedule... [B]oth management and technical arrogance brought about by the mindset caused by a period of spectacular successes. We in Congress, as well as NASA and the aerospace industry, must never again be*

[182]Richard P. Feynman, What Do You Care What Other People Think? 1988, W W Norton, p. 138.

[183]Report of the Presidential Commission on the Space Shuttle Challenger Accident, Chapter VIII: Pressures on the system, Washington, D.C., June 6th, 1986.

lulled into a sense of overconfidence that could contribute to such a tragedy. While history does not repeat itself, unfortunately people can repeat history".[184]

In August 1990, the U.S. General Accounting Office mentioned that "*NASA and the Air Force provide the majority of the contract dollars for the Thiokol Corporation [renamed Morton Thiokol Inc.]... Based on our interviews with Air Force and NASA officials, our study indicates that Thiokol will remain a viable part of the defense industrial base [Thiokol Corporation was also the manufacturer of the boosters for American ballistic missiles with nuclear warheads – Pershing, Peacekeeper/Trident, Poseidon, Minuteman*[185]*]... Air Force and NASA officials said that it really is not economically feasible to keep two sources in operation for these items... [Therefore] [p]urchases of Thiokol's solid rocket motors are planned through 1995 and beyond*".[186]

Ultimately, the combination of these facts explains the passive position of MTI management when, in July 1985, Boisjoly sent the report about shortcomings in the O-ring design. MTI executives realized that the budget of NASA and the available time were insufficient for serious improvement of the scores of solid rocket boosters that were already ordered—after the accident, it indeed took 32 months to redesign the solid rocket boosters.[187] It thus seemed impossible for NASA and MTI to halt the program and confess that there had been flaws in the design of the O-rings during many years before the first launch of the Shuttle in 1981. Moreover, NASA and MTI have been concealing this information from astronauts, the government and the public during two dozen launches. NASA was the major client for MTI with more than a billion dollar contract each year, and MTI management wished to maintain its contract. This explains why MTI management made the decision, in compliance with NASA requests, to keep the cost of launches down and to minimize delays, the later being interpreted by the media and politicians as due to technical shortcomings of the whole Shuttle program; NASA was anxious to demonstrate to Congress that the Shuttle could fly in any conditions with military staff and materials.

Because of the disaster, the USAF got resources from Congress to develop its Titan IV program, delivering military staff into space independently from the Shuttle program. After the disaster, MTI admitted guilt and legal liability for the disaster, paid out-of-court compensation to the families of astronauts, thus diverting the blame from NASA management… and received lucrative new contracts from NASA and USAF for the decades to come. After the commission hearing,

---

[184]Investigation of the Challenger accident: hearings before the Committee on Science and Technology, U.S. House of Representatives, Ninety-ninth Congress, June 15, 16, 23, 24, 1986, Vol.1, p. 53, p. 3.

[185]Thiokol Chemical Corporation profile, http://en.wikipedia.org/wiki/Thiokol.

[186]U.S. General Accounting office, Government Contracting: Review of Morton Thiokol Separation, NSIAD-90-220, August 1990, p. 3.

[187]Columbia Accident Investigation Board, CAIB Final Report, Volume 1, August 26, 2003, p. 25.

whistleblower Roger Boisjoly found himself shunned by colleagues and managers of MTI and resigned from the company.[188]

Richard Feynman summed it up as follows: "*Let us make recommendations to ensure that NASA officials deal in a world of reality in understanding technological weaknesses and imperfections well enough to be actively trying to eliminate them. They must live in reality in comparing the costs and utility of the Shuttle to other methods of entering space. And they must be realistic in making contracts, in estimating costs, and the difficulty of the projects. Only realistic flight schedules should be proposed, schedules that have a reasonable chance of being met. For a successful technology, reality must take precedence over public relations, for nature cannot be fooled*".[189]

Challenger Space Shuttle Disaster: Why Risks Were Concealed

- Unrealistic projections about the launch schedule and a **culture of continuously rushed organization**. NASA management's desire to demonstrate to Congress and the military that the Shuttle program could send any load to space in any weather conditions on a timely basis.
- **Habituation/wishful thinking/false reassurance/self-suggestion/self-deception** among NASA and MTI decision-makers about the supposedly minuscule probability of a failure of the Shuttle. This also led to an attitude of arrogance among NASA executives.
- MTI management's **fear of losing their main client** (NASA). General problem of incentives in risk management: if MTI had remained adamant and advised against the flight, how would the "success" of no disaster resulting from the flight cancellation be rewarded?
- The **reluctance of MTI management to confess their own mistakes** in the design of solid rocket boosters and in ignoring previous warnings (damage to the O-rings during previous launches).
- **"Success at any price" and "no bad news" culture**
- MTI management's **fear of being accused of incompetence**. This question was also connected to **national security secrecy** because MTI was the supplier of solid rocket boosters for several American ballistic missiles.

[188]Sidney Dekker, Second Victim: Error, Guilt, Trauma, and Resilience, CRC Press, 2013. p. 23.

[189]Report of the Presidential Commission on the Space Shuttle Challenger Accident, Appendix F—Personal observations on the reliability of the Shuttle by R.P. Feynman, June 6th, 1986; Washington, D.C.

2.1.5 Chernobyl Nuclear Disaster (USSR, 1986)

Although the Soviet socialist press blamed capitalists and their dangerous and irresponsible working practices during the Bhopal accident, the Soviet nuclear industry and top officials did not implement the evacuation lessons learned from this American-Indian disaster in their own practice—and it was not long before the USSR faced quite a similar event. Previously, Soviet nuclear industry executives had not recognized any parallels between the situation within their industry and the American problems revealed during the Three Mile Island accident: insufficient exchange of risk information about incidents and near-miss cases, ignorance of the importance of the human factor in operating a nuclear power plant (NPP), self-deception about the overall reliability of reactors in any situation, and overconfidence about the impossibility of a worst-case scenario actually happening.

On April 26, 1986 at 1:23 a.m., during an experiment with the emergency power supply system at the Chernobyl nuclear power plant, a power excursion occurred in the RBMK-1000 Reactor #4 that led the reactor to burn uncontrollably. The plant was located in the Ukrainian Soviet Socialist Republic, which at the time was part of the Soviet Union. It was 700 km away from Moscow, 320 km from Minsk, and 140 km from Kiev. Because the reactor did not have a containment dome, the explosion led to the release into the atmosphere of 7.7 tons of uranium oxide fuel, amounting to 4 % of the total contained in the reactor; 96 % of the fuel, or 185 tons of uranium, stayed in the reactor.[190] Huge regions of Belarus, Russia and Ukraine were contaminated,[191] and traces of chemical elements from Chernobyl NPP were later found in Northern and Western Europe. The accident resulted in the release of approximately 5200 PBq (1 PBq (Penta Becquerels) = 10^{15} disintegrations per second)[192] of radioactive substances into the atmosphere.[193] This was the first accident since the beginning of the nuclear age to be classified as a level 7 event—the maximum level according to the International Nuclear Event Scale. More than 116,000 people were evacuated from the 30 km zone around the NPP.[194] Two workers died after the explosion, and 28 firefighters died in the first three months following the accident. Estimates from various sources of the total number of victims of the Chernobyl accident remain

[190]Alexandr Borovoy, Evgeny Velihov. Experience of Chernobyl, National Research Center "Kurchatovsky Institute", Moscow, 2012, p. 25.

[191]Frequently Asked Chernobyl Questions, IAEA, http://www.iaea.org/newscenter/features/chernobyl-15/cherno-faq.shtml.

[192]PBq stands for Peta Bequerel, so 1 PBq = 10^{15} atomic disintegrations per second (1 million billion).

[193]Chernobyl Accident 1986, World Nuclear Association, updated April 2014.

[194]From April, 27 to May, 8, 1986, 99195 residents with a 30-km zone were evacuated. 17122 people were additionally evacuated from May, 14 till September 1986. Total number of evacuated residents was 116287. Source: 25 years of the Chernobyl accident (1986–2011). Results and Prospects overcoming its consequences in Russia, Ministry for Civil Defense, Emergencies and Disaster Management of the Russian Federation, Moscow, 2011, p. 20.

contradictory because of political indecisiveness, different scientific approaches and the unavailability of health statistics from Soviet officials. In 2005, the UN report "*Chernobyl's Legacy: Health, Environmental and Socio-Economic Impacts*" contained a statement from an international team of more than 100 scientists that up to about 4000 people could eventually die of radiation from the Chernobyl NPP accident.[195]

The financial cost of the Chernobyl disaster remains controversial too. Mikhail Gorbachev, General Secretary of the Communist Party of the Soviet Union from 1985 until 1991, cited that the Soviet Union spent 18 billion rubles[196] (approximately US$27 billion[197]) on dealing with the consequences of the disaster. The government budget of the USSR was around 360 billion rubles from 1985–1987,[198] and the GNP in that period was around 780–800 billion rubles; so the expenses for the response to Chernobyl were 5 % of the annual Soviet budget, or approximately 2 % of GNP. According to estimates from academician Valery Legasov, a key member of the government investigation committee on the Chernobyl disaster, the total damage caused by the Chernobyl accident was in fact 300 billion rubles in pre-1990 prices, or approximately US$450 billion (of 1990 US$). This amount exceeds the combined profits of all Soviet nuclear power plants for the duration of their existence.[199]

In a nutshell, the Chernobyl disaster is the combination of (i) a fundamental design mistake on the class of RBMK reactors (leading to instabilities in certain conditions) that were hidden due to a culture of secrecy and arrogance, and (ii) bypassing safety rules due to structural lack of communication between competent agencies as well as amateurship actions of the Chernobyl staff who switched off all alarms during a standard feasibility test.

[195]Chernobyl: the true scale of the accident, World Health Organization, 2009, http://www.who.int/mediacentre/news/releases/2005/pr38/en/.

[196]Interview with Mikhail Gorbachev, documentary "The Battle of Chernobyl", Director: Thomas Johnson, 2006.

[197]Official exchange rate of State Bank of the USSR by the end of 1986: Soviet ruble/US dollar—0.6783, Archive of Bank of Russia, http://cbr.ru/currency_base/OldDataFiles/USD.xls.

[198]Consolidated budget of the USSR and Russia, Information-Analytical Center "The budget system of the Russian Federation", http://www.budgetrf.ru/Publications/Magazines/Ve/1995/95-7illarionov/95-7illarionov020.htm.

[199]Valery Legasov, Problems of Safe Development of the Technosphere, Communist Journal, #8, 1987, pp. 92–101.

2.1.5.1 Risk Concealment Before the Disaster

Geopolitical Context and the Civil Nuclear Program Race

There was constant competition for innovation between the Soviet Union and the West, which manifested clearly in the development of nuclear weaponry and in space exploration. At the turn of the 1970s, the competition began in civil nuclear power. By 1972, the USSR was behind the USA and United Kingdom, which had constructed more than 50 reactors between them, while the Soviet Union had only 7.[200] The Soviet Union had not tried to develop nuclear power in the 1950s and 1960s, because the assumption within the powerful Soviet Planning Commission was that coal from the Donbass—the Donetsk coal basin, located in the Ukrainian Soviet Socialist Republic—could provide enough energy for the Western part of the USSR. But double-digit industrial growth and massive construction of civil infrastructure in the 1960s and 1970s provoked energy shortage within the western part of the Union, and new calculations showed that the Donbass would not have sufficient coal resources for long-term supply. Although the Soviet Union had access to massive coal deposits, these were located beyond the Ural Mountains and would therefore put huge pressure on the railways to transport coal to plants in the western part of the Union. One promising strategy to solve growing energy needs was the intensive development of civil nuclear energy in the heavily populated and highly industrialized western part of the Soviet Union. Moreover, after the 1973 Arab oil embargo, the price of oil increased by a factor 2.5 in six months, from US$4.90/barrel [US$22 in 2010 prices] to US$12 [US$53 in 2010 prices].[201] Oil exports from the gigantic oil and gas fields recently discovered in Western

[200]Nikolaii Karpan, Vengeance of peaceful atom, Dnepropetrovsk, 2006, p. 255.

[201]Andrew Scott Cooper, The Oil Kings: How the U.S., Iran, and Saudi Arabia Changed the Balance of Power in the Middle East, Simon and Schuster, New York, 2011, pp. 123, 149.

Siberia became very profitable for the USSR, because production costs there were very low, at only US$0.80 per barrel [US$3.50 in 2010 prices].[202] The export of hydrocarbons, principally oil, greatly increased the Soviet budget; so it was logical to focus on raising export revenue by reducing domestic oil and gas consumption for domestic electricity in parallel with large scale development of nuclear power plants for domestic uses. Nuclear power compares very favorably: burning just one ton of natural uranium oxide produces the energy equivalent of 16,000 tons of coal or 80,000 barrels of oil.[203]

The Politburo (the executive committee for the Communist Party of the Soviet Union) decided to invite the leading developers of the Soviet nuclear weapon (which worked within The Soviet Ministry of Medium Machine Building (Minsredmash)), to work on designing and building a new high-capacity reactor. Academician Anatoly Alexandrov, director of the Kurchatov Institute of Atomic Energy (subordinated to Minsredmash), which had researched the theoretical physics underlying the Soviet nuclear weapon program, was appointed as the scientific director of the new civil reactor project. Academician Alexandrov personally took part in the development of the Soviet nuclear submarine fleet, nuclear icebreakers and the civil Water-Water Energetic Reactor (VVER), a successful Soviet variant of the Western pressurized water reactor. Around 80 % of all operating civil nuclear reactors in the world are light water reactors: either pressurized water reactors (PWRs) or boiling water reactors (BWRs).[204] An immensely respected scientist, he was president of the Academy of Sciences of the USSR from 1975 until 1986.

Strengths and challenges during the development of RBMK reactor

The chief design engineer of the new reactor was Nikolay Dollezhal, another respected member of the Academy and director of the Scientific Research and Design Institute of Energy Technologies (NIKIET), which was responsible for the design of the Soviet nuclear submarine fleet, the first Soviet uranium-graphite channel water-cooled reactors and the VVER reactor. Academician Dollezhal recounted the history of the development of the new reactor—the high power channel reactor or RBMK, a water-cooled uranium-graphite channel reactor: "*In 1965, the design [of the RBMK] was sent to the Ministry. There were supporters and opponents of the reactor. The opponents considered that only VVERs should be developed… In the construction of [RBMK], we could use cooperative ties*

[202]Gennady Shmal, Energy heart of Russia. 60 years of West-Siberian oil and gas province, Drilling and Oil Magazine, Moscow, November 2013.

[203]Uranium Quick Facts, Environmental Science Division of Argonne National Laboratory for the United States Department of Energy, http://web.ead.anl.gov/uranium/guide/facts.

[204]Günter Kessler, Anke Veser, Franz-Hermann Schlüter, Wolfgang Raskob, Claudia Landman, Jürgen Päsler-Sauer, Safety Concepts of Light Water Reactors, Springer, p. 3.

between [existing] machinery plants, which were developed during the manufacture of the first industrial reactors [technologically speaking, RBMK reactors were an enlarged version of an existing military reactor developed for production of plutonium; so their construction required minimal restructuring of existing machinery plants, and RBMKs could use cheap natural uranium, while western analogs require more expensive enriched uranium]... It would allow us to cope with the task [of constructing each RBMK] in 5–6 years. As is known, the Americans construct large shell-type reactors [like PWRs and BWRs, which require sophisticated technological skills to manufacture the reactor vessel, which works under tremendous internal pressure] over 8–10 years; we [the Soviets] simply did not have the experience [to manufacture large reactors of this type], although we have a shell-type reactor of small capacity at Novovoronezhskay NPP... [The RBMK] did not require anything that would have been too far from the normal, no specialized engineering and manufacturing... Moreover, everything to do with the period [of construction] was important at that time: there was a difficult situation with power supply in the country [Politburo constantly put pressure on the developers in order to accelerate the pace of reactors' construction]... [Just for information,] the launch of the first VVER-1000 occurred only in 1979, but serial production of [this type of reactors began only in 1985, while in 1973 the first pilot RBMK-1000] was already online. By 1980, there was production of electricity from around ten serial RBMKs... The reactors were economical in terms of the cost of energy produced. Simple enough to run, of course, if [the operators] complied with all the requirements mentioned in the operating instructions".[205] Academician Alexandrov also confirmed that "*Soviet scientists were able to solve the problem of increasing the economic efficiency of nuclear power stations*".[206] In addition, the construction of the new reactor also implied the possibility of changing nuclear fuel without shutting down the reactor—unlike shell-type reactors, which require a compulsory shutdown—which made the RBMK very cost-effective in comparison with competitive reactor types.

Because the Soviet civil nuclear program originated from the military nuclear program, many approaches for designing and constructing civil reactors came from military experience. In the 1940s, USSR was eager to overtake Americans in the construction of nuclear bombs. In this race, drawing mainly from domestic assets and also from the German specialists they have been able to recruit, the Soviets were significantly constrained by a shortage of resources, both material and human, in comparison with the Americans who had invited the best nuclear physicists from around the world and had suffered no loss on their territory while the Soviets had to reconstruct after the most devastating war in the Union as well as Russian history with casualties mounting to more than 27 million people. Nevertheless, in 1949, the Soviets tested their first nuclear bomb (four years after the Americans). By 1953, the Union overtook the States in the development of

[205]Nikolay Dollezhal', At the root the man-made world, Moscow, 2010, 4th edition, pp. 160–162.

[206]Anatoly Aleksandrov, October and Physics, Pravda, November 10, 1967.

thermonuclear weapons. In 1954, Soviet nuclear scientists and engineers commissioned in Obninsk the first industrial civil nuclear reactor in the world: an uranium-graphite channel water-cooled reactor, it was a predecessor of the RBMKs. Academician Dollezhal outlined the experience of building this reactor as follows: "*In 1951, when designing of the reactor was in full swing … building the world's first nuclear power plant has already begun by laying the foundation of the plant … During experiments [i.e. during the simultaneous design and construction of the reactor], more and more new knowledge was revealed that was impossible to ignore. Not often, indeed, but still sometimes, there was the need to reconstruct already designed components and devices [of the reactor]… One thing is beyond doubt: if the construction [of the reactors] had been carried out 'by the rules', where construction was started [only] after the final completion of the design of the reactor, then the nuclear power plant would have been launched several years later*".[207] This approach was continued in the 1970s for the RBMKs: many design solutions for the new reactor were practically tested during operation of the first pilot model, while the construction of the first serial RBMKs was already launched. Minsredmash constructed and launched the first pilot RBMK reactor near Leningrad (now St. Petersburg) in 1973, while the foundations of Chernobyl NPP was already laid in 1970. Academician Valery Legasov, an executive at the Kurchatov Institute of Atomic Energy during the Chernobyl disaster, recalled: "*The first launch of the pilot RBMK reactor at Leningrad NPP already showed [that running an RBMK reactor safely] is quite a difficult task for the plant's operator. [There was] a problem with the instability of neutron fluxes and the challenges of managing them … It should be said that, of course, a positive coefficient of reactivity in this reactor appeared unexpected [for the developers] … We had to change the degree of nuclear fuel enrichment, and carry out a number of other technical measures in order to facilitate the operation of the reactor. Even after these measures, managing the reactor required tremendous attention from a plant's operator and it was always quite difficult*".[208] This shows that knowledge of serious intrinsic problems with the RBMK was present, but rampant misinformation and lack of communication nurtured the Chernobyl catastrophe.

RBMK Reactor Design and the SCRAM Effect

There is no single universally accepted version of the cause of the reactor excursion, but the majority of investigators mention two main causes: a combination of imperfections of the RBMK reactor design—especially a phenomenon known as

[207]Nikolay Dollezhal', At the root the man-made world, Moscow, 2010, 4th edition, p. 136.

[208]Valery Legasov, Record from cassettes, 1986–1988.

the "positive SCRAM effect"[209]—and unconscious mistakes by the executives and operators of Chernobyl NPP, who approved and conducted the experiment on Reactor #4. The reactor's developers blame the staff of the plant and vice versa. The "positive SCRAM effect" first came to light during operation of the first RBMK reactor near Leningrad, in an incident in November 1975. There was a reactor reactivity excursion after SCRAM (emergency shutdown), which overheated a small part of the core, rupturing a channel of the reactor. The investigation commission in 1976 concluded that some elements of the reactor core needed to be redesigned in order to reduce the void coefficient, change some features of the control rods and increase the speed of the SCRAM system.[210,211] But these recommendations were only implemented ten years later, after the Chernobyl disaster. The IAEA report after Chernobyl noted: "*The slow speed of the emergency protection system (the time for total insertion [of the control rods] into the core from the upper limit position is 18 s) and defects in the design of the rods (i.e. the positive reactivity excursion) resulted in a situation where, for a number of reactor operating modes, the emergency protection system not only did not function, but itself initiated a reactor runaway*".[212]

From 1976 onwards, Vladimir Volkov, head of the reliability and safety laboratory at the Kurchatov Institute, sent numerous memoranda to his supervisors about calculation errors in the design of the RBMK, and gave suggestions for their improvement. He mentioned the positive SCRAM effect, defined as a localized increase of activity in the bottom of the core of a nuclear reactor during emergency shutdown. But executives at the Kurchatov Institute and NIKIET did not pay serious attention to his warnings, or those of others.[213,214] Academician Valery Legasov concluded: "*I did not see in the Soviet Union a single collective body, which more or less competently put together and considered [i.e. made a systematic search of sources of problems and critical shortcomings of existing reactors, which could lead to accidents at nuclear plants, and assessed the probability of each]... [On the contrary] the struggle against [critical shortcomings of the reactors] was conducted as a separate struggle within each particular case: if there was failure of the steam generator at a plant – than it launched a decision-making process about changing the design of steam generators. And, of course, sooner or*

[209]The SCRAM system refers to the control rods that are inserted into a nuclear reactor core to suppress nuclear fission. A "positive SCRAM effect" is a localized increase of activity in the bottom of the core of a reactor during emergency shutdown with low power range: introducing graphite rods leads to decreased absorption of neutrons by the xenon in the core ("xenon poisoning") and accelerates the nuclear reaction.

[210]The Chernobyl Accident: Updating of INSAG-1, IAEA Publications, Vienna, 1992, pp. 47–48.

[211]Anatoly Dyatlov, Chernobyl. How it was, Nauchtekhlitizdat, Moscow, 2003, p. 153.

[212]The Chernobyl Accident: Updating of INSAG-1, IAEA Publications, Vienna, 1992, p. 39.

[213]Anatoly Dyatlov, Chernobyl. How it was, Nauchtekhlitizdat, Moscow, 2003, pp. 61–64, 91–93.

[214]Nikolaii Karpan, Vengeance of peaceful atom, Dnepropetrovsk, 2006, pp. 294–296, 399.

later, it led to improvements in the situation... [There would be an improvement of that exact shortcoming and things would then] calm down until the next case".[215]

The developers of the RBMK assumed that the positive SCRAM effect would only manifest in rare cases, and preferred to take organizational measures to ensure the safe operation of the reactor (clear instructions, staff training, etc.) rather than making technical changes to the reactor design.[216] They were confident that the high quality of education and self-discipline of the staff at military-prone Minsredmash would compensate for any technical disadvantages of the RBMK when it became operational. Moreover, due to Western sanctions against the USSR during the Cold War, Soviet scientists and engineers were not able to use American supercomputers in the 1960s and 1970s to calculate technical solutions for the reactors at the design stage without testing their assumptions on prototypes. An IAEA report in 1992 mentioned the following: "*There are a number of explanations for the poor quality of the calculation analysis of the safety of the design of [RBMK reactors]. These include the fact that, until recently, Soviet computer techniques were chronically outdated and the standard of computer codes was very low. Three-dimensional non-stationary neutron-thermal-hydraulic models are required in order to calculate the physical parameters of a RBMK reactor under different operating conditions. Such models first became available only shortly before the Chernobyl accident and were not really developed until after the accident... As a result of the misguided selection of the core's physical and design parameters by the designers, the RBMK-1000 reactor was a dynamically unstable system with regard to power and steam quality perturbations. The steam quality, in its turn, was dependent on many parameters characterizing the reactor state*".[217]

Economic and Political Pressure to Quickly Build Many RBMK Reactors

The Politburo put strong pressure on the RBMK developers to launch serial production of the reactors immediately in order to satisfy domestic electricity needs. So, in parallel with the test operation of the prototype RBMK in Leningrad, full-scale construction was initiated not only at the Chernobyl NPP near Kiev (in the Ukrainian Soviet Socialist Republic), but at Kursk and Smolensk in the Russian Soviet Socialistic Republic, and at the Ignalina NPP near Vilnius (in the Lithuanian Soviet Socialist Republic). The original RBMK design was not fundamentally redeveloped or revised for these serial units, and did not even include the improvements recommended after the 1975 accident at Leningrad NPP.[218] Academician Valery Legasov concluded that "*[The Soviet Union] built the world's first nuclear power plant [in Obninsk], but later we slowed down the development*

[215]Valery Legasov, Record from cassettes, 1986–1988.

[216]Nikolaii Karpan, Vengeance of peaceful atom, Dnepropetrovsk, 2006, pp. 294–296.

[217]The Chernobyl Accident: Updating of INSAG-1, IAEA Publications, Vienna, 1992, pp. 31, 59.

[218]Ibid, p. 87.

of this technology and the review of all safety issues associated with the operation of such plants [until changes in our energy supply strategy for the western part of the USSR, when we] began to hurry. Consequently this haste led to more units being built with limited funding. There was a need from the economy. Keeping the costs down began with [rejection to construct] containment [buildings over reactors, which would have increased the construction costs of Soviet NPPs by 30 % and lengthened the construction period for the plants]... We all began to show concern about the quality of education and training of the personnel responsible for the design, construction and operation of nuclear power plants, because the number of units increased dramatically, but the quality of the personnel involved in the process decreased [while developers of the RBMK expected that comprehensive organizational measures could compensate the technical shortcomings of the reactors]... There was a constant need for new buildings, new benches, new people for this job, because the number of units [reactors] increased. However this development was still not qualitative, [only] quantitative... [The problems of the Soviet nuclear energy, revealed in the Chernobyl disaster] generally originate from the organizational approach toward development, more rapid development, of new technology".[219] It was common practice for the Politburo to issue deadlines for the construction of nuclear plants according to the date of the next Congress of the Communist Party of the Soviet Union, with no regard for the availability of equipment for the plant or for the recommended schedule for proper construction. All this resulted in constant rush in the development of the Soviet civil nuclear industry in the 1970s and 1980s, repeating the practice that had prevailed during the nuclear arms race between the USSR and the West in the late 1940s and 1950s.

A Fatal Regulation Mistake

The Politburo issued another pivotal decision: responsibility for all the new RBMK NPPs would be transferred from the predominantly military Minsredmash to the civil Ministry of Energy and Electrification of the USSR. There were several explanations for this decision. The developers of the RBMK reactor—all nationally respected and honored scientists—convinced everybody, especially the senior executives of the Soviet Union, of the absolute safety of the RBMK reactor and the infallibility of Soviet nuclear technology.[220,221,222,223] Their overconfidence persuaded Politburo members and executives at the Ministry of Energy and

[219]Valery Legasov, Record from cassettes, 1986–1988.

[220]Anatoly Dyachenko, Experience of liquidation of Chernobyl disaster, Federal State Unitary Enterprise "Institute of Strategic Stability" of Rosatom, Moscow, 2004, http://www.iss-atom.ru/book-7/glav-2-3.htm.

[221]Ibid.

[222]Interview with Mikhail Gorbachev, documentary "The Battle of Chernobyl", Director: Thomas Johnson, 2006.

[223]Nikolaii Karpan, Vengeance of peaceful atom, Dnepropetrovsk, 2006, p. 401.

Electrification that it was safe to hand over the operation of nuclear power plants to personnel who had experience of running thermal power stations, but no education in nuclear science. For instance, the General Director of Chernobyl NPP, who was in place from the digging of the foundations of the plant in 1970, had training and experience as a turbine specialist and had worked on a coal power station before his appointment at Chernobyl. On one occasion before the accident, the director vividly revealed his wishful thinking about the harmlessness of nuclear reactors: "*What are you worried about, the nuclear reactor is a samovar [a traditional Russian pot used to heat water for tea]. It is much easier [to operate] than a thermal station, and we have experienced personnel – and nothing will happen*".[224] The background of the deputy of the Ministry of Energy and Electrification and department head, who was responsible for the construction of all nuclear plants in USSR, was in the building of hydropower plants. The head of the Ministry's unit, who was responsible for the exploitation of all nuclear plants in USSR, was a former executive of the State Planning Commission without any experience in nuclear industry.[225]

In addition, the plant's Chief Engineer was an electrician, who had worked previously on thermal power stations and the national electric grid. The Chief Engineer latter played a critical part in developing the plan for the experiment with the emergency power supply system on Reactor #4 that led to the disaster. This experiment was necessary because the emergency operating modes had not been properly tested prior to the RBMK being rolled out across the USSR.[226] It involved testing the potential rotational energy of the turbine during emergency shutdown of the reactor, in order to produce emergency electrical power for the water pumps (for a duration around 1 min) until the emergency diesel generators on the plant could be started up to full capacity.[227] The experiment was stipulated in the development project for the reactor, but the detailed steps involved were left to the personnel of each plant. In the case of Reactor #4, the plan for the experiment was not submitted to the developers because there was no requirement to obtain their approval for experiments.[228] The Chief Engineer of the plant—with no experience of running nuclear plants and little understanding of the risks involved in such a test to be performed on an RBMK[229]—decided to conduct it as a routine

[224]Valery Legasov Record from cassettes, 1986–1988.

[225]Grigori Medvedev, Chernobyl Notebook, New World Magazine, № 6, 1989.

[226]The Chernobyl Accident: Updating of INSAG-1, IAEA Publications, Vienna, 1992, p. 51.

[227]Günter Kessler, Anke Veser, Franz-Hermann Schlüter, Wolfgang Raskob, Claudia Landman, Jürgen Päsler-Sauer, Safety Concepts of Light Water Reactors, Springer, p. 173.

[228]The Chernobyl Accident: Updating of INSAG-1, IAEA Publications, Vienna, 1992, p. 52.

[229]"*The causes of the accident lie not in the programme [of the experiment] as such, but in the ignorance on the part of the programme developers of the characteristics of the behavior of the RBMK-1000 reactor under the planned operating conditions*", The Chernobyl Accident: Updating of INSAG-1, IAEA Publications, Vienna, 1992, p. 52.

electrotechnical test within a turbine-generator system, during a regular service break of Reactor #4 in April 1986, as part of compulsory measures stipulated by the reactor project.[230,231] Obviously, staff at the Ministry of Energy and Electrification was less competent in nuclear matters than the Minsredmash team who had developed the reactor, and did not have advanced knowledge of the physics involved.

Overall, there was a deficit of qualified personnel to run the constantly growing number of nuclear plants in the USSR. After the accident, Academician Alexandrov explained his position regarding the safety of the experiment: "*Nobody within our institute (the Kurchatov Institute) knew about the impending experience or participated in its preparation. Academician Dollezhal, Chief Design Engineer of the reactor, was also unaware of it. [Later, after the accident], when I was reading the plan of the experiment, I was shocked. Many actions of this plan led the reactor beyond its design state... Let's also ask who developed the plan. Executives of the NPP employed for this project – an organization that had no experience with the nuclear power plant [the organization was contractor of the Ministry of Energy and Electrification and specialized only in electric equipment]. Dilettantes can be well intentioned, but they could cause immense catastrophe – which happened at Chernobyl. The director of the station, without summoning even the deputy chief engineer of the plant who had education in physics, signed a contract with [the service contractor] to develop the plan of the experiment. The final version of the experiment was sent for consultation and testing by the Hydroproject Institute [the designers of the Chernobyl plant]... Members of the Institute, who had some experience with nuclear power plants, did not approve the plan and refused to endorse it... I often think now what would have happened if the Hydroproject Institute had informed us! However, the staff [of the Institute] could not even have imagined that the plant [staff] would dare to conduct the experiment. Minsredmash was not informed about the experiment because Chernobyl NPP had been transferred into the control of the Ministry of Energy... In Minsredmash ... were professionals with military-like discipline, who strictly followed the instructions, which in our case are extremely important... There are instructions, which must be followed by any NPP staff. This technical regulation is a guarantee of safety [of the plant]... [Furthermore], the experiment plan violated applicable instructions for operating nuclear power plants in twelve sections! We can say that the design of the reactor has flaws. However, the cause of the accident, after all – [was] a poorly prepared experiment, [in] flagrant violation of the instructions for NPP operations... I repeat, there are deficiencies in the reactor. Nowadays, these disadvantages are reduced. [Nevertheless], the problem is not the construction of the reactor. [Imagine] you are driving a car and turning the steering wheel in the wrong direction – and an accident takes place! Is it the fault of the engine? Or the designer of the car? Everyone will answer that it is the*

[230]Grigori Medvedev, Chernobyl Notebook, New World Magazine, № 6, 1989.

[231]Nikolaii Karpan, Vengeance of peaceful atom, Dnepropetrovsk, 2006, pp. 446, 451.

fault of the unskilled driver".[232] In 1999, Academician Dollezhal also stated his position: "*We left our chairs with Alexandrov [after the Chernobyl accident]. We, of course, are guilty [as developers of the reactor]. I have my version of the accident. First of all, the personnel was terrible; we were sending warnings in all instances but without results; we warned about the negligent regime of operation*".[233] Nikolay Fomin, Chief Engineer of Chernobyl nuclear power plant during the disaster, confirmed also his responsibility and that of his staff with respect to the accident: "*I was largely blamed. Not everything that has been said about me was fair, as I see it. Nevertheless, one thing I blame myself for: I have always believed that the key to the work of the nuclear industry is technology – but it turned out that the main thing is people. I underestimated their value*".[234]

Tragic Lack of Communication Between the Main Responsible Agencies

The situation was exacerbated by a total lack of communication about accidents between the military Minsredmash and the civil Ministry of Energy and Electrification, because of the culture of total secrecy[235,236,237] that developed within Soviet military nuclear programs during the Cold War. Consequently, neither the developers of the reactor nor Minsredmash officials informed personnel at other Soviet NPPs with RBMK reactors about the accident at the Leningrad NPP in 1975, or about technical imperfections of the reactor design.[238,239] Moreover, because developers did not eliminate defects revealed during the accident in 1975, the positive SCRAM effect within the RBMK series was observed again at Ignalina NPP, and during the launch of RBMK-1000 Reactor #4 at Chernobyl NPP in 1983.[240,241] The Chief Design Engineer for the RBMK reactors discussed the problem with his colleagues by correspondence,[242] stating that design changes would be made to correct the problem. But he made no such changes, and the

---

232Anatoly Dyatlov, Chernobyl. How it was, Nauchtekhlitizdat, Moscow, 2003, p. 134.

233Sergey Leskov, Smart guys, Vremya publisher, Moscow, 2011. p. 65.

234Mikhail Moshkin, Went down in the history by sentence, Moscow News, № 22, April 26, 2011.

235Anatoly Dyatlov, Chernobyl. How it was, Nauchtekhlitizdat, Moscow, 2003, p. 134.

236Anatoly Dyachenko, Experience of liquidation of Chernobyl disaster, Federal State Unitary Enterprise "Institute of Strategic Stability" of Rosatom, Moscow, 2004, http://www.iss-atom.ru/book-7/glav-2-3.htm.

237Unapprehended atom. Interview with Victor Bryukhanov, "Profile" Magazine, Moscow, № 29(477), 24.04.2006.

238The Chernobyl Accident: Updating of INSAG-1, IAEA Publications, Vienna, 1992, pp. 47–48.

239Unapprehended atom. Interview with Victor Bryukhanov, "Profile" Magazine, Moscow, № 29(477), 24.04.2006.

240The Chernobyl Accident: Updating of INSAG-1, IAEA Publications, Vienna, 1992, p. 43.

241Nikolaii Karpan, Vengeance of peaceful atom, Dnepropetrovsk, 2006, p. 290.

242Anatoly Dyatlov, Chernobyl. How it was, Nauchtekhlitizdat, Moscow, 2003, p. 136.

procedural measures he recommended for inclusion in plant operating instructions were not adopted.[243] At the Interdepartmental Science and Technology Council on Nuclear Power In December 1984, it was decided to postpone improvements of the RBMK—including the elimination of the positive SCRAM effect—for several years, until a period of planned reconstruction of the existing reactors.[244,245] Apparently, there was a widespread view that the conditions under which the effect would be important would never occur.[246] Personnel at the NPPs across the USSR were informed neither about these discussions within the development team, nor about near-miss cases of positive reactivity on other NPPs.

It important to ask the question: where were the regulators of the Soviet civil nuclear industry? Why did they allow 14 reactors with technical defects to go into operation? After the collapse of the Soviet Union, some supporters of Michael Gorbachev declassified secret shorthand records of Politburo meetings during the Chernobyl disaster. These records clearly show the complexity of the situation regarding the regulation of the developers of the RBMK, and the way shortcomings of the reactor were concealed to the Soviet government and the operators of the NPPs.

May 22, 1986. Mikhail Gorbachev declared: "*The Institute (the Kurchatov institute) was the only one [in the country] that was engaged in nuclear matters. It worked and nobody among us [the Politburo] knew what was going on. But it was only after Chernobyl that it was checked, 'exposed', and we saw a dangerous monopoly. The Director of the Institute, and President of the Academy of Sciences of the USSR [the collective Soviet body of advanced scientists from different fields, which could adequately assess any theoretical conclusions regarding the physics of the RBMK] is comrade academician Alexandrov in one person. He locked all things [regarding any criticism of nuclear matters] on himself… [There was] a 40-year friendly relationship [between the executives of the Kurchatov institute, NIKIET and Minsredmash] – and that's what happened*".[247]

June 5, 1986. Mikhail Gorbachev pronounced: "*[In this crisis situation, we have to eliminate the influence of] narrow departmental interests [when each ministry cares only about their field of responsibility at the expense of others]. Sometimes we hear [from ministers]: 'I have only my own object'. Everyone has its own object. No, we all have one object – Chernobyl!*".[248] Such non-coordination led to a situation where the Kurchatov institute, NIKIET,

[243]The Chernobyl Accident: Updating of INSAG-1, IAEA Publications, Vienna, 1992, pp. 44–45.

[244]Nikolaii Karpan, Vengeance of peaceful atom, Dnepropetrovsk, 2006, pp. 262, 398.

[245]The Chernobyl Accident: Updating of INSAG-1, IAEA Publications, Vienna, 1992, p. 31.

[246]Ibid, p. 13.

[247]Excerpts from transcripts of Politburo meetings concerning the Chernobyl disaster (April 1986–November 1989). The transcripts were published in the book "Within the Soviet Politburo… Records of Anatoly Chernyaev Vadim Medvedev, Georgy Shakhnazarov (1985–1991), The Gorbachev Foundation, Moscow, 2008, www.gorby.ru/userfiles/protokoly_politbyuro.pdf.

[248]Ibid.

Minsredmash and the Ministry of Energy and Electrification of the USSR did not transmit information about their problems or shortcomings, or critical information about the operation of the reactor to each other, and nobody understood the whole picture of risks associated with the RBMK; later there was similar non-coordination during response measures to the accident and Gorbachev called on ministers to enjoin them to cooperate.

The True Scale and Nature of the Faults Revealed in Meetings of Politburo in July 1986

July 3, 1986. Boris Sherbina (Deputy Chairman of the Soviet Council of Ministers—the Soviet government—and Head of the State Commission for liquidation of the consequences of the Chernobyl disaster) said: "*Evaluating the operational reliability of the RBMK reactor, a group of professionals working on behalf of the Commission concluded that its characteristics fall short of modern safety requirements... RBMK reactors are potentially dangerous... Apparently, all were under the impression that nuclear power plants were highly safe, as they were aggressively advertised to be... [Therefore] since 1983, the executive board of the Ministry of Energy and Electrification has never discussed the issues related to the safety of nuclear power plants... [Everybody] believed that the issue of civil nuclear safety was solved. A statement to this effect was in a Kurchatov Institute publication*".

Mikhail Gorbachev declared: "*Over the last 30 years, we have heard from you [scientists, experts, and ministers] that everything here [in the nuclear industry] is reliable. In addition you expect that we will look at you as to gods. From this, all went wrong [concerning the regulation of the Soviet nuclear industry]. It occurred because all ministries and research centers were out of control [of the Politburo and the Soviet government]. Finally it ended in failure... It was the responsibility of the staff [of Chernobyl NPP] that the accident took place [because the experiment on Reactor #4 was approved by executives of the plant, while it was not endorsed by the Science Director and the Chief Design Engineer of the reactor; operators chose to deviate from the program of the experiment; several instructions were violated during the experiment,* etc.*], nevertheless, the scale of the accident [was caused by] reactor physics [and is therefore the responsibility of the developers of the RBMK]... [The Politburo] did not receive information about what was happening in reality... All [nuclear-related] matters were classified and kept away from the reach of the Politburo. No representative of the [Communist] Party was allowed to meddle in this sphere. Moreover, [the Soviet] government had no power to determine which type of nuclear reactor [the country] should develop. Within the entire system [the nuclear energy industry], there was a spirit of servility, fawning, factionalism, persecution of dissidents [as in the case of Vladimir Volkov, the whistle blower from the Kurchatov Institute], window dressing, personal ties and different clans around different executives*".

A representative of the State Committee for supervision of the safe conduct in the nuclear industry of the USSR (Gosatomenergonadzor USSR) said: "*Everybody in the industry should be afraid of Gosatomnadzor!*[249] ... *It is impossible to ensure the complete safety of existing nuclear power units. However, if operators strictly follow standing orders and instructions [they could be operated safely]. With the approval of the reactor design, it was known that it would have 'positive void' and 'positive temperature' effects... Nevertheless, [Gosatomenergonadzor] never checked and studied the shortcomings [of the reactor] concerning the 'physics' and the degree of danger*". The majority of the staff of Gosatomenergonadzor USSR during the Chernobyl disaster were former specialists from Minsredmash; obviously, they did not want to criticize former superiors, and thus the government oversight over the industry was not independent. In addition, academician Valery Legasov stated after the disaster that most of those at the Kurchatov Institute (part of Minsredmash) tried not to ask executives of Minsredmash embarrassing questions because they were receiving bonuses from the ministry. The general opinion was the following: "*If I say anything about [the necessity of] the containment [vessels at Soviet NPPs], obviously, I will not receive a premium [from the Ministry]! If I express anything [against the mainstream within the industry and the opinion of the management], I will not be published and [my] dissertation will be not defended*"...[250]

... Mikhail Gorbachev [question to representatives of Minsredmash]: "*What can you say about the RBMK reactor?*"

Alexander Meshkov *(First Deputy Minister of Minsredmash)*: "*It is a proven reactor. However it does not have containment. If [the staff of the NPPs] comply strictly with the instructions, then it is safe*".

Mikhail Gorbachev: "*Meshkov continues to assure us that the reactor is safe... So, is it [still] possible to operate them and to construct more? ... All that we collected about Chernobyl by this time leads to one conclusion – the reactor should be decommissioned. It is dangerous. And you [to Meshkov] defend esprit de corps*".

Alexander Meshkov: "*No, I am advocating nuclear power*".

... Mikhail Gorbachev: "*What should be done by the Kurchatov Institute*?

---

[249]In IAEA's INSAG-7 report was noted that "*The USSR State Committee for the Supervision of Nuclear Power Safety was established only three years before the Chernobyl accident and, notwithstanding the safety culture concept, it could not be regarded as an independent body, since it was part of the same state authorities responsible for the construction of nuclear power plants and electricity generation. ... However, since the regulatory bodies have no legal basis, no economic methods of control, and no human and financial resources, and since it is very difficult to set up an institute of independent experts in this country, the system that existed and still exists is one consisting of many links providing step by step control and finicky supervision of nuclear power plants, rather than a full blooded regulatory system for the safe use of nuclear energy in the interests of the whole population*" [The Chernobyl Accident: Updating of INSAG-1, IAEA Publications, Vienna, 1992, p. 88].

[250]Valery Legasov, Record from cassettes, 1986–1988.

Anatoly Alexandrov (Science director of RBMK, director of the Kurchatov Institute, president of the Academy of Sciences of the USSR): "*I'm sure that we will not build RBMK reactors in future. Concerning improvement [of the RBMK series], the costs will not be compensated. [Nevertheless], I think that [positive reactivity] of the reactor can be eliminated [on existing reactors]. We have some ideas about possible solutions to this problem. This could be done in one or two years... The existing reactors can be made safe. I put my head on the block... that they can be improved. I beg you to release me from the duties of President of the Academy of Sciences and give me a chance to correct my mistake about the shortcomings of this reactor*".

... Mikhail Gorbachev speaks to representatives of Minsredmash: "*The reactor is unreliable. The reactor was transferred to the industry and [further] theoretical studies [of physics of the reactor] were suspended. Why was theoretical research not continued? ... Academician Alexandrov confirmed it [the unreliability of the reactor]. [Moreover] he missed something. He is [taking his mistakes] seriously, although he bears great responsibility for it [the defects of the reactor]. Whereas Meshkov lumps all accusations onto the operators [of Chernobyl NPP]*".

Alexander Meshkov: ... "*[However, we] will not allow one [RBMK] reactor to be built every year [to solve urgent domestic energy needs]. This involves – constant rush. Consequently, it leads to [low] quality of the equipment and [poor] safety measures*". ...

... Gennady Shasharin (First Deputy Minister of Energy and Electrification): "*The ministry believed that Chernobyl NPP was exemplary.*[251] *We appointed the best director to the plant ... [The main problem was] that we began to interact with nuclear energy on a first-name basis, but it [nuclear energy] requires respect. The staff [of Chernobyl NPP] did not know that the reactor can accelerate. Moreover, we [the Ministry of Energy and Electrification] did not know [about the 'positive SCRAM effect']. The staff [of Chernobyl NPP] is responsible for the accident. Nevertheless, I agree that the scale of the accident was caused by the physics of the reactor... Obviously, the first stages of Smolensk NPP, Kursk NPP and two reactors on Leningrad NPP should be closed. They are not subject for reconstruction... It is possible to get some units on these NPPs into shape. However, it will take one year. In addition, it will be very expensive*".

Mikhail Gorbachev: "*The statement of Shasharin about [immediate] decommissioning of [RBMK] NPPs is not serious*".

... Mikhail Solomentsev (member of Politburo) speaks to Gennady Shasharin: "*Did you know that the reactor was unreliable?*"

---

[251]In 1985, the plant was called best nuclear plant within the Ministry of Energy and Electrification of USSR [Nikolaii Karpan, Vengeance of peaceful atom, Dnepropetrovsk, 2006, pp. 410, 487]. In 1992 IAEA's report, it was mentioned that "*As a whole, the Chernobyl personnel in 1986 were characterized as a fairly typical, mature and stable group of specialists with qualifications regarded in the USSR as satisfactory. They were no better, but no worse, than the personnel at other nuclear plants*" [The Chernobyl Accident: Updating of INSAG-1, IAEA Publications, Vienna, 1992, p. 31].

Gennady Shasharin: "*Yes, I did. But it was not acknowledged on paper. There was a lot of resistance. Alexandrov was against it. The Academy of Sciences too. Minsredmash required [the Ministry of Energy and Electrification] to increase the production of energy in nuclear power plants by 2000*"...

... Victor Bryukhanov (Director of Chernobyl NPP): "*We did not know that something similar had happened at the Leningrad nuclear power plant in 1975*".

... Mikhail Gorbachev: "*The director of the plant ... was sure that nothing could happen [with an RBMK reactor] ... The Chief Engineer [of Chernobyl NPP] is an electrician. [His] main concern is supplying more electricity*"...

... Anatoly Mayorets (Minister of Energy and Electrification): "*This reactor does not ... and will never meet safety regulations even under ideal [conditions]. Sooner or later, it [catastrophe] could happen. Alexandrov says that [the RBMK] can be modified. In the meantime, what do we have to do [about electricity production for the Soviet economy]? ... It is necessary to bring all matters relating to the nuclear power plants together into one ministry. Moreover, we need to implement paramilitary discipline [within the new ministry]!*"

Vladimir Dolgikh (Soviet-Russian political figure and head of the Metallurgical Department of the Central Committee Secretariat): "*Radical reconstruction of the reactor makes it uneconomical. For many years, we were unaware what might happen. We stubbornly moved towards the accident. It was inevitable as a result of such behavior. A legend was created about the safety of nuclear power*".

Nikolay Ryzhkov (Chairman of the Soviet Council of Ministers: the equivalent of Prime Minister in Western countries and second in command within the Politburo after Mikhail Gorbachev): "*At the dawn of the nuclear industry, everything was conducted strictly and soundly. Gradually, the civil nuclear industry has gone beyond the boundaries [of Minsredmash], but [military] discipline 'has not fallen off' ... In addition, there was evidence of excessive authority in the hands of [the executives of Minsredmash] and Alexandrov. Things became less exacting and vigilant on all levels. After all, there was no single year without emergency situations at [Soviet nuclear] plants [according to data revealed at this meeting for the period 1981-1985, there were 1042 emergency showdowns among all nuclear reactors in the USSR, including 381 at RBMK reactors; at this time, there were 104 incidents at Chernobyl NPP]. There were no conclusions from the accident at Leningrad NPP. There were shortcomings ... and they were obscured, concealed in order to avoid publicity. Principalities took less responsibility. Without serious measures, we are not guaranteed against repetition [of the disaster]*".

Egor Ligachev (Politburo member): "*This is the lesson [about what happens when we have] a monopoly in science and production! We need to fully replace the structure of the nuclear industry. The current structure implies irresponsibility. [We are witnesses that] within the Ministry of Energy [and] at the Academy of Sciences, there has been extreme self-confidence*".

Mikhail Gorbachev: "*The accident could have been prevented. If there had been proper and timely information [about the features of the RBMK], then [the*

Politburo] could have taken action and we would have avoided this accident. However, we were faced with an extreme manifestation of irresponsibility".[252,253]

This meeting took place in July 1986; those present knew that a Soviet delegation had been invited to the International Atomic Energy Agency's Conference on Nuclear Power Performance and Safety in Vienna the following month, to present the main causes of the disaster to the international nuclear community. The Politburo discussed this problem too as follows.

Anatoly Mayorets (Minister of Energy and Electrification): "*Based on analysis of foreign sources, we can see that they have carried out a reconstruction of the accident at Chernobyl. So, do we submit a lie to IAEA [by attributing responsibility for the accident to the plant staff and denying the existence of defects in the reactor]?* "

... Valery Legasov (Deputy Director of the Kurchatov Institute and a key member of the government investigation committee of the Chernobyl disaster): "*The reactor does not meet safety requirements in critical areas*".

... Mikhail Gorbachev: "*We have suffered huge losses, not only economic, not only human. Huge political damage: there is some attempt to cast doubt on the level of our energy program. Throwing ideas around criticizing and dismissing the Soviet Union, Soviet science and technology, saying that our nuclear energy is ugly... In any case, we will not agree to ... hide the truth... [We have to] frankly inform the socialist countries, IAEA, the world public. All nations should be aware of the consequences of our actions and our response measures. Secrets will only bring losses for us. Openness will benefit us. We will lose if we do not say everything fully and clearly. Let's give the world as much information as possible [about the accident]. In any case, the actual situation is [already] known to the West*".[254,255]

It is also interesting to note that the information deficit was not solved by the layers of bureaucrats watching others, such as KGB agents. The opened KGB archives about the disaster confirm the existence of a lot of confusion among Politburo members due, as we have shown, to a lack of comprehensive situation assessment before and during the disaster from advanced Soviet intelligence, quite similarly to the situation during TMI—nobody in Washington understood the severity of the accident during the first few days. It is fair to conclude that intelligence services are powerless to mitigate industrial disasters probably because their goals are actually quite different: gathering risk information about technical

---

[252]Excerpts from transcripts of the meetings of Politburo concerning Chernobyl disaster (April 1986–November 1989). The transcripts was published in book "Within the Soviet Politburo... Records of Anatoly Chernyaev Vadim Medvedev, Georgy Shakhnazarov (1985–1991), The Gorbachev Foundation, Moscow, 2008, www.gorby.ru/userfiles/protokoly_politbyuro.pdf.

[253]Nikolaii Karpan, Vengeance of peaceful atom, Dnepropetrovsk, 2006, pp. 400–404.

[254]Excerpts from transcripts of the meetings of Politburo concerning Chernobyl disaster (April 1986–November 1989). The transcripts was published in book "Within the Soviet Politburo... Records of Anatoly Chernyaev Vadim Medvedev, Georgy Shakhnazarov (1985–1991), The Gorbachev Foundation, Moscow, 2008, ww.gorby.ru/userfiles/protokoly_politbyuro.pdf.

[255]Nikolaii Karpan, Vengeance of peaceful atom, Dnepropetrovsk, 2006, pp. 400–404.

problems of NPPs is a side job compared to antiterrorist surveillance. Other KGB reports and documents about the disaster are still classified and are likely to remain closed from public access until the decommissioning of the last RBMK reactor in 25–50 years.

Official Declarations Covering up the Truth

Nevertheless, in August 1986, Soviet officials headed by academician Valery Legasov declared to the IAEA that the major responsibility for the accident fell to the staff of Chernobyl NPP, not the developers of the RBMK. They even claimed the developers had informed the NPP operators about the positive SCRAM effect.[256] The official Soviet press also blamed mainly the operators, not the developers. In addition, there was a closed court hearing resulting in ten-year prison terms for key staff at the plant. In April 1988, on the second anniversary of the Chernobyl disaster, academician Valery Legasov committed suicide. Right before his suicide, he recorded on audio cassettes his confessions about previous concealment actions regarding flaws of RBMK and challenges concerning the development of the civil nuclear industry in USSR, excerpts of which we mentioned above.

After the collapse of the Soviet Union in 1991, the free press of Russia and Ukraine conducted several interviews with former personnel at the Chernobyl NPP, some of whom also published their own books. All of them asserted that they had not known about any technical shortcomings of RBMK reactors, and had believed that these reactors were absolutely safe. Moreover, they did not have any special instructions on how to handle RBMK reactors to avoid the positive SCRAM effect.[257,258,259] So, on April 26, 1986, when the operators of Chernobyl NPP pushed the emergency SCRAM button of Reactor #4 during the experiment, they were unaware of the existing technical shortcomings of this type of reactor and were convinced that an accident beyond the reactor design parameters could not happen.[260]

The truth of the matter was only revealed to the world community in 1992, when the IAEA published the INSAG-7 report—an updated version of their 1986 report, including new conclusions issued by the Soviet state committee for the supervision of safety in industry and nuclear power in 1991: "*The reactor designers were aware that the dangerous property of the reactor they had developed could be a cause of nuclear instability, but failed to estimate quantitatively its*

[256]The Chernobyl Accident: Updating of INSAG-1, IAEA Publications, Vienna, 1992, pp. 13, 30.

[257]Anatoly Dyatlov, Chernobyl. How it was, Nauchtekhlitizdat, Moscow, 2003, p. 102.

[258]Unapprehended atom. Interview with Victor Bryukhanov, "Profile" Magazine, Moscow, № 29(477), 24.04.2006.

[259]Vladimir Shunevich, Victor Bryuhanov: I was expelled from the party directly at a meeting of the Politburo, Fakty newspaper, Kiev, July 7,2012.

[260]Anatoly Dyachenko, Experience of liquidation of Chernobyl disaster, Federal State Unitary Enterprise "Institute of Strategic Stability" of Rosatom, Moscow, 2004, http://www.iss-atom.ru/book-7/glav-2-3.htm.

possible consequences and attempted to protect themselves by imposing operating limitations which, as it turned out, provided extremely poor protection... However, the defects identified in the reactor design and its unsatisfactory physical parameters have not been widely publicized among the scientific community and general public in the Soviet Union. They were also not included in the papers presented to the IAEA [in August 1986]... The design deficiencies and instability of the physical and thermal-hydraulic characteristics of the RBMK-1000 reactor had been theoretically and experimentally determined prior to the accident on 26 April 1986. However, no adequate remedial action was taken, firstly, to eliminate the defects and, secondly, to warn the personnel about the consequences of these dangerous characteristics and to provide them with appropriate training in the operation of the reactor, the parameters of which did not comply with the requirements of the technical documentation standards. The designers and authors of the standard operating procedures for the RBMK-1000 reactor did not inform the personnel about the very real danger of a number of reactor characteristics if certain possible personnel actions (including erroneous ones) were taken, because they failed to understand the possible cost of the consequences of personnel actions in operating such a reactor... The personnel violated the Operating Procedures... Some of these violations did not affect the initiation and development of the accident, others created favorable conditions for the manifestation of the negative design characteristics of the RBMK-1000 reactor. The violations were largely the result of the poor quality of the operating documentation and its contradictory nature caused by the poor quality of the RBMK-1000 reactor design".[261]

It is noteworthy that these imperfections in the design of RBMK reactors were eliminated shortly after the accident[262,263] and, for more than two decades, a dozen RBMK units have been operating without severe accidents in the ex-Soviet Union countries until their gradual decommission. After Chernobyl, only one new RBMK reactor was commissioned—at Smolensk NPP in 1990—while the remaining eight RBMK units under construction were cancelled. Nowadays, RBMKs represent only 3.4 % of all operating nuclear power plants in the world.[264]

2.1.5.2 Risk Concealment After the Disaster

Information Misrepresentation Due to Disbelief in the Improbable

Only many months after the disaster was the whole picture concerning operating risks of RBMK reactors recognized by all parties involved in the Soviet nuclear

---

[261]The Chernobyl Accident: Updating of INSAG-1, IAEA Publications, Vienna, 1992, pp. 31, 48, 87.

[262]Ibid, pp. 27, 48–49.

[263]Mikhail Moshkin, "Evgeny Adamov: 'The repetition of the Chernobyl scenario is impossible'", Moscow News, № 22, April 26, 2011.

[264]Günter Kessler, Anke Veser, Franz-Hermann Schlüter, Wolfgang Raskob, Claudia Landman, Jürgen Päsler-Sauer, Safety Concepts of Light Water Reactors, Springer, p. 3.

industry. Before that time, many of the people responsible were working without a full understanding of the real situation, which led to huge mistakes in their response in the first critical days and weeks after the accident. Immediately after the explosion, personnel at the plant did not believe that the worst-case scenario had actually happened. During the trial of the plant personnel who had conducted the experiment on that tragic night, they shared their perception of what happened in the first hours after the accident: "*Everyone was shocked... Total shock... Frankly speaking, I still believed that something had happened with the turbine... In spite of the night and poor lighting, it was clear enough. The roof and two walls of the reactor building were gone... This was Hiroshima... Walking around the reactor building, it became clear to me that the reactor was wrecked... [As Deputy Chief Engineer of the plant, with huge experience of running military and civil nuclear reactors], I'm probably guilty in that in my haste I did not explain to anyone [to the reactor operators in the control room, to the Chief Engineer and the Director] that the reactor had perished and that cooling [by pumping water into the destroyed reactor] was not necessary... I have been running uranium graphite reactors for 34 years, but never, never, have I known them to explode*".[265]

Several hours after the accident occurred, the Director of Chernobyl NPP arrived at the plant and saw that there had been an explosion, which had blown off the roof of Reactor #4 and caused a fire that was still burning. Obviously, the reactor was no longer there. Nevertheless, the director was given assurances from the operators who had been running the reactor during the accident that it was not damaged. He could not immediately verify this statement[266] and, during the first few hours, felt obliged to send encouraging reports to his superiors in Kiev and Moscow: "*The reactor is intact, continuing to pump water into the reactor, the radiation level is within the normal range*".[267] One hour after the accident, the chief of the Civil Defense Service reported to the director that radiation levels near the plant were 80,000 times the maximum acceptable level. However, the plant director did not believe him and ordered his arrest for spreading rumors and causing panic.[268] He continued to send reports to his superiors giving understated levels of radioactivity at the plant.[269] This misinformation of the authorities delayed

[265]Nikolaii Karpan, Vengeance of peaceful atom, Dnepropetrovsk, 2006, pp. 339–342.

[266]Seven hours after the accident (at 10:00 a.m. on 26 April 1986), one of the engineers of Chernobyl NPP explored the reactor room and found out that the reactor was demolished, but the Director of the plant did not believe his statement. It took a helicopter ride 12 h after the accident (around 3:00 p.m. on 26 April 1986) to establish the fact that Reactor #4 was destroyed and was throwing out radioactive material into the atmosphere (Alexandr Borovoy, Evgeny Velihov. Experience of Chernobyl, National Research Center "Kurchatovsky Institute", Moscow, 2012, p. 11).

[267]Alexandr Borovoy, Evgeny Velihov. Experience of Chernobyl, National Research Center "Kurchatovsky Institute", Moscow, 2012, p. 11; Grigori Medvedev, Chernobyl Notebook, New World Magazine, № 6, 1989.

[268]Documentary "Chernobyl. Chronicle of silence", Director: Irina Larina, 2006.

[269]Nikolaii Karpan, Vengeance of peaceful atom, Dnepropetrovsk, 2006, p. 427.

by more than 36 h the evacuation of the residents of Pripyat, a town of 47,000 inhabitants located near the nuclear power plant.[270] Twenty-five years later, Mikhail Gorbachev revealed: "*Regarding information [about real condition of damaged Chernobyl NPP], we tried to obtain it [immediately after the accident], but we could not do it. Even people, who, I believe, were honest and open with me (Academician Velikhov and many other young smart, intelligent, energetic academics) could not initially assess what happened [on the plant]*".[271] And because top-level managers in the Soviet central government were deceived by false reassurances from the plant, they did not understand the real scale of the accident, and could not take appropriate managerial decisions. The government commission, which arrived at the site of Chernobyl NPP from Moscow in the late evening on the day of the disaster, found that managers and all services of plants were demoralized. Later, the plant director said: "*People have been doing this [misrepresentation] with no malice. This was the practice within the industry: nothing bad to report. We always had to say – everything is going well*".[272] Ukrainian writer Boris Oleynik characterized the communication between industry executives and Communist Party officials in the following way: "*Talk more and prettier, in order to please your boss. You can act as you think best, even if your work is the polar opposite of your words*".[273] In a crisis situation, this common practice of passing on completely inaccurate information led, as already mentioned, to a critical delay in the evacuation of a city of some forty seven thousand people.

The behavior of the operators and of management of Chernobyl NPP—their certainty that a worst-case scenario was impossible, their unwillingness to believe extreme instrument readings, their misjudged if not consciously misleading status reports to superiors during the first and most important hours after the disaster—all these resemble the actions of the staff at the TMI-2 reactor of Three Mile Island in the US, seven years before Chernobyl. Unfortunately, openly published accounts of what happened at TMI-2—including the recommendations of protagonists and regulators regarding human factors, managerial decisions in a crisis situation and problems with timely and accurate communication about the plant status—were obviously not accessible to common Soviet nuclear specialists.

After the disaster, a middle manager at the Soviet Ministry of Energy and Electrification in the early 1980s confirmed the lack of communication about risks within the Soviet nuclear industry and the impossibility of accessing accident reports from abroad: "*In those years, information about any accidents and*

[270]The accident occurred at 1:24 a.m. 26 April 1986, but evacuation was started only at 2:00 p.m. 27 April. Central government officials arrived in Chernobyl by the end of 26 April, and recognized that the real picture differed completely from the one in reports.

[271]Online-interview with Mikhail Gorbachev, BBC Russian Service, March 8, 2002, http://news.bbc.co.uk/hi/russian/talking_point/newsid_1861000/1861942.stm.

[272]Maria Vasil', "Victor Bryuchanov, former director of Chernobyl NPP: 'If they could find to me apropriate criminal article, I think, they will execute me", Fakty newspaper, Kiev, October 18, 2000.

[273]Boris Oleynik, Literaturnaya Gazeta, Moscow, № 39 (5105), September 24, 1986.

malfunctions at NPPs was strongly filtered by the Ministry: the publicity of such information was allowed only if superiors found it necessary to publish. I well remember a landmark event of those years – the American nuclear power plant accident at Three Mile Island on March 28, 1979, which inflicted the first serious blow to the nuclear industry and dispelled the illusion of the safety of NPPs... [Nobody among my colleagues] had complete information about this accident. Details about the accident in Pennsylvania were published only as a restricted memo. However, it was common practice: only senior management had access to negative information, while subordinates [had to be] satisfied with truncated information, which did not contradict the official point of view about the complete safety of nuclear power plants".[274]

Refusal to Learn from Western Experience and Soviet Arrogance

Vladimir Asmolov was a middle manager at the Kurchatov Institute at the time of the Chernobyl accident (he is now an executive at Rosenergoatom, the operator of all Russian NPPs). He recalls: "*[We presented to superiors] the first documents about the structured protection [of Soviet NPPs] in 1982, but we received ... very simple feedback: 'In the West, they have oppression, capitalism, they do not think about the people and they have breakdowns in their reactors, as at Three Mile Island, for example. However, we have safe reactors, because they [the reactors] are Soviet!'*".[275] This arrogant dismissal of the experience of their colleagues from other industries and countries could naturally lead to the repetition of tough lessons. The case of Chernobyl clearly confirms this thesis—but unfortunately on a much bigger scale than Three Mile Island. On October 2, 1986, during a Politburo meeting, Mikhail Gorbachev accused Soviet nuclear scientists of not exchanging experience with international colleagues: "*It is a disgrace when we avoid participating in international scientific symposia about civil nuclear safety and we do not send our delegations to these for 10 years. [At these symposia], the experience of nuclear accidents occurring in the West has been summed up. What is it? Overconfidence, carelessness or the absence of a mechanism for regulating such participation? Chernobyl in this regard should provide a lesson – that people should not try to reinvent the wheel*".[276] Indeed, the preventive implementation of many recommendations from the TMI-2 accident could have reduced the magnitude of the disaster. If the Soviet nuclear industry had been able to take into account what was learnt in America—about the common practice of concealing minor shortcomings in the hardware, about the human factor during the operation

[274]Grigori Medvedev, Chernobyl Notebook, New World Magazine, №6, 1989.

[275]H. Seldon, Chernobyl in the mirror three generations, Druzhba Narodov, № 7, 2006, p. 169.

[276]Excerpts from transcripts of the meetings of Politburo concerning Chernobyl disaster (April 1986–November 1989). The transcripts was published in book "Within the Soviet Politburo... Records of Anatoly Chernyaev Vadim Medvedev, Georgy Shakhnazarov (1985–1991), The Gorbachev Foundation, Moscow, 2008, www.gorby.ru/userfiles/protokoly_politbyuro.pdf.

of reactors in emergency situations, and about the need to communicate correct information and organize timely evacuation in the first hours after an accident—the story of Chernobyl might have been very different.

In the days following the disaster, the Politburo could not get reliable information about the conditions at the plant and the possible consequences of the event in Chernobyl.[277] A commission of Soviet executive scientists was sent to the site of the accident but, at this stage, the real scale of the disaster was unclear.[278,279] Because of the lack of scientifically verified information about the accident in the first few days, the Politburo adopted a policy of keeping quiet and underplaying any possible threat. The central press issued a small statement in the evening of April 28, 1986: "*There was an incident at Chernobyl nuclear power plant. One of the reactors was damaged. Measures are being taken to eliminate the consequences of the incident. Necessary assistance is being given to victims. A government commission is investigating the accident*".[280] The IAEA was informed on the same day—only 60 h[281] after the disaster. Two days later, the Party leaders had the residents of Kiev—140 km from Chernobyl NPP—out on the streets for the First of May celebrations without any warning or precautionary measures. After the Soviet Union collapsed, documents were published demonstrating that the level of radiation on that day in Kiev exceeded normal background radiation levels by a factor of 125.[282]

Mismanagement of Information Communication to Soviet Citizens

The first official government statement to the people of the USSR about the situation at the plant was made by Michael Gorbachev on the central TV channel, a full 18 days after the accident.[283] Even at this stage, the scale of the accident was played down to avoid alarm. Meanwhile, the international media were in hysteria, publishing data from Scandinavian countries about the heightened level of radiation there without any information from the Soviet Union. Through various unofficial channels, information about the accident eventually began to trickle to Soviet

---

[277]Documentary "Chernobyl. Chronicle of silence", Director: Irina Larina, 2006.

[278]For example, it would take two years of intensive scientific research to draw the overall picture of the distribution of nuclear fuel lava in the reactor and nearly two decades to create a model of the flow disturbance—the formation and spread of nuclear lava (Alexandr Borovoy, Evgeny Velihov. Experience of Chernobyl, National Research Center "Kurchatovsky Institute", Moscow, 2012, p. 33).

[279]Documentary "The Battle of Chernobyl", Director: Thomas Johnson, 2006.

[280]ITAR-TASS announcement (April 28, 1986), BBC, http://www.bbc.co.uk/russian/specials/1550_chernobyl_chro/page9.shtml.

[281]Documentary "The Battle of Chernobyl", Director: Thomas Johnson, 2006.

[282]Alla Yaroshinskaya, Lies without borders, Rosbalt, May 8, 2013, http://www.rosbalt.ru/blogs/2013/05/08/1126494.html.

[283]Documentary "Chernobyl. Chronicle of silence", Director: Irina Larina, 2006.

citizens from abroad. This led to huge public disappointment in the ability of Mikhail Gorbachev and other communist leaders to deal adequately with the situation. Later, Mikhail Gorbachev stated that Chernobyl disaster was one of the triggers for the collapse of the Soviet Union.[284]

However, communist leaders still had a unique chance to convince their own citizens and the wider international community of their competence, power and ability to manage the country. In the first days after the accident, an operation unprecedented in world industrial history had been launched to liquidate the consequences of the Chernobyl disaster. For this mission, a group was established that included not only top scientists, but also military experts on chemical warfare, the air force, both military and civil engineers, civil defense forces, military logistics, controls and communications units, as well as coal miners to dig beneath the reactor and construct a core melt trap. Between 1986 and 1987, 240,000 military personnel and civilians took part in the liquidation. By the end of the operation, more than 600,000 people had been involved.[285] Finally, the emission of radioactive particles was halted and the temperature inside the reactor began to fall. By November 1986—only 7 months after the accident—a massive concrete cover known as the Sarcophagus had been constructed over Reactor #4, and most of the radioactive material released in the explosion had been collected and stored inside the reactors, or in special landfills.

Unfortunately, the strategy initially adopted by officials of misinforming their own citizens had created a situation where the Soviet population responded negatively and with distrust to any subsequent news about Chernobyl. As a result, the liquidation operation—which was one of the most remarkable and heroic achievements of the Soviet Union—was lost in the rhetoric of accusation. One devastating result of concealing the real information about the consequences of irradiation, and about the availability of effective antidotes, was psychological: a widespread phobia of radiation that seized thousands of Soviet citizens, leading to psychosomatic disorders, chronic heart disease and psychological problems. According to the Nuclear Safety Institute at the Russian Academy of Sciences, the death rate from such diseases was much higher than the death rate from radiation.[286]

---

[284]Mikhail Gorbachev, Turning Point at Chernobyl, Project Syndicate, April 14, 2006, http://www.project-syndicate.org/commentary/turning-point-at-chernobyl.

[285]Occupational Radiation Protection in Severe Accident Management. Interim Report, Organisation for Economic Co-operation and Development, Nuclear Energy Agency, January 7, 2014, pp. 65–66.

[286]25 years of the Chernobyl accident (1986–2011). Results and Prospects overcoming its consequences in Russia, Ministry for Civil Defense, Emergencies and Disaster Management of the Russian Federation, Moscow, 2011, p. 20.

Chernobyl Nuclear Disaster: Why Risks Were Concealed

- **Short-term profitability**, and the production of cheap nuclear energy in the Soviet Union, **took priority over the long-term resilience of the Soviet nuclear industry** and the protection of the environment.
- **A "rush culture"** was established by the Politburo in order to increase the speed of construction of nuclear power plants to meet urgent domestic energy needs. This environment of constant haste encouraged people to ignore possible measures to correct minor shortcomings of the reactor, which were perceived by developers as insignificant and unlikely to cause a serious problem in practice.
- The **nationalistic arrogance** of Soviet nuclear executives, and their **overconfidence** in the infallibility of Soviet nuclear technology.
- **Habituation/wishful thinking/self-suggestion/self-deception** among representatives of the Soviet civil nuclear industry about the minuscule probability of severe accidents at Soviet nuclear reactors: they totally **refused to believe that a serious disaster could happen**. As a result, the Politburo allowed the transfer of the control of nuclear power stations to the civil Ministry of Energy and Electrification, which was unprepared for such a complicated task.
- The Kurchatov institute, NIKIET, Minsredmash and the Ministry of Energy and Electrification **focused only on their narrow departmental interests,** which **prevented timely and adequate communication of risk information between different agencies.**
- **National security secrecy**. Before the accident, **operators at the plant did not receive any information about the accidents that had occurred previously** at other Soviet NPPs, or about international nuclear accidents.
- The developers of the RBMK reactor were **reluctant to confess their own mistakes** in the design of the reactor. They were **afraid of accusations of incompetence**. The prosecution of the plant's operators that would inevitably follow from such accusations was again **question of national security** because the developers of the RBMK reactor were members of the development team for the Soviet nuclear weapon. If they were found guilty, it would then cast doubts on the reliability of the Soviet civil nuclear program and of the Soviet nuclear weapons.
- It was common practice among Soviet bureaucrats to present themselves to superiors in the best possible light, which created an **organizational culture of "Success at Any Price" and "No Bad News" within the industry**. So the real defects of the RBMK were concealed from the Politburo - who also received misleading information about the real condition of the plant during the first hours after the accident, delaying the evacuation of the residents of Pripyat.

- The Politburo's delay in making any public announcement about the accident to the Soviet people and the international community was caused by the following factors: **uncertainty about the real scale of the disaster** in the first few weeks, the **absence of objective estimates of the possible consequences of the disaster**, and the **fear of panic** in the region of Kiev because, in the public perception, nuclear accidents and radiation constitute the most dangerous threats.

2.1.6 Exxon Valdez Oil Spill (USA, 1989)

On March 24, 1989 at 12:04 a.m., the oil tanker Exxon Valdez ran aground on Bligh Reef in Prince William Sound in Alaska (USA). The vessel was carrying approximately 1.2 million barrels of North Slope oil, which was loaded in port Valdez (40 km from the site of the accident). In the collision, eight of the ship's eleven cargo tanks were punctured, resulting in the leakage of around 250,000 barrels of oil during the first 3.5 h after the accident.[287] The total amount of leaked oil is estimated to be between 250,000 and 260,000 barrels.[288] The slow and inadequate response to the spill resulted in extensive oil contamination of 2000 km of pristine coastline on the Gulf of Alaska.

The main cause of the collision with Bligh Reef was the decision of Joe Hazelwood, the Exxon Valdez's captain, to deviate from the approved tanker route

[287]Final Report "SPILL: The wreck of the Exxon Valdez", Alaska Oil Spill Commission, State of Alaska, February 1990, p. 5.

[288]Ibid, p. iii.

in order to avoid colliding with small icebergs from the nearby Columbia glacier. However, the third mate failed to properly maneuver the ship and collided with Bligh Reef.[289] At the time of the accident, the captain may have been drunk[290] and the third mate was suffering from exhaustion.[291] The exceptional size of the oil spill was caused by the lack of oil spill response equipment and professional personnel at the Alyeska Pipeline Service Company, which was the responsible party in the first hours and days after any oil spill in the Valdez harbor and some areas of Prince William Sound. According to the Alyeska contingency plan, the oil spill response barge should have reached the area during the first 5 h after the accident but, in the case of Exxon Valdez, the spill response team arrived at the site only after 14 h.[292] The tanker was fully surrounded by containment booms only within 36 h after the accident. In the first 72 h, less than 3000 barrels of spilled oil were collected in spite of Alyeska's previous assurances that they could collect 100,000 barrels in less than in 48 h.[293] In addition, Alyeska, Exxon and the government of Alaska underestimated the possibility of a large oil spill in this area and, as a result, none of the parties had the essential amount of oil spill response equipment installed in the Valdez port area at the time of the accident. The clean-up operation in the summer of 1989 required 10,000 people, 1000 vessels, 38 oil skimmers and 72 aircraft. Over the 4 years following the accident, Exxon exerted huge efforts to clean the beaches of the Gulf of Alaska. Exxon's total expenses to deal with the consequences of the accident including penalties exceeded US$4.3 billion.[294]

2.1.6.1 Risk Concealment Before the Disaster

Pledge of Oil Companies to Protect the Pristine Environment of Alaska During the Exploitation of the Trans-Alaska Pipeline and Marine Oil Transportation

In 1968, the supergiant Prudhoe Bay Oil Field, the largest in North America with estimated resources of 25 billion barrels, was discovered on the Alaska North

---

[289]Exxon Valdez Oil Spill Trustee Council, Questions and Answers, http://www.evostc.state.ak.us/facts/qanda.cfm.

[290]Final Report "SPILL: The wreck of the Exxon Valdez", Alaska Oil Spill Commission, State of Alaska, February 1990, p. 13.

[291]Ibid, p. 11.

[292]Ibid, p. 17.

[293]Ibid, p. 49

H. Russel Holland, U.S. District Judge, Stanley Sporkin, U.S. District Judge, Exxon Valdez Oil Spill, Prince William Sound, Alaska, U.S. House of Representatives, April 8, 1991, p. 6 and Patrick Lee, Alaska Oil Co-op Accused of Plan to Ignore Spills: Energy: Rep. George Miller contends that before the Exxon Valdez disaster, the Alyeska pipeline consortium decided against cleaning up slicks. The group denies the charge, Los Angeles Times, April 10, 1991.

[294]Anne C. Mulkern, BP's oil spill bill could dwarf Exxon's Valdez tab, The New York Times, May 3, 2010.

Slope on the Arctic Ocean. The main problem facing oil companies exploiting this field would be the transportation of extracted oil. Due to the continuous ice and severe wind conditions of the Arctic Ocean, year-round oil export by icebreakers and oil tankers seemed risky and unpredictable. A trans-Canadian pipeline route was rejected because of U.S. energy security concerns. In the end, oil companies proposed the 1,287 km Trans-Alaska pipeline from the Arctic Ocean (Prudhoe Bay) to the Pacific Ocean, emerging at the year-round ice-free port of Valdez. For the construction and management of the Trans-Alaska Pipeline System (TAPS) pipeline, the Alyeska Pipeline Service Company (Alyeska) was established. Majority of shares in Alyeska were distributed between BP Pipelines (50.01 %), ARCO Pipeline Company (21.35 %) and Exxon Pipeline Company (20.34 %).[295] In the early 1970s, the idea of a Trans-Alaska pipeline met with opposition from environmentalists and Alaskan native tribes, but after the 1973 oil crisis, when oil prices dramatically increased from US$3 to US$12 per barrel, the pipeline development went ahead. The project was finished by July 1977. By the end of the 1980s, TAPS was carrying 25 % of US domestic oil production—around 2.2 MMbbl/day.[296]

Alyeska pledged to focus on the safe transportation of North Slope oil through the last American wilderness—and especially on the safe marine transportation from port Valdez through the pristine Prince William Sound. In March 1977, Alyeska promised regulators to deploy so many booms in the case of large oil spill that it "*would be like the Normandy invasion*".[297] The company demonstrated to officials the existence of booms stored along the U.S. West Coast (26.6 km of booms in total from sites spanning Prudhoe Bay to San Francisco) and assured that a sufficient number could be transported by air in case of emergency; however, response times were not indicated.[298] Alyeska's worst-case scenario was estimated at 200,000 barrels of oil spill, but the probability of such an event was calculated as once in every 241 years.[299] The company never believed the worst-case scenario could happen, declaring that the most likely spill volume for vessels operated by the Valdez terminal "*appears to be in the 1000 to 2000 barrel range*" and "*we feel Alyeska has adequately addressed the major issues*".[300] After the accident, Frank Iarossi, president of Exxon Shipping Co., said that "*there is no*

[295]Final Report "SPILL: The wreck of the Exxon Valdez", Alaska Oil Spill Commission, State of Alaska, Feb. 1990, p. 35.

[296]Oversight hearing before the subcommittee on Energy and Mineral Resources of the Committee on Natural Resources, U.S. House of Representatives, One Hundred Twelfth Congress, First Session, June 2, 2011.

[297]Patti Epler, Blueprint For Disaster: Despite Years of Warnings From Its Field Staffers About Alyeska's Poor Oil Spill Preparedness, The DEC Did Next To Nothing, Anchorage Daily News, Oct. 22, 1989, p. A1.

[298]Final Report "SPILL: The wreck of the Exxon Valdez", Alaska Oil Spill Commission, State of Alaska, Feb. 1990, p. 41.

[299]Ibid, p. 55.

[300]Ibid, p. 55.

doubt that all of these contingency plans and all of this planning and everything did not anticipate ever having to respond to a spill this big. I do not know why it didn't. I wasn't involved in the plan, but clearly no one ever anticipated trying to handle 250,000 barrels of oil on the water. This spill just overwhelmed everybody. No one was organized to control a spill of this magnitude".[301] With such optimistic risk assessment, Alyeska decided in 1977 that 5.5 km of oil spill booms were enough for Port Valdez.[302]

Weak State Government Control Over the Activities of the Main Taxpayer of Alaska's Budget

From the first months after the launch of the Valdez terminal, Alyeska neglected the maintenance of the oil spill response equipment and team. Thus, in December 1977, a representative of the Alaska Department of Environmental Conservation (ADEC) found out that 137 pieces of oil spill response equipment were broken or missing from a compulsory list of 170 items at the Valdez terminal.[303] Alyeska was the major taxpayer of the State of Alaska and had a strong influence on the State decision-making process for decades. In spite of a considerable increase in the budgets of both the State of Alaska and ADEC, which went from US$125 million in 1969 to US$2 billion in 1989 for the former, staff complained that ADEC's Prince William Sound District Office "*has been under-budgeted and under-staffed to adequately inspect the terminal and keep in touch with their day-to-day operations*".[304] Valdez oil terminal was just one of 93 onshore oil terminals under the oversight of ADEC. The regulator was responsible for more than 400 facilities (tankers, barges, and drilling platforms). In 1988, when ADEC asked for an additional US$0.5 million to hire inspectors to review contingency plans and inspect facilities, the department received only US$0.15 million.[305] Over many years, ADEC Prince William Sound District Office representatives pointed out problems at the terminal, including outdated oil spill recovery equipment, reduced training programs and questionable equipment reliability. They criticized the practice of staff-only drills, which showed that "*Alyeska's spill response activities have regressed to a dangerous level*". The reaction of the president of Alyeska to these revealed shortcomings illustrates well the real power of Alyeska within the State of Alaska: instead of making safety improvements, he changed the procedures for

---

[301]Exxon Shipping Company vs Grant Baker, The Supreme Court of the United States, Joint Appendix Volume Two, Oct. 29, 2007, p. 1237.

[302]Final Report "SPILL: The wreck of the Exxon Valdez", Alaska Oil Spill Commission, State of Alaska, Feb. 1990, p. 41.

[303]Ibid, p. 45.

[304]Final Report "SPILL: The wreck of the Exxon Valdez", Alaska Oil Spill Commission, State of Alaska, Feb. 1990, p. 50.

[305]Harold R. Linstone, Multiple Perspectives on the Alaska Oil Spill, Prepared for the Alaska Oil Spill Commission, Aug. 22, 1989, p. 22.

the access of ADEC to the terminal! He decided that ADEC officials should give "*preferably one day's notice*" before inspections, he refused to allow them to bring video cameras and he assigned a designated Alyeska representative "*who will accompany them at all times during their stay on the terminal to answer any questions or address any concerns they may have at the time*".[306]

This dangerous lack of maintenance continued for a decade. Thus, Jim Woodle, U.S. Coast Guard commander at the Marine Safety Office of the Port of Valdez, testified that the equipment was not in good shape. He said five booms were physically in the inventory—although the contingency plan required all booms to be situated on the oil spill barge for quick response. Moreover, Alyeska reduced the number of response team members from 18 to 10 or 8 people. Woodle asked to test all booms and inflate them, but the Alyeska oil spill response team said that "*they didn't have 1) the capability of activating all five at one time from the standpoint of manpower; 2) they weren't sure that three of them could operate. They basically kept two available for drill purposes, and the other three had never been used*". In 1984, Woodle wrote a letter to the president of Alyeska concerning oil spill recovery: "*Due to reduction in manning, age of equipment, limited training opportunities, and lack of experienced coordination personnel, serious doubt exists that Alyeska would be able to contain and clean-up effectively a medium or large size oil spill*".[307]

Concealment of Alyeska's Inability to Respond Adequately to Large Oil Spills

Apparently, the management of Alyeska did not react adequately to such information over the following few years. The reasons were as follows. Firstly, the statistics of oil spills in Prince William Sound made a large oil spill seem unlikely. For instance, from 1977 to 1989, 8700 oil tanker transits occurred, with only 400 small oil spills, the majority of which were located in the Port of Valdez during oil tanker loading.[308] The largest oil spill happened when the Thompson Pass oil tanker leaked 1700 barrels (150 times less then Exxon Valdez) within the terminal area.[309] Secondly, in 1982, the Exxon Company stated that "*for most tanker spills, the response plan outlined in the Alyeska plan will suffice. However, in the event of a major spill by an Exxon owned and operated vessel, it is anticipated that the Exxon Company, U.S.A. Oil Spill Response Team... would be activated to manage*

[306]Final Report "SPILL: The wreck of the Exxon Valdez", Alaska Oil Spill Commission, State of Alaska, Feb. 1990, p. 50.

[307]Ibid, p. 49.

[308]Patti Epler, Blueprint For Disaster: Despite Years of Warnings From Its Field Staffers About Alyeska's Poor Oil Spill Preparedness, The DEC Did Next To Nothing, Anchorage Daily News, Oct. 22, 1989, p. A1.

[309]Mary Evans, Effects of U. S. Coast Guard Enforcement Performance on Oil Tanker Safety, Prepared for the Alaska Oil Spill Commission, Dec. 4, 1989, p. 23.

the spill response".[310] Thirdly, the oil companies that owned Alyeska came to a mutual agreement that Alyeska's area of responsibility should be limited to "*Valdez Arm and Valdez Narrows only. Further efforts in the Price William Sound would be limited to use of dispersants and any additional efforts would be the responsibility of the spiller*".[311] However, the authorities, including ADEC representatives and the U.S. Coast Guard, were not informed about these internal organizational issues between Alyeska and its owners. This was a violation of Alyeska's agreements with both the United States and the State of Alaska. In 1973, in exchange for the right to build the Trans-Alaska pipeline on public lands, Alyeska signed an agreement where, in the section on oil spill contingency plans, they promised to control and clean up any oil spill: "*If during any phase of the construction, operation, maintenance or termination of the Pipeline, any oil or other pollutant should be discharged from the Pipeline System, the control and total removal, disposal and clearing up of such oil and other pollutant, wherever found shall be the responsibility of Permittees, regardless of fault... Full scale, company-wide field exercises will be held at least once per year to insure overall readiness for response to large scale oil spills... Alyeska will direct cleanup operations of spills resulting from operation involving tankers carrying or destined to carry crude oil transported though the Trans-Alaska Pipeline System, occurring at Valdez terminal, in Port Valdez, Valdez arm or Prince William sound*".[312] Because of the agreement, it was widely anticipated within Alaska that Alyeska would immediately react to any oil spill in the first days after an accident while the spiller was launching an independent oil spill response plan.

On 6–7 April 1988, ten months before the accident, Theo L. Polasek, vice president of operations of Alyeska, made a presentation to an internal operation subcommittee comprising representatives from BP Pipelines (Alaska), ARCO Pipeline Company, Exxon Pipeline Company and other companies participating in the Alyeska consortium. In the presentation, entitled "*Alyeska Response Capability to Spills at midpoint of Prince William Sound*", Polasek stated that an "*immediate, fast response to midpoint of Prince William Sound [is] not possible with [the] present equipment complement*".[313,314] One of the members of the operations subcommittee, who represented BP Pipelines (Alaska), proposed an "*acceptable compromise*", which stated that "*[the] current stockpile of clean up*

[310]Final Report "SPILL: The wreck of the Exxon Valdez", Alaska Oil Spill Commission, State of Alaska, Feb. 1990, p. 49.

[311]H. Russel Holland, U.S. District Judge, Stanley Sporkin, U.S. District Judge, Exxon Valdez Oil Spill, Prince William Sound, Alaska, U.S. House of Representatives, Apr. 8, 1991, p. 10.

[312]Ibid, pp. 3, 5.

[313]Ibid, p. 8.

[314]Patrick Lee, Alaska Oil Co-op Accused of Plan to Ignore Spills: Energy: Rep. George Miller contends that before the Exxon Valdez disaster, the Alyeska pipeline consortium decided against cleaning up slicks. The group denies the charge, Los Angeles Times, Apr. 10, 1991.

equipment is adequate" for spills at the terminal, but "*should be maintained to the highest state of readiness*". It was offered to use dispersants "*on a widespread basis*" for possible spills occurring in Prince William Sound.[315] Ultimately, Alyeska's proprietors secretly decided that Alyeska would not respond to an oil spill in Prince William Sound in the way prescribed in the contingency plan.[316] Apparently, information about the real condition of the Alyeska oil spill response team, and about the decision not to respond to an oil spill in Prince William Sound, was withheld not only from regulators and the government of Alaska, but even from management of oil companies included in the Alyeska consortium. For example, after the Exxon Valdez accident, Don Cornett, CEO of Exxon Company in Alaska and spokesman on the accident, said that he was not aware of any cutbacks in Alyeska's response team,[317] which suggests that he was probably also unaware of the inadequate amount of equipment for first response actions and of its poor condition. The same lack of awareness of the risks prevailed among local communities: the majority of them believed in Alyeska's promise, issued during discussions about the Trans-Alaska Pipeline System (TAPS) project, that "*the contingency plan which will be drawn up will detail methods for dealing promptly and effectively with any oil spill which may occur, so that its effect on the environment will be minimal... operations at Port Valdez and in Prince William Sound [will be] the safest in the [w]orld*".[318]

Nobody Among Oil Executives Understood the Whole Picture of Risks Associated with Transportation of Oil Through Alaskan Waters

There is no evidence that executives of Exxon Shipping Co. (the oil transportation subdivision of Exxon Company) were informed about the real level of Alyeska's response capability, either by Alyeska representatives or by the Exxon Pipeline Company. In the 1980s, Exxon Shipping Co. practiced the risky strategy of reducing the running costs of its oil tankers because the industry was in depression, with a third of the world's supertankers out of business. In the early 1980s, the government of Alaska tried to impose additional requirements on oil transportation, with wide-ranging authority over the design, equipment, navigation, operation, certification, inspection, financial responsibility, oil spill liability, cleanup capability and responsibility of oil tankers entering Alaskan waters.[319] However, in 1984, the

[315]H. Russel Holland, U.S. District Judge, Stanley Sporkin, U.S. District Judge, Exxon Valdez Oil Spill, Prince William Sound, Alaska, U.S. House of Representatives, Apr. 8, 1991, p. 9.

[316]Ibid, p. 9.

[317]Exxon Reduced Its Staff of Oil Spill Experts, AP, Mar. 30, 1989.

[318]Harold R. Linstone, Multiple Perspectives on the Alaska Oil Spill, Prepared for the Alaska Oil Spill Commission, August 22, 1989, p. 19.

[319]Final Report "SPILL: The wreck of the Exxon Valdez", Alaska Oil Spill Commission, State of Alaska, February 1990, p. 44.

industry organized a lawsuit (Chevron v. Hammond), which claimed that Alaska's new oil transportation laws and regulations were unconstitutional.[320] This enabled oil companies to reduce the cost of tankers and to continue using single-hulled tankers instead of safer but more expensive double-hulled tankers. Exxon Shipping Co. also cut down expenses on tanker staff. By 1989, the required number of crewmembers shrunk by a factor 18 as a result of the automation of oil tankers and their transformation to diesel propulsion.[321] Thus, on the Exxon Valdez in 1989, there were only 0.35 crewmembers per million gallons of oil. This meant that the tanker's crew had little time for rest, and felt constantly tired because of the excessive workload.[322] In addition, Exxon Shipping Co. did not implement the agreed system of health checks on all crewmembers before shifts, and made no effort to prevent alcohol and other restricted substances. After the accident, Exxon officials confirmed that they were aware that Captain Hazelwood had gone through a program of alcohol detoxification, but allowed him to command the Exxon Valdez nonetheless.[323] However, Exxon Shipping Co. emphasized that Captain Hazelwood concealed from his supervisors that he kept drinking while on duty. Exxon Valdez crew members, who later admitted drinking with Hazelwood aboard, knew they had violated company rules and had concealed it from the company's management.[324]

As a result, nobody within the Alyeska consortium of oil companies, the hierarchy of Alyeska itself, the Exxon Shipping Co., or the regulators of the State of Alaska understood the whole picture of risks, namely that oil was being shipped in tankers with chronically fatigued and in some cases alcoholic staff, in areas that were dangerously vulnerable to large oil spills because of a total lack of oil spill response equipment and reliable personnel.

Exxon Valdez Oil Spill: Why Risks Were Concealed

- **Short-term profitability won priority over the long-term sustainability** of the Trans-Alaska Pipeline System and over environmental protection.
- **Habituation/wishful thinking/overconfidence/self-suggestion/self-deception** among representatives of the Alyeska consortium about the low probability of a severe oil spill in Prince William Sound after more than

[320]Ibid, p. 45.

[321]Ibid, p. 11.

[322]National Transportation Safety Board, Safety Recommendation, Washington, D.C., September 18, 1990.

[323]O.C. Ferrell, John Fraedrich, The Wreck of the Exxon Valdez, Business Ethics: Ethical Decision Making and Cases, Third Edition, Boston: Houghton Mifflin, 1997.

[324]Grant Baker VS Exxon Mobil Corporation, Supreme Court of the United States, No. 07-276, p. 19.

a decade of intensive shipping of supertankers. This led the consortium being reluctant to admit the importance of readiness in the case of a large oil spill and to pay for a high-capacity oil spill response team.

- **Lack of consideration of scenarios** that could lead to large oil spills, such as a super-tanker collision: only past spills that had occurred were considered as representative of possible future events. This is well-known as historical sampling bias.
- **Cozy relationships between the Alyeska consortium and representatives of the State of Alaska**, who allowed Alyeska to exert a strong influence on state government decisions concerning the regulations of the consortium's activity, the funding of the state government environmental regulator (ADEC) or heeding its warnings. This helped the Alyeska consortium to conceal for years and with impunity the risks resulting from the inadequately prepared oil spill response.
- A **fragmented perception of risks** (i.e., the **absence of the whole picture of risks**) among decision-makers of the stakeholders led companies to resist revealing their own risks to members of the oil spill response team. Ultimately, nobody understood the risks existing in other involved organizations.
- A **permanent rush culture** among the crew of Exxon Valdez, because of unrealistic projections about the shipping schedule, which compelled the crew to conceal their chronic fatigue from employers. **Crew members were also afraid to lose their jobs** during the depression occurring in the oil supertanker market.

2.1.7 Ufa Train Disaster (USSR, 1989)

On the night of June 3–4, 1989, about 50 km from the city of Ufa in the Bashkiria region of the Ural Mountains, the Western Siberia/Ural/Volga natural gas liquids pipeline ruptured, causing the build-up of a potentially explosive hydrocarbon-air mixture. At 1:15 a.m., two passenger trains came into the zone of gas contamination, passing in opposite directions with a total of 37 railroad cars carrying 1284 passengers and 86 crew members. Apparently, a spark from a susceptor on one of the electric locomotives ignited the lethal gas mixture, causing an explosion in which 575 people perished and 623 were injured.[325] The explosion, equivalent to 300 tons of TNT, became the most deadly railway accident in the history of the Soviet Union and of the Russian Federation. The leader of the USSR, Mikhail Gorbachev, stated that this disaster was "*caused by mismanagement, irresponsibility, [and] disorganization*".[326]

---

[325]Alik Shakirov, In Bashkiria will hold a memorial ceremony dedicated to the 18th anniversary of the tragedy at the station Ulu Telyak, RIA Novosti, May 31, 2007.

[326]Bill Keller, 500 on 2 Trains Reported Killed By Soviet Gas Pipeline Explosion, The New York Times, June 5, 1989.

2.1.7.1 Risk Concealment Before the Disaster

Contrast Between the "Crude Oil Pipeline" of Official Documents Versus the "Natural Gas Liquids Pipeline" in Reality

In December 1980, the Minister of Petroleum of the USSR sent a letter to the Council of Ministers of the USSR, pointing out the serious deficit in raw materials for the Soviet petrochemical industry in the Volga region. He proposed reallocating a surplus of assorted petroleum gas from the gigantic oil fields of Western Siberia by constructing an 1852 km Western Siberia/Ural/Volga natural gas liquids (NGL) pipeline, to transport a mixture of methane, propane, butane and pentane. He also mentioned that, because there were as yet no rules and regulations for the proposal of such large NGL pipelines in the USSR, the design stage of the pipeline alone would take more than two years. To speed up the construction of the pipeline, he offered to use standard blueprints for oil and gas pipelines and begin construction immediately in parallel with the design of the pipeline.[327] We have already seen how the Soviet civil nuclear industry adopted the practice of simultaneously designing and building sophisticated technological facilities, leading to the tragic consequences revealed in 1986 at Chernobyl. In the case of Soviet petroleum, the consequences of this practice became clear in 1989.

Less than a month later, in January 1981, the Council of Ministers of the USSR issued permission to the Soviet Ministry of Petroleum for the construction of a "crude oil pipeline", while everybody in the industry was aware that this "crude oil pipeline" would in fact be the proposed NGL pipeline. The development was launched immediately. Because the Ministry had available the documentation for standard crude oil pipelines with a diameter of 720 mm, this diameter was chosen

[327]Determination of session of Supreme Court of USSR under the chairmanship of Judge V.I. Cherkasov, Dec. 26, 1991, pp. 8–11.

for the NGL pipeline. In making this decision, Ministry executives were ignoring the main condition for safe NGL transportation through pipelines, which requires that the diameter of such a pipeline should not exceed 400 mm including the bold wall of the pipe. At that time, nobody in the world was operating NGL pipelines wider than this diameter of 400 mm[328]: this requirement is dictated by the physical features of the hydrocarbon mix in NGLs, which reduces the temperature of the pipe. Moreover, to provide anticorrosive insulation, the constructors used polymer films designed for oil pipelines,[329] not suitable for NGL transportation characterized by continuous temperature changes. The combination of the thin wall of the pipe and of the temperature changes as the mixture flowed through it made the pipeline a highly dangerous structure. And in the case of a break, faster leakage of hydrocarbons from the wrecked segment of a wider-diameter pipeline could greatly increase the magnitude of the disaster. The experience obtained from safe operation of NGL pipelines showed that reducing the diameter of the pipeline to below 400 mm, and constructing several parallel pipelines with lower capacity in each line, constitute efficient preventive measures against giant leaks. During his testimony before the Supreme Court of the USSR, the chief engineer of the project admitted that, with the normal sequence of research, design and construction, the development of the pipeline would have taken 5 to 6 years. But in the design of this pipeline, there had been violations of the normal technological process: the blueprints were provided directly from the Ministry, the design of the pipeline was changed four times and all works were carried out in a rush in order to launch within four and a half years. Other witnesses also confirmed that there had been undue haste, compromising the quality of construction.[330]

By 1984, the "crude oil pipeline" was in the final stage of construction, and the Ministry of Petroleum proposed the urgent re-commissioning of the pipeline from oil to NLG.[331] Naturally, the USSR had a state regulatory body for the construction industry, responsible for checking project documentation to ensure the safety of buildings and prevent the development of facilities that would violate the rights of individuals or other organizations. But—based on existing construction norms and rules—the regulator rejected re-commissioning of the existing 720 mm pipeline. Obviously, when the pipeline had originally been proposed to transport oil, the pipe diameter and the route were judged to be acceptable; but the new project would involve transporting a far more flammable mixture through highly populated areas of the Ural Mountains. The safe transportation of NGLs through a 720 mm diameter pipeline demanded a total replacement of the type of pipes used. Nevertheless, the Soviet Ministry of Petroleum lobbied the Council of Ministers of the USSR to waive the requirement to assess the new project; and ultimately the

---

[328]A.Usoltsev, S. Shkaev Where will pipeline explode? Soviet Russia, October 17, 1990.

[329]Sergei Kudryashov, History of one disaster, Kommersant, №126, July 8, 1995.

[330]Determination of session of Supreme Court of USSR under the chairmanship of Judge V.I. Cherkasov, Dec. 26, 1991, pp. 8–11.

[331]Shamil Rahmatullin, Large pipe at the cost of life, The Chemical Journal, August 2011, pp. 36–38.

pipeline was re-commissioned according to the parameters mentioned above.[332,333] A French international expert in NGL pipeline construction warned Soviet petroleum officials at the design stage that the proposed pipeline would be dangerous to operate. This cautioning warning was ignored by Soviet petroleum representatives, and was not revealed to the Council of Ministers of the USSR, pipeline operators or railway representatives.[334]

Because of the tormented landscape of the Ural mountains, and in order to reduce costs and give easy access to maintenance using nearby transport infrastructure, the pipeline was constructed dangerously close to the railway: for 273 km, they were less than 1 km apart. Moreover, the pipeline crossed the bed of the railroad, which included the high traffic Trans-Siberian railway, in 14 places.[335] In this connection, another French expert warned construction team managers that heavy freight trains could generate intensive vibration and that the pipeline would thus require special joints to cope with the impact of this vibration on the pipes; but builders replied that "*all necessary safety measures are stipulated in the design of the pipeline in Moscow and that the joints are not necessary*"[336]

Massive Cost Reduction on Safety Matters During Construction and Exploitation of the Pipeline

Furthermore, in May 1984, executives of the Soviet Ministry of Petroleum canceled the installation of an automatic telemetry system for real-time control of possible leaks from the pipeline.[337] The Supreme Court enquiry did not find an adequate explanation of this decision, but some witnesses testified that there had been a shortage of investment, others about the lack of import equipment or service contractors qualified to install the system.[338] Nevertheless, regular helicopter sorties to check for possible high concentrations of methane in the atmosphere near the pipeline, and squads of trackmen with gas leakage detectors, worked effectively during the first years after the launch of the pipeline in October 1985.

---

[332]Alexey Skripov, Asha explosion. Why the largest in the history of the country's rail disaster occurred, Rossiyskaya Gazeta—Week—Ural, June 11, 2009.

[333]Shamil Rahmatullin, Large pipe at the cost of life, The Chemical Journal, Aug. 2011, pp. 36–38.

[334]Personal communication with executive representative of association of relatives of people, which were perished and injured in the accident.

[335]History of NGLs pipeline "Western Siberia—Ural—Volga", JSC "Yamal-Volga", http://yamal-povolzhye.ru/project-history.

[336]I sounded the alarm, Literaturnaya Gazeta, #24, June 14, 1989.

[337]The torch of death, 18 years ago there was an accident in Bashkortostan, which world did not face before, MediaKorSet (Ufa), Jun. 3, 2007.

[338]Minutes of session of the Supreme Court of the USSR under the chairmanship of Judge V.I. Cherkasov, Dec. 26, 1991, pp. 8–11.

Constructors of the pipeline were aware of the risks posed by possible NGL leakage to human habitation areas. Therefore, in September 1985, builders returned to one segment of the pipeline to construct a bypass around the village of Sredniy Kazayak, which was less than 1 km from the pipeline. According to the project schedule for the pipeline, the residents of the village should have been relocated elsewhere, but in 1985 the village was still inhabited. (There were 35 places along the pipeline where the pipes were very close to populated localities[339]). By the end of October 1985, the bypass was built and connected to the main pipeline by special valves. During the construction process, a powerful excavator caused considerable mechanical damage to the pipe close to the valves, which became the main cause of the NGL leakage in 1989. Moreover the soil in the area was rocky, but there was no special protection—like, for instance, a cushion of sand—where the vulnerable pipeline ran among rocks. In addition, nobody from the construction and maintenance crews checked the condition of the pipe—by ultrasonic scanning or even visual checks—before resuming the flow of hydrocarbons through the bypass section.[340] The pipeline operator was not aware of these hidden defects, but had a very low opinion of the quality of the pipeline construction and design in general: "*When in 1987 the pipeline was transferred from builders to us [the pipeline operator], we conducted an investigation of the condition of the pipeline and recognized that the pipeline was not fit for operation. We drew up a statement with remarks [to the Ministry], but nobody wanted to listen to us – we were forced to accept the transfer of the pipeline into full operating regime*".[341] The initiators and builders of the pipeline received government awards for developing the project in record time.

In the next four years, more than 50 incidents occurred over the whole length of the pipeline, fortunately with no casualties. Nevertheless, because of pressure from executives of the Soviet Ministry of Petroleum to keep costs down, the helicopter sorties stopped and the teams of trackmen with gas leakage detectors were disbanded. Until this point, up to 15–20 workers—on horseback because of the steep slopes of the Ural mountains—had been conducting regular monitoring of the pipeline, and the condition of the pump equipment, on the segment where the accident took place. But after the budget cuts, the maintenance team had to rely on information from locals about leaks on the pipeline! Four days before the disaster, the maintenance team published a warning in a local newspaper about possible leaks, with a request to be immediately informed about them if they were to be noticed by local inhabitants. The warning included the following: "*…before the arrival of representatives of the pipeline, there is necessity to cordon off [the*

[339]A. Usoltsev, S. Shkaev Where will pipeline explode? Soviet Russia, October 17, 1990.

[340]Alexey Skripov, Asha explosion. Why the largest in the history of the country's rail disaster occurred, Rossiyskaya Gazeta—Week—Ural, June 11, 2009.

[341]A. Usoltsev, S. Shkaev Where will pipeline explode? Soviet Russia, Oct. 17, 1990.

hazardous area] and prohibit the movement of equipment and people...".[342] But inexplicably, the administration of the pipeline did not inform railway officials or local rail traffic controllers of possible leaks in the area, or of the dismissal of the monitoring teams.[343] Railway representatives were of course aware of the existence of the pipeline near their lines, but they assumed that it was perfectly safe because originally it had been an oil pipeline.[344] They were not told about the lack of specialist equipment or regular surveillance by the NGL pipeline operators to detect leaks, but would presumably not have realized such measures were necessary.

Failure of Inter-organization Risk Transmission Led to Catastrophe

A critical difference between NGL pipelines and conventional oil and gas pipelines is the fact that the hydrocarbon mixture stays in a liquid state only if there is sufficient pressure within the pipeline – in this case at least 10 atmospheres. Below this pressure, it will revert to the more unstable gaseous state. So, for safe transportation, operators should use a pressure of around 84 atmospheres. But in this instance, the operators generally maintained only 36–38 atmospheres, because of fears that the pipeline would not sustain such high pressure due to the thin wall of the pipes, the low quality of construction and certain features of the NGL mix.[345] Several hours before the disaster, operators received a call from one of their NGL consumers, the Minnibaevsky Gas Processing Plant, that the plant had detected reduced pressure within the pipeline and that the NGL delivery rate had gone down. The pipeline control room was located 250 km from the wrecked segment—and as we have seen, the staff did not have the resources to immediately verify the pressure drop, because there was no telemetry system and the pipeline monitoring squads had been disbanded. To make matters worse, the conversation between the operator and the refinery took place just before a shift change in the pipeline control room, and the outgoing operator was in a hurry to catch the bus home; so all he said to the next operator was that the pressure had dropped and would need to be increased.[346] Because a constantly high pressure had to be maintained within the system, the new operator just turned up the NLG flow at the

[342]Advertisement page, Light of October [local newspaper, which was distributed within Iglinskiy district of Bashkiria, where the accident took place], № 65, May 30, 1989, p. 4.

[343]Igor Makarov, 575 dead, 623 injured … Can we forget about it? 20 years from the tragedy at Asha, ChelNovosti Information agency, June 6, 2009.

[344]Personal communication with executive representative of association of relatives of people, which were perished and injured in the accident.

[345]Personal communication with executive representative of association of relatives of people, which were perished and injured in the accident.

[346]Personal communication with executive representative of association of relatives of people, which were perished and injured in the accident.

nearest compressor station to get the pressure back to normal.[347] Reduced pressure in a given section of the pipeline was common practice, and a regular occurrence for the operators: usually the pressure had been cut intentionally because of maintenance works on the pipeline. So, on the night of the disaster, the operator simply assumed that the drop was not dangerous. The investigation after the accident concluded that a 1.7 m crack had developed in the pipeline only 20–40 min before the explosion, at the exact point of the bypass construction in 1985.[348] Nevertheless, some locomotive drivers and local residents later testified that they had noticed the smell of gas for 20–25 days before the explosion.[349,350] This would mean that the pipeline lost integrity at least three weeks before the disaster but, in the absence of regular monitoring, the leak was not identified. On the night of the disaster, when the operator increased the pressure in the system, it provoked a more serious rupture of the pipe at the already weakened joint—and the massive gas release that ensued caused the explosion.

A few hours before the disaster, the driver of a freight train informed a traffic controller that there were serious gas levels in that area, but the controller was unwilling to stop the trains. Obviously, several factors influenced the decision not to close the line despite gas warnings. Firstly, there had been no information about previous and potential leaks from the pipeline administration, and the railway dispatchers had no direct contact line with the pipeline control room. Secondly, the timing of the gas warnings in the night from Saturday to Sunday made it difficult to launch a prompt and detailed investigation of the cause of the gas smell by railway workers. Finally, because railways were the major means of transport for industrial goods and passengers alike in the Soviet Union, local traffic controllers had to follow an intensive train schedule—more than 100 trains a day in both directions; they had no authority to decide, without consulting supervisors, to halt a key section of the Trans-Siberian railway. In the hour before the accident, nine freight trains passed the contaminated section. Consequently, passenger trains were given the green light to enter what was to become the disaster zone.

2.1.7.2 Risk Concealment After the Disaster

After the disaster, the pipeline was finally shut down and abandoned. Remarkably, immediately after the disaster, the pipeline's designer issued a special order prohibiting the construction of NGL pipelines with diameters greater than 400 mm

[347]William R.Doerner, Communism: Soviet Union Hard Lessons and Unhappy Citizens, Time magazine, June 19, 1989.

[348]Minutes of session of the Supreme Court of the USSR under the chairmanship of Judge V.I. Cherkasov, Dec. 26, 1991, pp. 8–11.

[349]The torch of death, 18 years ago there was an accident in Bashkortostan, which world did not face before, MediaKorSet (Ufa), June 3, 2007.

[350]Alexey Skripov, Asha explosion. Why the largest in the history of the country's rail disaster occurred, Rossiyskaya Gazeta—Week—Ural, June 11, 2009.

and without a leakage telemetric system. And Soviet railway management issued permission for locomotive drivers and dispatchers to suspend traffic if gas contamination was suspected.

The investigation that followed the disaster was biased. From the beginning, prosecutors focused mainly on scapegoating among subordinates, especially the members of the bypass team who had carried out such a poor quality job in 1985. These were the executives of the Soviet Ministry of Petroleum who had ordered costs to be cut and safety rules to be violated at all stages of the project's development: lobbying for the construction of a dangerous piece of industrial infrastructure within highly populated areas; promoting an environment of total rush and the constant raising of productivity targets during construction of the pipeline, at the expense of construction quality and safety[351]; canceling the telemetry system, the helicopter surveillance and the manual monitoring of leaks; and neglecting to inform other organizations operating in the immediate vicinity about the absence of leakage detection equipment on the pipeline. Ministry executives were ultimately charged, but they were amnestied during the preliminary investigation because they were highly respected captains of industry—with numerous Soviet state awards (including for the fast construction of this pipeline) and merits. The investigation dragged on for six years, and the court hearing eventually took place after the collapse of the Soviet Union—when Russian society paid little attention to the lenient sentences handed down to the defendants, because there were too many other serious challenges facing the newly independent Russia.

Neither Soviet government officials nor petroleum executives published any investigation reports describing the causes of the disaster. This led to a situation where executives of the Russian oil, gas and petrochemical industries did not learn lessons out of it. Thus, more than twenty years after the disaster, Russian oil and petrochemical lobbies are proposing the construction of new NGL pipelines with a diameter of more than 400 mm, instead of the more expensive option of doubling the lines to keep the diameter below the 400 mm diameter known to be safer.[352,353] Moreover, one of the consortiums bidding to develop a NGL pipeline plans to transport a mixture with up to 27 % ethane—even though such a high proportion of ethane reduces the temperature of the NGL mixture and thus of the pipe walls to −66 °C. This makes the pipeline dangerously unreliable due to the fragility of metal at such extremely low temperatures.[354,355] In the absence of officially

[351]Minutes of session of the Supreme Court of the USSR under the chairmanship of Judge V.I. Cherkasov, Dec. 26, 1991, pp. 8–11.

[352]Shamil Rahmatullin, Large pipe at the cost of life, The Chemical Journal, Aug. 2011, pp. 36–38.

[353]Routes of NGLs pipeline "Western Siberia—Ural—Volga", JSC "Yamal-Volga", http://yamal-povolzhye.ru/products-pipeline-route.

[354]Shamil Rahmatullin, Large pipe at the cost of life, The Chemical Journal, Aug. 2011, pp. 36–38.

[355]Alexey Skripov, Asha explosion. Why the largest in the history of the country's rail disaster occurred, Rossiyskaya Gazeta—Week—Ural, 11 June, 2009.

recognized findings or recommendations from the inquiry after the Ufa disaster, and with inadequate implementation of that previous experience, a similar disaster could well occur again.

UFA Train Disaster: Why Risks Were Concealed

- **Short-term tasks** (timeline, productivity, carrier opportunities and awards) **took priority over long-term consequences** (quality of works, safety and reliability of the pipeline in the long term).
- There was a **rush work culture** prevailing during development and construction of the pipeline.
- **Executives of the Soviet Ministry of Petroleum were reluctant of to admit their own mistakes** during the redesigning and lobbying for the pipeline, or their negligence during its subsequent operation (long-term spending cuts on the maintenance of the pipeline; cancellation of the telemetry system for real-time monitoring of leaks; scrapping the helicopter and ground monitoring teams; poor quality of reconstruction works, and so on).
- The **lack or absence of communication between representatives of the pipeline, Soviet railways, and local residents**, in spite of the fact that the pipeline had 50 leakage incidents over 3 years and was constructed parallel to the railway for a length of more than 270 km. In addition, **nobody could imagine that such a catastrophic event could ever take place**.
- The fact that railway traffic controllers had **no authority** to preventively halt traffic on any section of the Trans-Siberian railway during the detailed investigation of the causes of the gas smell.

2.1.8 Sayano-Shushenskaya Hydropower Station Disaster (Russia, 2009)

The Sayano-Shushenskaya Hydropower Station (SSHPS), on the Yenisei River in south-central Siberia, is the largest hydroelectric power plant and the largest power producing facility in Russia in terms of its installed capacity (6400 MW). The station produces 2 % of all Russian electricity, and 15 % of the country's hydroelectricity. In 2009, the station was the sixth largest hydroelectric plant in the world, exceeded in average annual power generation only by Three Gorges in China, Itaipu in Brazil/Paraguay, Guri in Venezuela, Tucuruí in Brazil, and Churchill Falls in Canada. On August 17, 2009, the rotor of SSHPS Turbine 2 shot out. This flooded the turbine hall of the station, damaged nine of SSHPS's ten turbines and killed 75 station workers. After the disaster, the Minister of Emergency Situations for the Russian Federation evaluated the event as "*the biggest man-made emergency situation [in Russia] in the past 25 years [after Chernobyl] – for its scale of destruction, for the scale of losses it entails for our energy industry and our economy*".[356] Recovery costs after the accident came to over US$1.5 billion and the reconstruction of the station took more than 5 years.

2.1.8.1 Risk Concealment Before the Disaster

The Problems Coming from Simultaneous Design and Construction of Highly Sophisticated Energy Infrastructure (Common Soviet Practice, as We Have Seen in Previous Cases)

In 1962, one year after the Soviet Union launched the first human being into space, the Communist Party set an ambitious new goal for Soviet engineers—to construct the largest hydropower plant in the world at that time on the powerful Yenisei River in Siberia, to provide extremely cheap electricity for large non-ferrous metal plants. In 1963, an initial design for the plant was developed. In 1968, construction started on a unique arch-gravity dam, 245 m high. It was ten years before the first turbine of SSHPS began to generate electricity, and another ten years before the whole project was completed. The erection of the dam and the completion of the station took such a long time because of Siberia's extreme continental climate: during the winter months, temperatures can fall to −44 °C. These conditions required special treatments to enable the solidification of massive amounts of concrete, and innovative approaches to many other areas of design and construction, which had never been tested in other projects before the installation of SSHPS.

The design of this hydropower station proceeded in parallel with its construction in a manner similar to the design/construction of several of the major Soviet energy projects we have already seen: the first Soviet civil nuclear plant in Obninsk in the 1950s, the RBMK reactor series in the 1970s and the Western

[356] Joe P. Hasler, Investigating Russia's Biggest Dam Explosion: What Went Wrong, Popular Mechanics, February 2, 2010.

Siberia/Ural/Volga natural gas liquids pipeline in the 1980s. As in these cases, the motive of parallel design and construction was to accelerate the commissioning of SSHPS to solve an energy shortage for the heavily industrialized economy of the USSR: the cost of construction had to be minimized, and the USSR lacked the automated computational capabilities to calculate design solutions for such a complex technical project, so the developers had no choice but to test many of their ideas on real operating facilities.

In 2000, before a full-fledged operating permit was issued for SSHPS, a report noted: "*The essential disadvantages of organizing the construction [of the station] include the fact that the general scheme of construction was not finally adopted before construction was started and for [both] objective and subjective reasons, the station's design was changed in the middle of construction... [This] caused several negative consequences (incidents) which were eliminated during [the] operation [of the station in the 1990s]... The existing expertise for [the] construction [of hydropower stations in the USSR], coupled with a lack of funding, did not allow [developers to carry out] a full program of preparatory works [or] ensure the readiness of construction phases stipulated in the design, resulting in a significant lengthening of the construction period. The actual duration of the preparatory period was 12 years (against 5 years provided for the design), and the total duration of construction of the station – 27 years (against 9 years)*".[357]

Moreover, the turbines were commissioned while the construction of the dam body was still incomplete. For example, the Politburo and the State Planning Commission set a deadline for Turbine 1 to be commissioned by December 1978, despite the fact that the builders were behind schedule on the paving by 0.9 million m^3 of concrete: by the deadline, they had laid only 3.2 million m^3 while the design required 4.1 million m^3 to be in place before Turbine 1 could be fully commissioned. This decision to launch Turbine 1 before the dam was fully developed led to changes in the water flow circuit during flooding in 1979, as a result of which the passage of floodwater could not be fully controlled, and the station building and turbine warehouse were flooded. And the subsequent launching of other turbines with the body of the dam still incomplete led to cracking processes in the concrete of some of the dam pillars, and the decompaction of the bedrock foundation, resulting in increased water seepage through the body of the dam and partial degradation of the concrete in these zones. The consequences of these violations of the station's design were subsequently eliminated by the operating personnel when the station was running.[358]

In addition, during the construction of the station in the 1960s and 1970s, important safety features for the turbines were not included in the master plan. The absence of these elements predetermined the enormous scale of the accident in 2009 when, due to the failure of just one turbine, the other nine were flooded and damaged. For example, blueprints for the turbines had initially included penstock

[357]The act of technical investigation of the accident at the Sayano-Shushenskaya HPP, Rostechnadzor, October 3, 2009, pp. 32–33.

[358]The Ibid, pp. 32–33.

butterfly valves, which would shut off water flow through the turbines in case of an emergency. If the turbines at SSHPS had been fitted with such valves, the scale of the accident could have been limited to the destruction of Turbine 2 and water would not have uncontrollably flooded other turbines. However, penstock butterfly valves were eliminated from the station's master plans: it was simply beyond the limits of Soviet technology at that time to produce such valves on a scale gigantic enough for the tallest dam in the USSR, and strong enough to withstand the tremendous water pressures involved. Some other safety elements specified in the initial plans (e.g., a shore spillway) were also dispensed with during construction to cut costs and save time; but the projected parameters of power generation and load on the equipment—which had only been envisaged together with these safety elements—were not changed.

Finally, in 1988, the station got its permit for trial operation. Obtaining a fully-fledged operating permit was postponed because technical shortcomings had emerged in the design of the station, which would require further improvements to resolve. Then in 1991, the Soviet Union collapsed; for the next decade, the new Russian government, which now owned SSHPS, did not have the budget to invest in the station to eliminate these imperfections in the dam and the turbines. From 1988 until 2009, the station had no severe accidents; nevertheless, there were a significant number of minor turbine incidents, and minor breaches in the body of the dam when water seepage went beyond the design specifications.[359,360]

Lack of Communication About Minor Incidents in the Soviet Electro-Energetics Industry in the 1980s as a Key Cause of the Accident at SSHPS in 2009

On July 9, 1983, there was an incident with Turbine 1 at Nurek hydropower station, in what was then the Soviet Socialist Republic of Tajikistan. At 304 m high, the Nurek dam was the tallest in the world at that time. The radial vibration of the turbine bearing led to metal fatigue in the stud-bolts of the turbine cap, and 50 out of a total of 72 bolts finally broke off from the cap. Fortunately, the station staff quickly detected the water flowing out of the turbine shaft into the generator hall, and used the penstock butterfly valve to shut down Turbine 1 with no consequences to the other turbines. Despite the prompt and effective resolution of the problem, information about this incident was not widely distributed by the Soviet Ministry of Energy and Electrification among engineers and managers of other Soviet hydroelectric stations. The event was only mentioned in a classified annual

359 N.Vulfovich, L. Gordon, N. Stefanenko, Arch-gravity dam of Sayano-Shushenskaya HPP. Technical evaluation according to field observations, St. Petersburg, 2012, pp. 45, 53–55, 60.

360 Dissenting opinion of R.M. Haziahmetov (member of investigation commision of Rostechnadzor) regarding the Act of technical investigation of the accident at the Sayano-Shushenskaya HPP, Destruction of Tubine 2 of Sayano-Shushenskaya Hydropower Station: causes and lessons, Volume III, Hydrotechnical Construction, Moscow, 2013, p. 273.

review of accidents and other disturbances at Soviet power stations and electric networks for 1983.[361] This review was available to executives and special engineering staff of Soviet hydropower stations—although not to the public—but information about the incident was scant and the majority in the industry paid little attention to it.

After the incident, staff at Nurek hydropower station carried out obligatory six-monthly tests on the condition of stud-bolts by ultrasound; between 1983 and 2009, 154 stud-bolts on the turbine caps failed the test.[362] Nevertheless, no special ministerial requirement was issued to make this practice mandatory at other stations, including SSHPS.[363] There are several reasons why obligatory ultrasound testing of the stud-bolts on turbine caps was not implemented on all Soviet hydropower stations. Firstly, because Nurek Turbine 1—like all the turbines of the station—was fitted with a penstock butterfly valve, the water flow was cut off as soon as the turbine cap started to tear away from the body of the turbine; so the event remained a minor incident at one of many Soviet power stations, rather than a major nationwide disaster. But because the incident was perceived as relatively unimportant, nobody took it as a diagnostic of a more systemic problem. Secondly, the specialist engineers on any particular hydroelectric facility tend to see their site as absolutely unique: a unique master plan is drawn up for each station, taking into account the specific natural features of the area in which the station will be located, and consequently specific technical solutions will be proposed for the generation of electricity with a given dam height and river flow rate. Although hardware vendors also develop unique equipment for different hydroelectric stations, their experience shows that they often use generally the same technical solution for different stations. Nevertheless, hydropower engineering specialists hold to the belief that their site is unique. This assumption means there is generally far less communication about risks—and even actual incidents—between different hydroelectric station operators than between the operators of thermal power plants, which are built to a broadly unified master plan for a whole plant series. In the case of the Nurek hydropower station, neither the turbine manufacturer from Kharkov nor the Ministry of Energy and Electrification of the USSR could imagine that similar processes of turbine bearing vibration and turbine cap stud-bolt fatigue would also be observed on other equipment produced by other manufacturers; in the case of SSHPS, for example, the turbines were designed and manufactured in Leningrad. So, the Kharkov turbine manufacturers

---

[361]Rostehnadzor: the accident at the Sayano-Shushenskaya HPP is not unique, in 1983 was a similar situation at Nurek HPP, Interfax, Oct. 3, 2009; Review about accidents and other disturbances on power stations and electric networks of USSR energy system for 1983, Soyuztechenergo, Moscow, 1984.

[362]S. Pryganov, Analysis of possible accidents on hydropower stations and response measures, Destruction of Tubine 2 of Sayano-Shushenskaya Hydropower Station: causes and lessons, Volume I, Hydrotechnical Construction, Moscow, 2013, p. 327.

[363]Ivan Sliva, Sayano-Shushenskaya HPP: conclusions have been drawn, RusHydro Herald, #5, May 2011, p. 3.

made changes according to the repair checklist for its equipment; and when the Ministry were informed by Kharkov about the changes, they passed this information to other producers but did not require them to change their regulations on repairing their turbines. Finally, the operators at SSHPS received no information about the incident at Nurek beyond a brief paragraph in the report we have mentioned.

After the disintegration of the Soviet Union, risk information exchange between stations from different republics was dramatically reduced, because the responsibility for running the electric utilities of each republic was transferred from the Soviet Ministry of Energy and Electrification to the governments of the 15 republics.[364] Moreover, there was very little international collaboration between hydroelectric industries, because different countries used completely dissimilar equipment and the Soviet and Western energy systems operated in very different ways—so Russian specialists did not receive detailed information about an accident at Manitoba Hydro's Grand Rapids hydropower station in Canada in 1992, when turbine failure led to the flooding of the turbine building. The problem was subsequently traced to the stud-bolts of the turbine cap, which had apparently failed.[365] Russian hydroelectric specialists only became aware of the Canadian accident in 2011, during the investigation of the accident at SSHPS.

Soviet Electro-Energetics in the Post-Soviet Russian Market-Oriented Economy

During communist rule, a single technological electro-energetics complex called the Unified Energy System of the USSR was developed across the entire Soviet Union, which covered 12 time zones from the Pacific to the Atlantic. It enabled the transmission of electricity to be organized across the largest country in the world from more than 1000 power stations. A reliable electricity supply was ensured by coordinating the operations of all these stations within a single nationwide technological complex. The efficiency of the Unified Energy System was achieved by optimizing the modes of operation of different stations and by the construction of trunk transmission lines, which reduced production costs and ensured low tariffs for both the industrial and domestic sectors.

In 1992, a year after the collapse of the Soviet Union, a joint stock company was formed called the Unified Energy System of Russia (RAO UES), in which the Russian Federation—represented by the Federal Agency for Federal Property Management—had a majority of shares. RAO UES became the legal successor

[364] S. Pryganov, Analysis of possible accidents on hydropower stations and response measures, Destruction of Tubine 2 of Sayano-Shushenskaya Hydropower Station: causes and lessons, Volume I, Hydrotechnical Construction, Moscow, 2013, p. 331.

[365] Fabian Acker, Fatal failures: Siberia's hydro disaster, The Institution of Engineering and Technology Magazine, July 11, 2011.

on Russian Federation territory of the Unified Energy System of the USSR, taking control of 72 regional grids—which comprised 70 % of the country's installed electric capacity including the majority of its hydropower stations, 96 % of its high-voltage grids and over 70 % of its transmission lines. The only sites not included in RAO UES were those that had already been transferred to the control of regional authorities or privatized. Around a decade into the post-Soviet era, Russian electro-energetics was generally working well without severe accidents or massive blackouts, in spite of very tough economic conditions for industrial consumers of electricity during the country's dramatic transition from a planned economy to a market economy. At that time, the industry did not receive real money for generated electricity—only 15 % of the total revenue was in cash—and used a system of netting, bill obligations and barter. The absence of severe accidents on electric facilities during this period suggests that Soviet engineers in general had laid solid foundations for the resilience of the nationwide electrical system, even under extreme conditions, for decades into the future.

The new Russian liberal government that replaced the Politburo was strongly oriented towards financial efficiency. This led to a situation where financiers and managers—loyal to the new anticommunist government but with no experience in electro-energetics—became executive managers of RAO UES, and began to implement massive savings by cutting back on capital renewals and investment on safety. In the early 2000s, less than 25 % of RAO UES board members had qualifications in the field of electricity, and most were not competent to manage a potentially dangerous high-tech power generation grid like the Unified Energy System of Russia.[366] In addition, the Russian liberal government was constrained by the terms of the International Monetary Fund and the World Bank, which in exchange for loans required the reform of Russian monopolies in natural gas production, railways and electro-energetics.[367] International financial institutions expected the dissolution of former Soviet unified energy and transport complexes through the promotion of free-market relations between different parts of the complexes. With regard to electro-energetics, this involved the separation of the Unified Energy System of Russia into numerous mutually independent companies for the production, transmission, distribution and sale of electricity. This led to the disintegration of centralized supervisory control over Russian electro-energetics. Thus, the departments responsible for unified technical development, and industry-wide technical inspections of all electric power plants and infrastructure, were eliminated. Research work on maintenance and repair operations, and even the

[366]Platonov V.V., Russian electroenergetics: reform and development, Materials of public seminar "Economic Problems and Energy Complex", Institute of Economic Forecasting of the Russian Academy of Sciences, Dec. 18, 2009, pp. 4, 26.

[367]Platonov V.V., Russian electroenergetics: reform and development, Materials of public seminar "Economic Problems and Energy Complex", Institute of Economic Forecasting of the Russian Academy of Sciences, Dec. 18, 2009, p. 12.

development of specialized equipment, was stopped.[368] From 2003, the system for reporting on equipment reliability and on emergencies occurring on Russian electrical installations ceased to function. Professor Vasily Platonov analyzed the disintegration of the Unified Energy System of Russia between 1998 and 2005 and came to the conclusion that profits amounting to US$39.5 billion were not invested in modernization and repair of electric equipment, while cash accumulated in the accounts of the vast number of independent companies created during the reforms of Russian electro-energetics.[369] Consequently, decades of investment shortage and the elimination of professional staff from all levels of the industry resulted in severe accidents in several Russian regions: in the winter of 2000/2001, a massive black-out occurred in the Far East of Russia and Eastern Siberia; and in 2003, the Ural interregional accident took place.

In spite of the fact that unified Russian electro-energetics infrastructures were clearly deteriorating during these monopoly break-up reforms, an executive of RAO UES, during a discussion of the massive Northeast blackout in the United States and Canada in August 2003—when 50 million people were affected for a period of 44 h—arrogantly claimed: "*For us [Russians, who have a united and centralized electricity network] it is impossible. Obviously, Americans have systemic problems [with their nonintegrated national electricity network]*".[370,371] These statements were shown to be vacuous in May 2005, when the massive Central Russia blackout initiated 35-h power cuts for more than 6.5 million people—half of Moscow, Tula, Kaluga, Ryazan and other regions—resulting from an unskilled response by dispatchers to an accident at the Chagino substation. The substation was equipped with six high-to-low transformers, three of which were built in 1958 and had not been maintained adequately after the collapse of the Soviet Union. An investigation concluded that the equipment at the substation had badly deteriorated—90 % of the equipment was still in operation after replacement age.[372] Other causes of the blackout were attributed to the disintegration of the formerly unified Moscow region energy system. It had been divided into several separate units, each of which was involved in generating, transporting, distributing or selling electricity without strong and efficient coordination between its own dispatchers and those of other units. The Moscow blackout revealed the urgent need for massive investment in Russian electro-energetics, in order to compensate for decades of shortage of capital investment to replace the equipment. If Russia wanted fast economic growth in the mid 2000s, there was an urgent need to

[368]Victor Kudryavy, Systemic causes of accidents, Hydrotechnical Construction, Moscow, #2, 2013.

[369]Vasily Platonov, Analysis of development strategies and problems of reforming Russian electroenergetics, Novocherkassk, 2006.

[370]Eugene Arsyukhin, Who is responsible for the accident? Rossiyskaya Gazeta, May 27, 2005.

[371]Vasily Platonov, Electricity crisis in Russia on American maner, Industrial Vedomosti, 2005, № 4–5.

[372]Report about investigation of accident in RAO UES May 25, 2005, Joint Stock Company of Power And Electrification RAO "UES Of Russia", Moscow, June 2005.

get new electric capacity into operation and modernize the existing stations and infrastructure. President Vladimir Putin and his government agreed with the proposal of the liberal, pro-market senior management of RAO UES that they should continue to dismantle the system into generation, transportation, distribution and electricity sale units, which could compete with one another in a free market. The idea of dividing the formerly Unified Energy System of Russia into privatized sections was based on the assumption that these smaller units could be easily controlled and managed by invited domestic and foreign investors. The rationale was that these new private owners would invest billions in Russian electro-energetics, instead of the government having to find the budget for the whole Unified Energy System of Russia. However, it became clear from the outset that these investors were focused on getting short-term return on their investment by raising electricity prices, rather than on the public-service priorities of the Unified Energy System of the previous USSR—whose aim had been to provide a reliable electro-energetics network in a gigantic country with a tough climate, while keeping tariffs low for consumers.

By July 2008, the Unified Energy System of Russia had been dismantled. In its place were six wholesale thermal power generation companies; 14 regional thermal power generation companies; the hydropower giant RusHydro – which operated 53 hydropower stations, including SSHPS, and became the world's third largest hydroelectric power producer and the largest power-generating company in Russia; the Federal Grid Company; the System Operator of the Centralized Dispatching Administration; and other companies. The former chief engineer of RAO UES (1993–1996) and Deputy Minister of Energy of Russia (1996–2003) estimated that after the reorganization, the number of electricity sale companies quadrupled and exceeded 320; and the total number of power grid companies reached an astronomical level at around 3600.[373] Only about US$36 billion of private investment was attracted[374] in spite of forecasts by the management of RAO UES before the reorganization of a potential $79 billion of investment.[375] Between 2008 and 2012, only 16.1 GW of new capacity was installed, instead of the 21.8 GW that RAO UES management had estimated before reorganization.[376] During this period, the world economic crisis, climaxing with the collapse of Lehman Brothers investment bank and the bailout by the US government of AIG, the largest insurance company in the world, in September 2008,[377] led to a suspension of any significant private investments; and the Russian government through state-owned companies was forced to become once again the major investor in Russian electro-energetics after the apparent failure of RAO UES's reforms.

---

[373]Victor Kudryavy, Mister kilowatt, The Soviet Russia, Nov. 28, 2013.

[374]Chubais agrees with the Rostekhnadzor's investigation conclusion concerning accident at SSHHPP—comment the former head of RAO UES of Russia, Interfax, Oct. 3, 2009.

[375]Neil Buckle, Russia seeks $79 bn electricity funding, Financial Times, Sep. 4, 2006.

[376]Neil Buckle, Russia seeks $79 bn electricity funding, Financial Times, Sep. 4, 2006.

[377]Didier Sornette and Peter Cauwels, 1980–2008: The Illusion of the Perpetual Money Machine and what it bodes for the future, Risks 2, 103–131 (2014) (http://ssrn.com/abstract=2191509).

Ten years of reforms within the industry had seen electricity tariffs increase by a factor of ten, from US$0.01/kWh in 1998 to US$0.1/kWh in 2008, and the reliability of the nationwide power supply had gone down as a result of the disintegration of the unified system.[378] Moreover, electricity prices in Russia reached a level 1.5 times higher than those in China and the United States—which made several Russian industries, which had been dependent on cheap power supply, less competitive on the international market.[379] This rise in tariffs was influenced by the need for the industry to attract/recoup investment. Because most of the units of the former RAO UES were now public companies, their managers began to focus on maximizing profits and financial efficiency, instead of the reliability of the sites and infrastructure they were supposed to be running. These became the priorities for the partly government-owned RusHydro—and at SSHPS in particular, according to one member of the Russian parliamentary commission, which investigated the accident at SSHPS: "*The operation of the station was subordinated to the main task – to generate profit... Therefore, financiers and economists were the main force in RusHydro and, perhaps, they had influence or put pressure on engineering services. It is hard to explain in any other way, why in spite of the fact that the technical lifecycle of Turbine 2 was practically expired, a new turbine had not been ordered and a special plan for the safe operation of the obsolete turbine was not even developed*".[380]

Tragic Consequences of the Ultra-Liberal Reforms of RAO UES on the Safe Operation of SSHPS

SSHPS had been producing hydroelectric power on a provisional operating permit from 1988 until 2000, because of several technical shortcomings that the Soviet Ministry of Energy and Electrification expected to eliminate during the 1990s. Nevertheless, in the wake of the collapse of the Soviet Union and the complicated financial situation of RAO UES, nobody could afford to plough hundreds of millions into constructing the shore spillway for the dam at SSHPS or changing the station's still unperfected turbines. During the 1990s, only minor, inexpensive and urgent maintenance works were carried out. When RAO UES top management began to discuss the reorganization of the Unified Energy System, managers came to the conclusion that the largest hydropower station in Russia could not be recognized as an asset for a prospective public company like RusHydro, given its

---

[378] Platonov V.V., Russian electroenergetics: reform and development, Materials of public seminar "Economic Problems and Energy Complex", Institute of Economic Forecasting of the Russian Academy of Sciences, Dec. 18, 2009, p. 55.

[379] Platonov V.V., Russian electroenergetics: reform and development, Materials of public seminar "Economic Problems and Energy Complex", Institute of Economic Forecasting of the Russian Academy of Sciences, Dec. 18, 2009, p. 78.

[380] Victor Khamraev, Responsibility for the accident on was laid on principle, Kommersant, Dec. 22, 2009, 239 (4294).

provisional operating permit. Consequently in May 2000, RAO UES executives issued a fully-fledged operating permit for SSHPS despite the existence of unsolved technical shortcomings at the station in previous years. The forms accompanying the permit mentioned many shortcomings of the station, and the management of the station and RAO UES received recommendations regarding these imperfections. For instance, several problems with the turbines were emphasized: "*In the initial period of operation of [SSHPS], some design flaws were identified in several parts of the turbines [abnormal vibration of turbines during different operating regimes], which were partially eliminated by operating staff and manufacturers. Nowadays, works to improve the reliability of individual components of the turbines are continuing, in particular, the station's staff [are eliminating] cracks on the blades of the turbines... More than 20 years... after the start of electricity production at SSHPP, therefore, there is the necessity to replace obsolete equipment and facilities ([the automatic process control system], the turbine impellers,* [etc.]*)... After 50 thousand hours of operating time [of the turbines], the volume of repairs has increased significantly. Annually, such repairs are performed on 4–5 turbines [SSHPS has 10 turbines in total]; [such activity] requires large labor costs and an increase of turbine downtime due to repair... Replacement of turbines is required*".[381]

In spite of these recommendations for major corrections to the flaws of the turbines and other imperfections, there was little serious investment from RAO UES at the beginning of the 2000s. The focus of top managers was on demonstrating the financial efficiency of the assets in order to attract potential investors and raise capitalization value: for ten years from 1998 to 2008 the capitalization of RAO UES rose from US$12 to $50 billion, while the degree of asset deterioration on RAO UES sites increased from 50 % in 1999 to 59 % in 2006; this measure had been 43 % in 1995.[382,383,384] Increasing the flow of water in SSHPS's reservoir, instead of dumping water vainly through the spillway embedded in the body of the dam, made for greater profitability because it raised electricity production on the station. Furthermore, dumping water was even more dangerous for the body of the SSHPS dam, because the idle discharge of water could destroy the base of the tailrace of the dam due to the absence of a shore spillway at that time. Thus, in 2006, the station generated record electricity outputs due to the heightened water inflow to the Yenisei River. In 2006, the net profit of RusHydro was US$47 million, and it continued to rise in the following years: in 2008, it increased more than 12 times

[381]The act of technical investigation of the accident at the Sayano-Shushenskaya HPP, Rostechnadzor, Oct. 3, 2009, pp. 30–34.

[382]Profile of the company. History transformations, RAO "UES of Russia", http://www.rao-ees.ru/ru/info/history/show.cgi?prof.htm.

[383]Anatoly Kuzovkin, Assets of Russian energy complex: how much eaten, how much is left, Industrial Vedomosti, Sep. 2001, № 16–17 (27–28).

[384]Accounts Chamber of the Russian Federation: the degree of asset deterioration of RAO UES is 59 %, Finam, Dec. 17, 2007.

to $605 million.[385] On July 3, 2009—45 days before the disaster—the station issued a press release celebrating new record levels of electricity production due to a higher than average inflow of water to the river: "*In June, the historical maximum output of hydroelectric power generation on SSHPS was recorded... More than 100 million kWh per day were produced*".[386] This record was reached with obsolete turbines, in spite of the recommendations issued in 2000 to replace them. At the time of the disaster, Turbine 2 had been in operation for 29 years and 10 months. The maximum period of operation during which the manufacturer guaranteed full compliance with design specifications was 30 years; but naturally this guarantee assumed timely and comprehensive routine maintenance of the turbine, provided by station personnel or special repair contractors. However, the fact is that at the moment of the accident, no plan was in place for Turbine 2, which would soon come to its 30-year safe operating limit, whether to extend its operation into the following decade or withdraw the turbine from service and replace it. Incidentally, it should also be noted that the station never worked at its full capacity (6400 MW), because the existing electrical network to the station was not able to take a load of more than 4000 MW.

Running obsolete turbines with known deficiencies and giving priority to economic concerns should have been compensated by serious attention to thorough equipment monitoring and the complex maintenance of the turbines. However, the reorganization strategy of RAO UES stipulated that the repair departments in all power stations had to be independent business units with a profit-based approach, and that all maintenance contracts should take place on a tender basis. This meant that the repair staff of SSHPS, who had decades of experience and knowledge, had to be transferred into a separate company. This company, Sayano-Shushensky HydroRepair—100 % of which was owned by SSHPS—now had to compete with other repair organizations on tenders and provide the lowest possible prices for maintenance of the unique equipment of the station. The separation of repair departments from Russian power stations was influenced by the prevailing free-market ideology among executives of RAO UES, which assumed that competitive markets would always be more effective.

Another reason for the transfer of repair staff from station to a separate company stems from the fact that, during Soviet times, there were several huge nationwide repair holdings, which had multi-station repair experience and were far more effective than the repair staff at any ordinary station in carrying out major overhauls; local staff focused on ongoing minor repairs. However, these holdings collapsed during the 1990s due to privatization and the total lack of funds at power stations to pay for outsourced repairs—so by the time RAO UES was reorganized, the most advanced repair staff in the industry were within the stations rather than with any of the external repair holdings.

---

[385] Annual Report of Joint-Stock Company "RusHydro", 2008, p. 57.

[386] On Sayano-Shushenskaya hydropower plant recorded maximum power output, press release of SSHPS, July 2, 2009, http://www.sshges.rushydro.ru/press/news/33083.html.

In the case of SSHPS—which is located in a remote mono-industrial city in the Siberian taiga near the Mongolian border, 3500 km from Moscow and 600 km from Novosibirsk—the only qualified service staff available to repair the turbines were former and current personnel of the station. In order to comply with the formalities of tenders dictated by the reform of RAO UES, the management of SSHPS founded a company called HydroRepair, which began to "compete" with Sayano-Shushensky HydroRepair for the station's tenders. Among the founders of HydroRepair were the CEO of SSHPS, the station's chief engineer (who formulated the tender requirements, and after repairs made formal acceptance of the work carried out), the chief financial officer and other executives of the station.[387] This looks like a clear example of a conflict of interests with signs of corruption. But SSHPS management were forced to go through the motions of running both Sayano-Shushensky HydroRepair and HydroRepair in order to prevent inexperienced companies, with no competence in repairing the station's sophisticated turbines, from winning repair contracts. Such companies could easily win contracts simply by offering the lowest price on maintenance work: Russian contracting legislation is flawed in that it requires the buyer to choose the lowest bid on tenders, without taking into consideration the experience of the bidding organizations or the quality of their previous performance (moreover, according to Russian legislation concerning turbine repair works, service companies are not required to obtain licenses, therefore, a company without any experience in repairing turbines can participate in the tenders).

The tender system had another grave shortcoming. The high-quality repair of sophisticated and unique equipment during its life cycle requires a constant accumulation and transfer of knowledge about previous repairs—so that long-term contracts, which guarantee ongoing work if the repair company fulfils its obligations, are beneficial and even necessary for safe operation. But after the reform of RAO UES, stations had to make contracts only for precisely defined maintenance tasks; along with the requirement to give work to the lowest bidder, this resulted in a high turnover of contractors. Therefore, in response to the demand of RAO UES for the withdrawal of repair units from the station staff, SSHPS managers chose what seemed for them to be the option that would minimize damage to the station—they founded and established the HydroRepair company, transferred the maintenance personnel from the station to the company and reduced the likelihood of tenders being won by unscrupulous competitors.

In 2005, the turnover of Sayano-Shushensky HydroRepair was around US$10 million, but the company began to lose contracts to the management-affiliated HydroRepair. Between 2005 and 2008, the total turnover of HydroRepair reached US$30 million. By 2009, the company was winning the majority of the repair contracts for SSHPS; and in particular, it was HydroRepair who repaired Turbine 2

[387]Ibid.

from January until March 2009.[388,389] Only five months after this medium scale repair, it was damage to this turbine that became the main cause of the disaster at SSHPS. Nowadays, it is hard to determine the quality of the repair work that was performed on Turbine 2, because of the total destruction of the turbine. Moreover, after the accident, investigators did not look into the legality of contracts with HydroRepair and there was no investigation of the quality of the repairs carried out on Turbine 2 early in 2009. All opinions expressed after the accident about this repair can only be considered as personal assessments by experts and officials, which are not supported by judicial decisions. And these opinions are radically different. For example, a senior representative of the plant designers concluded that the repair of Turbine 2 was made properly according to all existing standards[390]; on the other hand, an executive member of the Russian parliament's investigation commission, who was on the board of directors at RAO UES between 1997 and 1998 and has been on board of RusHydro since 2013, declared: "*They [the staff of HydroRepair] repaired [the turbines], but did not in reality make full-scale repairs, in spite of documentation which they provided describing fully-fledged repairs... We had assumptions that the management of the hydropower station had affiliated companies. This is true. Some repairs were not made at all [however, funds for these repairs were received by HydroRepair]*".[391] Later Vladimir Putin, during meetings about the accident, commented on the situation: "*It would be irresponsible and even criminal to save money on safety or entrust repairs to companies that are 'affiliated' with anyone, but especially with the management of facilities... The audit conducted by the Energy Ministry in some state-funded companies ... show[s] that many senior officials of these state-funded companies are also involved with commercial companies. [T]here was a conflict of interests, meaning that officials of state-funded companies should not use their position to conduct other commercial activity in the interests of private, generally speaking, family businesses... [W]e must fundamentally improve technological discipline in industry. Performance at sophisticated technological facilities is ... bad. Technological discipline is very low*".[392,393,394] Shortly after these harsh

[388]Irina Tumakova, Sergey Teplyakov, Firm-disaster, Izvestia, September 22, 2009.

[389]Elena Mazneva, Anna Peretolchina, Repair on trust, Vedomosti, September 21, 2009.

[390]Boris Yurkevich, About causes of the accident on Turbine 2 of The Sayano-Shushenskaya Hydropower Station, Destruction of Tubine 2 of Sayano-Shushenskaya Hydropower Station: causes and lessons, Volume I, Hydrotechnical Construction, Moscow, 2013, p. 459.

[391]Repairs on the 2nd turbine of SSHPS were not made in full—the commission of the State Duma, Interfax-Russia, October 30, 2009.

[392]Prime Minister Vladimir Putin chaired a meeting at the National Crisis Management Centre of the Ministry for Emergencies held to discuss the relief efforts at the Sayano-Shushenskaya power plant, September 21, 2009, http://archive.premier.gov.ru/eng/events/news/5032/.

[393]Vladimir Putin meets with Deputy Prime Minister Igor Sechin, December 30, 2011, http://archive.premier.gov.ru/eng/events/news/17554.

[394]Prime Minister Vladimir Putin chaired a meeting of the Government Presidium, August 20, 2009, http://archive.premier.gov.ru/eng/events/news/4757/.

assessments following the disaster, some senior managers of RusHydro were forced to resign, and just over a month later RusHydro—whose main shareholder is the Russian government—appointed a new CEO.

The Problems of Assessing the Real Characteristics of the Vibrations in Turbine 2

Operation of the repaired Turbine 2 was resumed on March 16, 2009 without detectable abnormal vibrations during the following 35 days. According to RusHydro comments given after the disaster, the sophisticated tender scheme for repairs on SSHPS was disclosed by management in April 2009, and consequently HydroRepair lost all contacts.[395] Nevertheless, the founders of HydroRepair all remained in their positions; this was explained by the fact that in previous years there had been no cause for complaint about the quality of repair work produced by HydroRepair.[396] But during spring 2009, when the station began to generate electricity under additional pressure caused by the spring flood and high reservoir levels, increased vibration levels were registered in all the turbines.[397,398] From April 21, staff at the station began to detect abnormally high vibrations in Turbine 2, based on data from one of sensors installed inside the turbine. By August 17, according to this sensor, the vibration amplitude of the bearing of Turbine 2 rose to 840 μm, more than five times the maximum permitted safe level of 160 μm.[399] Station safety instructions dictate that, in case of any sudden increase of vibration of the turbine bearing over 160 μm, the chief engineer must be consulted and the turbine must be unloaded or stopped immediately.[400] Nevertheless, SSHPS executives—including the founders of HydroRepair—took no action to investigate the vibration, eliminate this technical failure during the months before the disaster or order an emergency stoppage of Turbine 2.

There are several explanations for this inaction.

Firstly, SSHPS executives were not able to properly assess the risks involved in operating the turbines at the station, and the operators did not even consider that a

---

[395] Anastasia Lyrchikova, RusHydro has been uncovered machinations on SSHPS, Reuters Russia and CIS, Sep. 18, 2009.

[396] Personal communication with RusHydro's executives.

[397] Dissenting opinion of R.M. Haziahmetov (member of investigation commision of Rostechnadzor) regarding the Act of technical investigation of the accident at the Sayano-Shushenskaya HPP, Destruction of Tubine 2 of Sayano-Shushenskaya Hydropower Station: causes and lessons, Volume III, Hydrotechnical Construction, Moscow, 2013, p. 276.

[398] N. Baykov, Analysis of the circumstances of the accident at the Sayano-Shushenskaya HPP, Destruction of Tubine 2 of Sayano-Shushenskaya Hydropower Station: causes and lessons, Volume I, Hydrotechnical Construction, Moscow, 2013, p. 158.

[399] The act of technical investigation of the accident at the Sayano-Shushenskaya HPP, Rostechnadzor, Oct. 3, 2009, p. 66.

[400] Ibid, p. 75.

serious accident could occur. In fact, engineers had been recording minor vibrations in the station turbines over decades—especially in Turbine 2 during the first years after the station was launched in 1979, when the turbine was equipped with a poor-quality interim rotor.[401,402,403,404] In spite of that, there had been no serious turbine accidents. This is a clear example of habituation—the unwillingness to believe that the worst could happen, and the growth of a misplaced confidence in the soundness of the system.

Secondly, during routine maintenance on the turbines of SSHPS, the condition of the turbine cap stud-bolts was checked visually, without ultrasonic scanning. We have seen that similar accidents had occurred at Nurek hydropower station in 1983 and Grand Rapids in 1992, but that information about these accidents was not widely distributed among executives and engineers of the station for the reasons we described—so nobody thought that unusual vibrations could lead to such catastrophic consequences.[405] After the disaster, laboratory tests revealed that the average degree of fatigue in the stud-bolts was about 60–65 %, and that the majority of them had fatigue cracks.[406] But the laboratory did not estimate when the fatigue had started to develop, and thus could not establish whether it dated from 1979–1983, when a poor-quality interim rotor was installed on the turbine causing excessive vibration, or from the more recent period when HydroRepair had begun to service the turbine. The scanning of stud-bolts on other turbines did not reveal the same massive levels of fatigue seen in those of Turbine 2.

Thirdly, the systems in place to track and monitor the functioning of the turbines at SSHPS were not being used effectively. In 1999, the former director of SSHPS, who had been in post from 1977 and remained until 2001, wrote in a monograph about the development of the station: "*[The laboratory of technical diagnostics at SSHPS] conducted a study aimed to develop a system for permanent tracking of the mechanical state of critical components of the turbines. This*

[401]Valentine Bryzgalov, Monograph "From the experience of establishment and development of Krasnoyarsk and Sayano-Shushenskaya HPSs", Krasnoyarsk, Surikov Publisher, 1999, p. 541.

[402]Vladimir Demchenko, Andrew Krassikov, Sergey Teplyakov, Irina Tumakova. Was Turbine #2 on SSHPS shaking during 10 years? Izvestia, September 14, 2009.

[403]F. Kogan, Abnormal operating conditions and reliability of modern hydro turbines, Destruction of Tubine 2 of Sayano-Shushenskaya Hydropower Station: causes and lessons, Volume I, Hydrotechnical Construction, Moscow, 2013, p. 49.

[404]N. Baykov, Analysis of the circumstances of the accident at the Sayano-Shushenskaya HPP, Destruction of Tubine 2 of Sayano-Shushenskaya Hydropower Station: causes and lessons, Volume I, Hydrotechnical Construction, Moscow, 2013, p. 153.

[405]Rostehnadzor: the accident at the Sayano-Shushenskaya HPP is not unique, in 1983 was a similar situation at Nurek HPP, Interfax, Oct. 3, 2009 and review about accidents and other disturbances on power stations and electric networks of USSR energy system for 1983, Soyuztechenergo, Moscow, 1984.

[406]B. Skorobogatykh, N.Shepilov, S.Kunavin, V.Ushakov, Investigation of the metal and the nature of damage studs of turbine cover of Turbine 2 of Sayano-Shushenskaya Hydropower Station, Destruction of Tubine 2 of Sayano-Shushenskaya Hydropower Station: causes and lessons, Volume I, Hydrotechnical Construction, Moscow, 2013, p. 373.

system should provide information about malfunction and, eventually, provide recommendations for halting a [problem] turbine. Unfortunately, up to now in most cases, the vigilance of operators in the control room and the dial indicator remain the main monitoring instruments of the mechanical condition of turbines at the majority of HPSs in Russia. [The laboratory of technical diagnostics at SSHPS has made] repeated attempts to install and use modern instruments of vibration control to continuously monitor the mechanical state of turbines in order to receive a warning signal in case of off-limit vibration. However, the low level of accuracy, narrowness of frequency band, absence of full-scale spectrum of the diagnosed signal, and low reliability of the monitoring equipment did not lead to a practical positive result".[407] By the late 2000s, the station was equipped with more than 11,000 sensors for controlling all aspects of operation.[408] In March 2009, a new vibration monitoring system was installed on SSHPS's turbines on a trial basis, but it was working only as an internal information system: there were no specifications from the government regulation body, or requirements from the turbine manufacturer, as to how such systems should be used with the turbines. After the accident, an investigation by the Federal Service for Ecological, Technological and Nuclear Supervision of the Russian Federation (Rostechnadzor) concluded that "*The continuous vibration monitoring system installed on Turbine 2 in 2009 was not put into operation and the station's operating personnel and management did not take into account [the data it provided] during decision making*".[409] In other words, this system was collecting information about the vibrations—we know about this because investigators demonstrated an array of recorded vibration data regarding the turbines at SSHPS—but information from the detectors was not recognized by the technical staff as a reliable basis for decision-making. This was because there were four sensors attached to Turbine 2, but only the one for the turbine bearing indicated abnormal vibration levels (up to five times the approved limit), while the others registered acceptable levels of vibration.[410,411] Technical staff had been used to high vibration levels on Turbine 2 for decades, and as we have seen were unaware of the accidents on the Nurek and Grand Rapid stations; so they evaluated the abnormal vibration of the turbine bearing on Turbine 2 during the summer of 2009 as incorrect/inexact, because the

---

[407] Valentine Bryzgalov, Monograph "From the experience of establishment and development of Krasnoyarsk and Sayano-Shushenskaya HPSs", Krasnoyarsk, Surikov Publisher, 1999, p. 541.

[408] L.Godron, Assessment of condition of dam of the Sayano-Shushenskaya Hydropower Station before and after accident, Destruction of Tubine 2 of Sayano-Shushenskaya Hydropower Station: causes and lessons, Volume I, Hydrotechnical Construction, Moscow, 2013, p. 206.

[409] The act of technical investigation of the accident at the Sayano-Shushenskaya HPP, Rostechnadzor, Oct. 3, 2009, p. 66.

[410] Dissenting opinion of R.M. Haziahmetov (member of investigation commision of Rostechnadzor) regarding the Act of technical investigation of the accident at the Sayano-Shushenskaya HPP, Destruction of Tubine 2 of Sayano-Shushenskaya Hydropower Station: causes and lessons, Volume III, Hydrotechnical Construction, Moscow, 2013, p. 271.

[411] Victor Kudryavy, Systemic causes of accidents, Hydrotechnical Construction, Moscow, #2, 2013.

other sensors on Turbine 2 did not demonstrate serious deviation from their approved limits. Moreover, even when Turbine 2 was temporarily suspended during the summer months of 2009, the sensor still showed that the vibration of the suspended turbine bearing was 160 μm.[412] Believing that the sensor was recording the vibration level incorrectly, station managers neither logged this "faulty sensor" in their records nor carried out tests on the actual level of the vibration with portable diagnostic equipment. They perceived it as an unimportant defect, which they were not obliged to report to RusHydro headquarters.

After the disaster, during restoration work on SSHPS in 2010–2011, an advanced vibration monitoring system was installed on the turbines. Nevertheless, 15 cases of faulty sensors within Turbines 3, 4, 5, and 6 were registered. The main cause of incorrect turbine vibration measurements was "*multidirectional dynamic impulses of forces from the flow of water coming off the rotor blades*".[413] However, during the investigation of the accident, a laboratory examination of the assumed faulty sensor was carried out and its results were presented to the court—it turned out that the sensor was fully functional.[414] But why it showed abnormal vibration when other sensors indicated the turbine's vibration to be within the normal range, will never be known, because the turbine was completely shattered in the accident and all electrical circuits were also destroyed in the ensuing flooding. This means that after the event, nobody can draw any conclusions about the real vibration characteristics of the bearing of Turbine 2 on the day of the accident. It would be unreasonable to claim that the possibility of a sensor fault implies that there was no vibration in the turbine before the disaster. But there is further compelling evidence against such a claim: the head of Rostechnadzor revealed that "*seismologists recorded abnormal vibration at Sayano-Shushenskaya Hydropower Station 15-45 minutes before the accident*".[415,416] Therefore, a combination of evidence allowed investigators to conclude that "*the cause of the destruction of the turbine cap stud-bolts was fatigue cracks in the body of the stud-bolts. The origin and intensive development of [these] cracks resulted from actions and efforts*

[412]Dissenting opinion of R.M. Haziahmetov (member of investigation commision of Rostechnadzor) regarding the Act of technical investigation of the accident at the Sayano-Shushenskaya HPP, Destruction of Tubine 2 of Sayano-Shushenskaya Hydropower Station: causes and lessons, Volume III, Hydrotechnical Construction, Moscow, 2013, pp. 271–278.

[413]Victor Kudryavy, Systemic causes of accidents, Hydrotechnical Construction, Moscow, #2, 2013.

[414]From the sentence of Sayano-Gorsky District Court, Respublic of Khakasia, Russia, Dec. 24, 2014.

[415]A. Abakumov, M. Abdulin. D. Soloviev, V. Ustinenko, Numerical simulation of a possible accident scenario on Sayano-Shushenskaya Hydropower Station, Destruction of Tubine 2 of Sayano-Shushenskaya Hydropower Station: causes and lessons, Volume I, Hydrotechnical Construction, Moscow, 2013, p. 151.

[416]Dmitry Malkov, The Hydro power station was closed for construction, Kommersant, Aug. 28, 2009, № 158 (4213).

influenced by the horizontal vibration of the turbine bearing".[417] In December 2014, a court supported this accusation.

A fourth probable reason for the unwillingness to shut down Turbine 2 relates to another hydropower station in the region. In 1988 SSHPS, with its operating output of 6,400 MW, had been combined with the smaller Bratsk HPS for power regulation under the supervision of the System Operator of the Unified Energy System (SO UES) within the whole Siberia region. Bratsk HPS was located 700 km from SSHPS on the Angara River, and had an output of 1,400 MW. At midnight on August 17, 2009, there was a fire affecting the communication channels of Bratsk HPS, which led to loss of control over the station by the dispatcher of the Siberian branch of SO UES. Therefore, the operator ordered SSHPS to launch all available turbines in order to compensate for any possible suspension of output from Bratsk HPS. At 3:14 a.m. on August 17, 2009, Turbine 2 was resumed and was operating automatically as part of the "*regulator group for active and reactive power*" of the Siberian branch of SO UES. On the day of the disaster, the reservoir level at the SSHPS dam was 212 m instead of the optimum level of 197 m. The turbines of SSHPS only had a narrow range of adjustment to their output if the reservoir level was higher than 197 m. This meant that the turbines could operate safely with the reservoir at 212 m only from 0 to 265 MW and from 570 MW to 640 MW. Therefore, during load changes with this heightened reservoir level, the turbines would pass through a "*not recommended for use*" zone between 265 and 570 MW, during which there would be transient hydrodynamic processes, pressure fluctuations and high vibration. From the resumption of operation during the incident at Bratsk HPS, the load regime of Turbine 2 was changed twelve times and the turbine passed six times through the "*not recommended for use*" zone; in the longer period from March 2009, Turbine 2 passed through this zone 210 times—regulations allowed it to pass through the zone not more than 750 times annually—and was there for 2520 s. Thirteen minutes before the disaster, when the dispatcher of the Siberian branch of SO UES gave the order to reduce the output of Turbine 2 from 600 MW to 475 MW, the vibration amplitude of the turbine bearing rose by 240 μm according to the detector, from 600 μm to 840 μm. It is likely that the station's chief engineer and operator staff did not stop Turbine 2 in spite of the obvious extreme vibration within the machine, because of the continuing incident at Bratsk HPS, which was fully eliminated only by 2:03 p.m. on August 17, 2009—roughly 6 h after the disaster at SSHPS.[418]

Finally, SSHPS executives were obviously afraid to shut down Turbine 2 over many months because of the potential scandal with the questionable tender practice and inevitable questions about the quality of the repairs over the previous three years. Five months earlier, SSHPS executives had lost an affiliated business

[417]Andrei Mitrofanov, former chief engineer of SSHPP, considers that the accident could be repeated, Khakasia News Agency, Dec. 14, 2013.

[418]The act of technical investigation of the accident at the Sayano-Shushenskaya HPP, Rostechnadzor, Oct. 3, 2009, pp. 55–76.

with US$30 million turnover due to suspicions about conflicts of interests and evidence of corruption; they did not wish to give any occasion for more inconvenient scrutiny about the business activities they were engaged in on top of their managerial duties. The main reason for such concealment lays in avoiding potential criminal charges against the management of SSHPS for using questionable repair tender schemes as well as in steering clear of questions about the quality of repair of the station's equipment. Long before the disaster, RusHydro had a very long chain of communication of risk information. Typically, RusHydro's top management actually received information about technical incidents—even insignificant ones—not from the managers of hydropower stations but from the RusHydro security services, which monitored the station environment independently from the station's personnel.[419] Thus, a typical reply from stations to any question from Moscow about their reliability was "*everything is under control*".[420] In this case, RusHydro security services also missed the existence of abnormal vibrations in the plant. This is reminiscent of the KGB falling short of recognizing the importance of the SCRAM effect in RBMK nuclear plants.

2.1.8.2 Risk Concealment After the Disaster

After the disaster, Rostechnadzor issued an investigation report about the technical causes of the disaster. Rostechnadzor's findings were selected as the basis for a criminal investigation by Russian prosecutors towards the management and staff of SSHPS. Finally in December 2014, more than five years after the accident, the court found the director of SSHPS, the station's chief engineer and other technical managers of the station guilty of violations of safety regulations at work—in particular through their disregard of the signs of excessive vibration within Turbine 2—causing the death of more than two persons. The main defendants were given six-year prison sentences. Nevertheless, the accused staff of the station did not accept the judgment, and filed an appeal declaring that the main causes of the accident were the imperfection of the turbine design and the poor quality of production of Turbine 2 at the manufacturer's Leningrad plant, and wider shortcomings in the design of the station.[421] As we mentioned earlier, the prosecutors did not conduct a detailed investigation of HydroRepair's tender practice and the evident conflict of interests involved; nevertheless, most of the founders of HydroRepair were ultimately found guilty by the court in their principal roles as executives of SSHPS, but not as executives of the dubious HydroRepair.

The lack of interest from state prosecutors towards HydroRepair was hardly surprising: if they revealed this scheme to the public, the Russian government would have to admit their own mistake in allowing the reform of Russian electro-energetics

[419] Personal communication with RusHydro's executives.

[420] Personal communication with RusHydro's executives.

[421] Dmitry Malkov, SSHPS kept the last word, Kommersant, Dec. 2, 2014.

according to ultra-liberal conceptions of a free-market economy. So, the managers who were ultimately responsible for the proper operation of the turbines were punished in any case, but without public disclosure of the corruptive tactics of the reformed Russian electro-energetics industry. The prosecutors also neglected to investigate the possible guilt of the top management of RAO UES, who had implemented an apparently misguided reorganization of Russian electro-energetics, damaging the interests of national energy security, with the approval of the Russian government. This omission is all the more glaring in that some RAO UES executives, as well as some former government officials, were mentioned in Rostechnadzor's report as persons "*who contributed to the occurrence of the accident*".[422] After the disaster, the Russian parliament's investigation commission stated: "*Most of the causes of the disaster are systemic and multifactorial, influenced by indigenous deficiencies of the existing organizational scheme and functioning of domestic electro-energetics... During radical changes of property relations and principles of conduction of the sector [i.e. during the reforms of RAO UES], comprehensive conditions for ensuring technological safety were not formed*".[423]

Sayano-Shushenskaya Hydropower Station Accident: Why Risks Were Concealed

- The Politburo and State Planning Commission **focused on the short-term reduction of safety costs** through the redesign of the station, **and demanded constant rush** during the construction phase, because they wanted to accelerate the introduction of new energy facilities to meet the needs of the national economy.
- There was a general **reluctance** within the Soviet and Russian electro-energetics industry **to investigate in detail the causes of previous accidents/near-miss cases,** or to transmit the results among decision-makers, so that the remedies learned from the experience of previous accidents on other electro-energetical facilities could be implemented across the industry.
- After the collapse of the Soviet Union, the liberal-oriented government **gave priority to short-term financial results** in the operation of electro-energetical facilities, and to indicators of market capitalization, over the long-term reliability of Russian electro-energetics.
- **Habituation/wishful thinking/overconfidence/self-suggestion/self-deception:** engineers and management at the station believed that a severe turbine accident was highly unlikely, because of the station's 30-year history of generally safe operations.

---

[422] The act of technical investigation of the accident at the Sayano-Shushenskaya HPP, Rostechnadzor, Oct. 3, 2009, pp. 111–113.

[423] Victor Kudryavy, Mister kilowatt, The Soviet Russia, Nov. 28, 2013.

- SSHPS managers were afraid of **potential criminal charges** for using questionable repair tender schemes, which they felt obliged to implement after the misguided reorganization imposed by RAO UES. They were also afraid to seem incompetent in the eyes of RusHydro's superiors.
- **The Russian government was unwilling to admit its own mistakes in pushing through unreasoned free-market reforms of Russian electro-energetics,** or to admit the evident failure of its reorganization of RAO UES. This led to a situation where only SSHPS staff faced criminal charges after the accident.

2.1.9 Deepwater Horizon Oil Spill (USA, 2010)

From January to April 2010, floating in the Gulf of Mexico 66 km from the coast of Louisiana State, the Deepwater Horizon oil platform was drilling the Macondo exploratory well. The total depth of the well was 6500 m: 1500 m below sea level and 4000 m beneath the seafloor in Block 252 of the Mississippi Canyon. The proven reserves of the field were 110 million barrels[424]; the potential income from extraction of this amount of oil was approximately US$10 billion. The platform was owned by Transocean Ltd., the largest offshore drilling operator in the world. BP (formerly known as British Petroleum) leased the rig for exploration of the Macondo field. Halliburton Company, one of the world's largest oilfields services companies, was engaged as the cementing contractor.

On April 20th, 2010 at 9:45 p.m. US Central Time, a blowout of oil, gas and concrete from the well occurred on the Deepwater Horizon platform, causing an explosion and a fire that sunk the platform. There were 126 crewmembers on the rig during the accident; 11 people perished and 17 were injured. The rest of the crew survived unharmed, but the accident led to oil being discharged from the well for 87 days—for a total of 3.19 million barrels.[425] This was the third largest oil spill in the history of the oil industry, after the Kuwaiti oil fires in 1991 where the approximate discharge was 10 million barrels and the blowout at Lakeview Gusher Number One oil well in Kern County, California, which was out of control for nine months in 1910–1911 and led to the release of approximately 9 million barrels. BP was forced to cover all expenses incurred in shutting down the deepwater leak and

[424]Macondo: The Gulf Oil Disaster, Chief Counsel's Report, National Commission on the BP Deepwater Horizon Oil Spill and Offshore Drill, Government Printing Office, 2011, p. 25.

[425]On January 15, 2015, a US federal judge ruled that BP spilled 3.19 million barrels of oil into the Gulf of Mexico during the 2010 Deepwater Horizon disaster. The US government has argued for 4.2 million and BP for 2.45 million. Six experts testified that as few as 2.4 millions barrels of oil, and as many as 6 million, escaped during the 86 day long accident. BP and the US government agreed that 810,000 barrels were captured. The importance of this ruling lies in the fine facing BP, calculated at US$4300 per barrel. Source: http://scim.ag/_BPruling and News in brief, Science 347 (6220), p. 356 (2015).

in cleaning up the American part of the Gulf of Mexico coastline—an area where 14 million inhabitants reside—contaminated by spilled oil. In addition, they paid compensation to the fishing and coastal tourism industries in the area and a fine issued by the U.S. government. BP's total losses from the accident were estimated at US$46 billion (US$28 billion was spent on the accident and $18 billion on additional government fines and penalties[426]) and by June 2010, BP's stock market value had fallen by US$70 billion.[427] Because of the disaster, the U.S. government suspended any deepwater offshore activity in the United States for 6 months. In the middle of June 2010, the President of the United States Barack Obama declared: "*this oil spill is the worst environmental disaster America has ever faced*". More than 47,000 people and 7000 vessels[428] took part in the response to the spill.

In January 2011, the National Commission on the BP Deepwater Horizon Oil Spill and Offshore Drilling (hereafter the National Commission) stated in a report to the President of the United States that "*The explosive loss of the Macondo well could have been prevented. The immediate causes of the Macondo well blowout can be traced to a series of identifiable mistakes made by BP, Halliburton, and Transocean that reveal such systematic failures in risk management that they place in doubt the safety culture of the entire industry. Fundamental reform will be needed in both the structure of those in charge of regulatory oversight and their internal decision-making process to ensure their political autonomy, technical expertise, and their full consideration of environmental protection concerns*".[429]

[426]Bradley Olson, Margaret Cronin Fisk, Worst Case' BP Ruling to Force Billions More in Payout, Bloomberg, Sep. 4, 2014.

[427]Steve Hargreaves, BP's $70 billion whipping, CNN Money, June 2, 2010.

[428]Jonathan L. Ramseur, Curry L. Hagerty, Deepwater Horizon Oil Spill: Recent Activities and Ongoing Developments, Congressional Research Service, January 31, 2013, p. 2.

[429]National Commission on the BP Deepwater Horizon Oil Spill and Offshore Drilling. Report to the President, Jan. 2011, p. vii.

2.1.9.1 Risk Concealment Before the Disaster

Geological and Regulatory Contexts

In 1947, in Louisiana State, a first well was drilled by a fixed platform, which was located offshore, out of sight of land.[430] After the 1973 oil crisis, which led to a dramatic increase in oil prices, oil companies intensified offshore drilling. In 1978, Shell Oil Company's Cognac production platform launched drilling at a depth of 1000 ft (304 m) underwater. In 2006, Chevron, Devon Energy and Statoil drilled the Jack 2 exploratory well, 7000 ft (2133 m) underwater,[431] reaching a total depth of 28,125 ft (8572 m). In 2009, BP, working from the Deepwater Horizon platform, discovered the gigantic Tiber Oil Field, with resources between 4 and 6 billion barrels of oil at a total depth of 35,056 ft (10,685 m) and under 4130 ft (1258 m) of water.[432] In 2011, 30 % of U.S. crude oil production was extracted from the Gulf of Mexico.[433]

In 1982, the Minerals Management Service (MMS) under the U.S. Department of the Interior was established to regulate such intensive offshore drilling. Due to the widespread idea that government oversight of private enterprise should be kept to a minimum, active lobbying from the industry and cuts in public funding, the budget of MMS dropped from US$250 million in 1984 to less than US$200 million in 2009 (representing less than $100 million in 1984 dollar value due to inflation), even though oil companies progressed considerably in the development of deepwater drilling over this period.[434] The regulator had no budget for hiring advanced specialists who understood innovations in the field, and MMS came to rely on the expertise of deepwater operators and contractors. Moreover, by 2009, the number of unannounced MMS inspections of offshore oil infrastructure reached a negligibly low level compared with the 1980s.[435] The impotence of the US regulator led to a situation where innovations in the safety requirements for offshore drilling, which were widely implemented as compulsory measures in other countries after accidents, were left to the discretion of U.S. offshore drilling operators.[436] For example, in Norway and in Brazil, all deepwater blowout preventers have an acoustics trigger for remote emergency shutdown of a well—these triggers cost over US$0.5 million apiece—but in the USA the use of such devices

[430]Ibid, p. 23.

[431]Chevron Announces Record Setting Well Test at Jack, Press Release, Chevron, San Ramon, Calif., Sep. 5, 2006.

[432]BP Announces Giant Oil Discovery In The Gulf Of Mexico, Press release, BP, Sep. 2, 2009.

[433]The U.S. Energy Information Administration, 2011.

[434]National Commission on the BP Deepwater Horizon Oil Spill and Offshore Drilling. Report to the President, Jan. 2011, p. 73.

[435]Ibid, p. 75.

[436]Ibid, p. 72.

was optional.[437] In addition, BP did not have a contingency plan for any emergencies arising while drilling the Macondo well, because such plans were not obligatory under U.S. deepwater drilling legislation.[438]

When BP filed the plans for the drilling of the Macondo exploratory well to MMS in 2009, the probability of an oil spill in this area was assessed as low ("*[it is] unlikely that an accidental surface or subsurface oil spill would occur from the proposed activities*").[439] This was despite the fact that, since 2001, according to U.S. officials, there had been 948 fires and explosions on offshore oil platforms in the Gulf of Mexico, many of which were associated with the drilling of exploratory wells, where the risk of blowouts was extreme.[440,441] Moreover, the U.S. Department of the Interior exempted BP from a detailed evaluation of the environmental impact of the Macondo well after concluding that a massive oil spill was unlikely[442]—in spite of previous MMS study findings that 50 % of tested blowout preventers failed to cut through the pipe and halt the flow of oil during emergencies.[443,444] In fact anonymous representatives of the Bureau of Ocean Energy Management, Regulation and Enforcement—the federal agency that regulates offshore drilling—had recognized "*that the designs of blowout preventers were not adequate and that new requirements were needed, along with tougher government inspections*".[445] After the accident, new BP CEO Robert Dudley said that BP had never anticipated such a tremendous spill: "*we've been drilling in the Gulf of Mexico, in the deep water for 20 years now. You just never see an accident like this*".[446] However, in 1979, there had been a blowout on the Mexican Ixtoc I oil rig in the south-western part of the Gulf of Mexico, which was unable to shut down for 10 months at a depth of just 50 m, and which resulted in 3 million barrels of oil being discharged.

---

[437] Tom Doggett, Timothy Gardner, U.S. mulls requiring remote shutoffs for oil rigs, Reuters, May 3, 2010.

[438] Vladimir Milov, Oversept the accident, Gazeta.RU, June 16, 2010, http://www.gazeta.ru/column/milov/3385462.shtml.

[439] Cain Burdeau, Holbrook Mohr, BP downplayed possibility of major oil spill, Associated Press, May 1, 2010.

[440] Vladimir Milov, Oversept the accident, Gazeta.RU, June 16, 2010, http://www.gazeta.ru/column/milov/3385462.shtml.

[441] National Commission on the BP Deepwater Horizon Oil Spill and Offshore Drilling. Report to the President, Jan. 2011, p. ix.

[442] Juliet Eilperin, U.S. exempted BP's Gulf of Mexico drilling from environmental impact study, Washington Post, May 5, 2010.

[443] Mini Shear Study, WEST Engineering Services, Study for U.S.Minerals Management Service, Requisition No. 2, 1011–1003, Dec. 2002, p. 13.

[444] Allan Chernoff, Blowout preventers—disasters waiting to happen? CNN, June 10, 2010.

[445] Clifford Krauss, Henry Fountain, Report on Oil Spill Pinpoints Failure of Blowout Preventer, The New York Times, Mar. 23, 2011.

[446] America Speaks to BP, Full Transcript: Bob Dudley Interview, Public Broadcasting Service (PBS), July 1, 2010, http://www.pbs.org/newshour/bb/environment-july-dec10-dudleyfull_07-01.

The Minerals Management Service (MMS) has also been accused of being corrupted by oil companies in return for money, sex favors and drug.[447] After the disaster, MMS was dismantled and replaced by two separate organizations (Bureau of Ocean Energy Management (BOEM) and Bureau of Safety and Environmental Enforcement (BSEE)).

Business Pressure and Miscommunications Between BP, Halliburton and Transocean

The platform started to drill the Macondo well in February 2010, aiming to finish the job in 51 days with a budget of US$96.2 million.[448] However, with drilling still incomplete following delays and over expenditure, BP managers urged the staff of Transocean and Halliburton to operate faster[449] because BP expenses on leasing the platform exceeded US$1 million a day: by the disaster date, the delay added to 43 days and BP were already more than US$58 million over budget.[450] Some BP engineers considered that "*this has been [a] nightmare well, which has everyone all over the place*"[451]; nevertheless, by the middle of April 2010, the well was successfully drilled.

On the morning of the accident, the cementing engineer of Halliburton sent an e-mail to his colleague in Houston: "*We have completed the job and it went well*"[452] and a BP engineer informed onshore colleagues: "*just wanted to let everyone know the cement job went well. Pressures stayed low... The Halliburton cement team ... did a great job*". The reply from BP executives was encouraging: "*Great job guys*!".[453] The quality of the cement job is critical for the safe exploitation of deepwater wells: according to an MMS study, cementing was the single most significant factor in 18 of 39 well blowouts in the Gulf of Mexico over a 14-year period.[454] To save money and time, BP managers had reduced the number of centralizers, which distribute cement evenly in a well, from 21 to 6. Transocean's rig crew and several BP's representatives were unaware that

---

[447]The Friday Podcast: Sex, Drugs and Regulation, NPR, June 11, 2010, http://www.npr.org/blogs/money/2010/06/11/127772998/the-friday-podcast-sex-drugs-and-regulation; Investigative Report " OIG Investigations of MMS Employees", United States Department of the Interior, Office of Inspector General, Sept. 9, 2008, https://www.doioig.gov/sites/doioig.gov/files/RIKinvestigation.pdf.

[448]National Commission on the BP Deepwater Horizon Oil Spill and Offshore Drilling. Report to the President, Jan. 2011, p. 2.

[449]Deepwater Horizon's Blowout, Part 1, CBS, 60 Minutes, August 22, 2010, http://www.cbsnews.com/video/watch/?id=6795538n.

[450]National Commission on the BP Deepwater Horizon Oil Spill and Offshore Drilling. Report to the President, Jan. 2011, p. 2.

[451]Ibid, p. 2.

[452]Ibid, p. 102.

[453]Ibid, p. 4.

[454]Christina Ingersoll, Richard M. Locke, Cate Reavis, BP and the Deepwater Horizon Disaster of 2010, MIT Sloan School of Management, Apr. 3, 2012, p. 15.

Halliburton had run three laboratory tests of cement stability for the Macondo well between February and April 2010, all of which had failed.[455] The BP team was relying on the good quality of Halliburton's cement to compensate for previous BP cost-reduction measures: BP managers even canceled the final acoustic test of the cement job on the morning of the disaster day, thinking that they had saved $128,000 in doing so.[456]

After the disaster, the National Commission found out that managers of Halliburton "*did not comment on the evidence of the cement slurry's instability, and there is no evidence that BP examined the foam stability data in the report at all… Documents identified after the blowout reveal that Halliburton personnel had also conducted another foam stability test earlier in February. The earlier test had been conducted under slightly different conditions than the later one and had failed more severely. It appears that Halliburton never reported the results of the earlier February test to BP… Halliburton conducted another round of tests in mid-April, just before pumping the final cement job. By then, the BP team had given Halliburton more accurate information about the temperatures and pressures at the bottom of the Macondo well, and Halliburton had progressed further with its cementing plan. Using this information, the laboratory personnel conducted several tests, including a foam stability test, starting on approximately April 13. The first test Halliburton conducted showed once again that the cement slurry would be unstable. The Commission does not believe that Halliburton ever reported this information to BP… It appears that Halliburton personnel responded instead by modifying the test conditions – specifically, the pre-testing conditioning time – and thereby achieving an arguably successful test result… In fact, it appears that Halliburton did not even have testing results in its possession showing the Macondo slurry was stable until after the job had been pumped. It is difficult to imagine a clearer failure of management or communication… BP's fundamental mistake was its failure – notwithstanding the inherent uncertainty of cementing and the many specific risk factors surrounding the cement job at Macondo – to exercise special caution before relying on the primary cement as a barrier to hydrocarbon flow… BP, Transocean, and Halliburton failed to communicate adequately. Information appears to have been excessively compartmentalized at Macondo as a result of poor communication. BP did not share important information with its contractors, or sometimes internally even with members of its own team. Contractors did not share important information with BP or each other. As a result, individuals often found themselves making critical decisions without a full appreciation for the context in which they were being made (or even without recognition that the decisions were critical)*".[457]

A year after the disaster, in April 2011, BP filed a lawsuit against Halliburton accusing it to have intentionally destroyed the evidence related to Halliburton's

---

[455]National Commission on the BP Deepwater Horizon Oil Spill and Offshore Drilling. Report to the President, Jan. 2011, pp. 117, 123, 224.

[456]Ibid, p. 4.

[457]National Commission on the BP Deepwater Horizon Oil Spill and Offshore Drilling. Report to the President, Jan. 2011, p. 123.

non privileged cement testing, in part because it wanted to eliminate any risk that this evidence would be used against it in any trial that would attempt to determine the adequacy of Halliburton's cement job on the Macondo well. In return, Halliburton blame BP for reducing the number of centralisers, a course of action that allowed hydrocarbons to escape through channels that formed in the cement liner.[458] In 2013, in its coverage of the trial, Dow Jones Newswires published the following statement: "*Halliburton had concealed and failed to disclose evidence 'as part of an effort by upper management to ratify and conceal Halliburton's pre-blowout callous disregard for safety*'".[459] Moreover, during the trial, former Halliburton lab manager testified "*that a company official asked him not to record the results of a cement stability test related to BP's blown out Macondo well*". This allowed BP to declare: "*Plaintiffs respectfully submit that Halliburton willfully and intentionally concealed and/or otherwise failed to preserve and/or timely produce and disclose material evidence and/or potentially relevant evidence to the parties and to the court in advance of trial... The cumulative effect of Halliburton's pattern of destruction and spoliation of evidence has been to deprive the court and the parties of significant post-incident evidence relevant to the inherent quality and performance of the cement Halliburton provided for the job at the Macondo well, and the role of that Halliburton slurry design as a cause of the events of April 20, 2010*".[460] Tommy Roth, Halliburton vice president, noticed that "*these tests weren't authorized and he didn't know about them*" and that he was aware "*in April 2011 that a Halliburton employee had conducted unauthorized tests and in 2012 that the results were discarded*".[461]

After the cement job on the morning of April 20, 2010, the staff of the platform made several tests of the well's integrity (positive- and negative-pressure tests). The positive-pressure test passed successfully, but the negative-pressure test showed contradictory information. Operators interpreted this data as the "*bladder effect*" and concluded that the negative-pressure test was successful.[462] "*Many BP and Halliburton employees were aware of the difficulty of the primary cement job. But those issues were for the most part not communicated to the rig crew that conducted the negative-pressure test and monitored the well*".[463] So nobody from the rig crew immediately informed decision makers, either on the platform or at onshore headquarters, about the apparently contradictory condition of the well. On being questioned by a Transocean executive "*You got everything under control*

[458]BP accuses Halliburton of hiding evidence, Al Jazeera, Dec. 6, 2011.

[459]Alison Sider, BP Asks Court to Sanction Halliburton in Deepwater Horizon Trial, Dow Jones Newswires, Mar. 22, 2013.

[460]Harry R. Weber, Plaintiffs: Halliburton's reckless behavior marring Gulf spill trial, Fuelfix, Mar. 21, 2013.

[461]Allen Johnson Jr., Margaret Cronin Fisk Halliburton, Official 'Surprised' by Unauthorized Tests, Bloomberg News, Mar. 12, 2013.

[462]Ibid, pp. 6, 107.

[463]Ibid, p. 123.

here?" the drilling master said "*Yes, sir*"; to the question "*How did your negative test go*?" a rig crew member answered "*It went good*".[464] For everybody on the platform, the blowout that followed was unexpected. For a number of technical reasons, the blowout preventer on the Macondo well failed to cut the pipe, and a flow of oil and gas began to surge from the well.

BP managers arrived on the day of the disaster to celebrate the safety award earned by Transocean for the previous year and all staff members were eager to party. The staff did not wait long enough for the cement to dry and they did not see the blowout coming through a series of faults due to the haste to finish the well. They died for it.

The National Commission stated that "*each of the mistakes made on the rig and onshore by industry and government increased the risk of a well blowout ... the cumulative risk that resulted from these decisions and actions was both unreasonably large and avoidable*".[465]

Lack of Learning from Earlier Disasters

It is remarkable that, in 1988, two decades before this disaster, the Piper Alpha offshore platform in the North Sea was destroyed because of failure of communication as well as blatant disrespect for safety rules.[466] In this case, the problem that led to the explosion arose between two repair shifts operating within the existing "permit-to-work system", when the second team was not informed that the first had removed a pressure safety valve for routine maintenance.[467] This absence of information about a minor maintenance process had major consequences, when the unwittingly dangerous actions of the second shift caused a leakage of condensate—which exploded, causing a massive fire in which 167 crew members perished. Dr. M. Elisabeth Paté-Cornell, in a discussion of the lessons to be learnt from the Piper Alpha disaster in 1993, wrote: "*The culture of any industry that discourages internal disclosure and communication of bad news leads to ignoring small incidents and near-misses as long as they do not result in full-scale accidents. In such an environment, the fact that a severe accident did not occur seems to be sufficient proof that the system works and that 'an inch is as good as a mile'. The possibility that several minor problems could occur at the same time does not seem to be considered. Consequently, small, isolated incidents are seldom*

[464] Ibid, pp. 5–7.

[465] Ibid, p. 115.

[466] Months before the explosion, Jean Laherrère, then assistant to the VP E&P of Total, visited the Piper Alpha platform and discovered that all safety rules were ignored in the production process. But on the day following the explosion, a manager of the UK department of Trade and Industry announced in the media that all safety rules had been respected! (private communication, Dec. 18, 2014).

[467] M. Elisabeth Paté-Cornell, Learning from the Piper Alpha Accident: A Postmortem Analysis of Technical and Organizational Factors, Risk Analysis, 1993, 13(2), 226.

discussed openly since they would constitute a black mark for the personnel involved. Therefore, the same problems are likely to recur elsewhere. In fact, even when an accident does occur, appropriate measures to avoid its recurrence are not necessarily taken. The permit-to-work system, for example, had failed before, in particular on Piper Alpha in 1987, when a worker was killed... The accident was the result of a breakdown of communications in the permit-to-work system and an error in the shift handovers. In spite of memos and warnings to other [offshore installation managers], the lesson was not learned on Piper Alpha itself'.[468] After the explosion, the safety rules were updated, showing that they were previously insufficient and inadequately enforced. In their turn, BP, Transocean and Halliburton also failed to learn the serious lesson of Piper Alpha.

2.1.9.2 Risk Concealment After the Disaster

Over 87 days, the wellhead discharged a total of 4.9 million barrels of oil; on April 22, the daily discharge was 62,200 barrels and on July 14 it was still 52,700 barrels.[469] However, in the first few days, BP and the U.S. Coast Guard hesitated in concluding that there was an oil leakage: they assumed that the blowout preventer had shut down the well properly and that the oil slick consisted predominantly of 700,000 gallons[470] of diesel fuel from the sunken platform.[471] Coast Guard Admiral Mary Landry told correspondents "*We are only seeing minor sheening on the water... We do not see a major spill emanating from this incident*".[472] After remotely operated submarines dived to the wellhead, could not manually stop the blowout preventer and found oil leaking from the end of the riser, BP declared a leakage of 1000 barrels per day[473]—2 % of the real discharge. Later, the U.S. Coast Guard and the National Oceanic and Atmospheric Administration estimated the leakage at between 5000 and 8000 barrels per day[474,475]—still just 10–15 %

---

[468]Ibid, p. 231.

[469]National Commission on the BP Deepwater Horizon Oil Spill and Offshore Drilling. Report to the President, Jan. 2011, p. 167.

[470]To put in perspective, a barrel of oil is equivalent to 42 US gallons or 159 liters and a US gallon is 3.79 liters; thus, the 700,000 gallons of diesel fuel amounted to approximately 16700 barrels, a very small quantity compared to the total spill.

[471]National Commission on the BP Deepwater Horizon Oil Spill and Offshore Drilling. Report to the President, Jan. 2011, p. 130.

[472]Ibid, p. 130.

[473]Oil rig wreck leaks into Gulf of Mexico, Associated Press, Apr. 24, 2010.

[474]National Commission on the BP Deepwater Horizon Oil Spill and Offshore Drilling. Report to the President, Jan. 2011, p. 133.

[475]Congressional Republicans seize on oil spill crisis to attack Obama, Michael D. Shear, Washington Post, June 4, 2010.

the actual discharge from the well—and these estimates became official during the following four weeks.[476] Ultimately, in the first few days—the most important time in any crisis—this underestimation of the seriousness of the spill affected not only public perception, but also the response inside the crisis team, which led to a delay in preventive measures for oil dispersion and collection far away from the coastline. For example, Louisiana State declared a state of emergency only on the ninth day after the accident. Surprisingly, American oil-spill removal organizations were not able to supply enough containment booms for such a large oil spill. After the well was finally capped, the federal government released a report entitled "*BP Deepwater Horizon Oil Budget: What Happened to the Oil?*" in which they assumed that 75 % of the spilled oil was "*gone*".[477] However, the public and the media were skeptical about these estimates: "*From the start of the disaster, the government has badly underestimated the amount of oil spewing from the runaway well. That poor track record makes people understandably skeptical of [the Oil Budget] report*".[478]

The Center for Public Integrity revealed that, in the hours after the Deepwater Horizon oil rig caught on fire, the US Coast Guard failed to follow its own internal firefighting procedures.[479] They did not call for an expert to assist them and poured 6000 tons of salt water per hour on the rig to attempt to extinguish the fire, while it is well known that fires involving hydrocarbon fuels should be quenched with foam and not water. As a consequence, the ballasts of the platform were filled with water causing it to sink and, dramatically, the riser piper to rupture. And the riser pipe did not start leaking until after the rig sank. According to Jean Laherrère, a retired geologist and oil engineer from Total who is known for his work on risk management of the oil industry, the lack of competence and communication concerning fires of hydrocarbon substances led to the wrong decision with horrendous consequences.[480] From a technical point of view, letting the fires burn (until extinction with the right foam[481]) and keeping the platform afloat would have avoided the marine pollution and would have allowed to shut off and secure the riser. Could it have been that some decision makers found more appealing to sink a very visible and inconvenient blaze from the media and public? Moreover, the discussion of the U.S. Coast Guard actions would have been also problematic, given the political and public will to pass on the blame to BP, Transocean, and Halliburton, with good reasons.

---

[476] Flow Rate Group Provides Preliminary Best Estimate Of Oil Flowing from BP Oil Well, Press Release, U.S. Department of the Interior, May 27, 2010.

[477] National Commission on the BP Deepwater Horizon Oil Spill and Offshore Drilling. Report to the President, Jan. 2011, p. 167.

[478] Ibid, p. 169.

[479] Matthew McDermott, Improper Coast Guard Firefighting Responsible For Sinking Deepwater Horizon? Maybe, New York, NY, July 28, 2010 (http://assirou.net/?p=1905).

[480] Jean Laherrère, private communication, Dec. 18, 2014.

[481] Fire fighting foam, http://en.wikipedia.org/wiki/Fire_fighting_foam.

Deepwater Horizon Oil Spill: Why Risks Were Concealed

- **Habituation/wishful thinking/false reassurance/self-deception** among representatives of the Minerals Management Service, BP, Halliburton Company, and Transocean Ltd. in assuming **that a massive well blowout on American deepwater oil fields was unlikely**.
- The **nationalistic arrogance of American regulators and oil companies**: they ignored international experience of previous disasters on deepwater drilling platforms, and assumed that they could neglect advanced oil drilling requirements because the Americans were pioneers in deepwater oil drilling and had the most skilled staff.
- **Deliberate lobbying by the American oil industry to persuade government to deregulate the sector and massively reduce the budget of the regulators**: unattractive wages, lack of skilled staff, inadequately qualified government officers, and so on. This led to a situation where regulators began to rely on information concerning new technologies from, and on the experience of, oil companies and their contractors, the very entities they were supposed to independently assess and regulate; as a result, **regulators failed to identify systemic failures in risk management, which the industry was trying to hide from regulators and the public**.
- **Fragmentary risk perception (failure to see the whole picture of risks) and lack of communication among representatives of the different organizations** working on the project about the real risks involved.
- **Rush during drilling because of delays in the schedule and cost overrun**, which encouraged those involved to ignore warnings and conceal information from other participants about defects during the cementing job in order to save time and money.

2.1.10 Raspadskaya Coal Mine Burnout (Russia, 2010)

In 2006, the Raspadskaya coal mine together with other mines managed by the open joint stock company (OJS) "Raspadskaya" had 781 million tons of coal reserves, but only 22 million tons were extracted by 2008.[482] The company still had more than 750 million tons of coal reserves—assets that could lead to substantial profits for their owners for decades to come. The coal company was very profitable: in 2009, its profitability was an incredible 51 %, while Gazprom, the Russian natural gas giant, exhibited a profitability of 36 %, and Lukoil, the largest Russian private oil company, of only around 17 %.[483]

---

[482]"Raspadskaya" is a leader of the coal industry in Russia, Mining Magazine, Mar. 2008, p. 4.

[483]Natalia Telegina, Victor Dyatlikovich, Why mines explode, Russian Reporter, May 20, 2010, № 19 (147).

The Raspadskaya coal mine itself was the largest underground mine in Russia with reserves of 450 million tons of coking coal: the mine produced up to 20 % of the coking coal in Russia and was among the top ten coking coal producers in the world. It was located in the Kuznetsk coal basin (Kuzbass) in the southern part of Western Siberia. On the night of 8 to 9 May 2010, two blowouts occurred at the mine. As a result, 91 people were killed and 94 injured. The blowouts ignited a huge underground fire, which continued to burn for years, destroying 300 km of coal roadways and making it one of the worst in the history of coal mining worldwide.

2.1.10.1 Risk Concealment Before the Disaster

Flawed Regulation of Coalfield Exploitation with High Concentrations of Methane

Kuzbass produces 56 % of Russia's coal and about 80 % of the coking coal for ferrous metallurgy.[484] The coal beds of Kuzbass contain high concentrations of methane, which has led to a tremendous number of methane-related blowouts since the Soviet industrialization of Siberia. The privatization of coal mining in Russia in the 1990s deteriorated the situation regarding production safety. Most of the current Russian underground coalmine projects were designed in the 1960s and 1970s, when the productivity of coal-plow machines was modest. During coal plowing in some Kuzbass mines, as much as 15–30 m^3 of methane could be released for every ton of coal. Soviet designers calculated the requirements for methane-ventilation equipment in mines according to the capacity of the coal-plow machines at the time.

[484]The economic potential of the Kemerovo region, Administation of Keverovo region, April 16, 2010, http://www.ako.ru/Ekonomik/potens.asp?n=1.

During the collapse of the Soviet Union and the privatization that followed in the 1990s, neither the Russian government nor the new private mine owners had the investment resources to make significant upgrades to the mines overall design; but the productivity of the modern coal-plow machines had dramatically increased since the 1970s. This led to a situation where the productivity of the mines increased manifold and the length of coal roadways expanded to hundreds of kilometers, but the capacity of ventilation systems for the timely elimination of methane and the anti-coal dust equipment in the mines did not keep pace. In addition, private owners did not have the resources to ensure preventive degassing of the methane from the belts prospective coal belts by the process of expensive drilling of special degassing wells to depths of up to 500 m: each well could cost up to $1 million and hundreds of wells are required for each coalfield. Over a period of up to five years, these wells allow for the concentration of methane within a prospective coal belt to be seriously reduced.[485,486] Such methods are used in the United States and Australia, where it is strictly prohibited to produce coal if a coal belt releases more than 9 m^3 of methane per ton of coal.[487] However the new Russian private owners, instead of investing in expensive methane venting equipment and a comprehensive system for degassing the coal roadways of prospective coal belts, usually preferred to pay compensations to the families of coal miners killed in methane blowouts—just US$35,000 per death.[488] Sergey Slastunov, a professor at the Russian mining university, commented on the situation: "*Nowadays, Russia has 50 mines, in which the concentration of methane has increased, but in our country, no one spends on degassing: this process takes at least three years. Therefore, the owners of the mines cannot wait – they need to quickly make a profit, and for them it is cheaper to bury the miners, rather than to wait a few years until the methane is eliminated*".[489]

The government's approach to overseeing the mining industry was based on the assumption that private owners would not violate safety procedures because of their own interest in the long-term productivity of mines, which would bring stable profitability and growing business capitalization value. However, experience showed that this assumption did not uphold in practice. Private owners focused only on short-term profitability instead of long-term benefits, as in many other cases involving various industries reported in this book. At the same time, accusations of extortion and corruption by some representatives of the state regulator in the coal industry led to the deregulation of government control. Any immediate

---

[485]Press-conference "How to prevent explosions in the mines? The opinions of scientists", RIA Novosti, Moscow, May 13, 2010, http://ria.ru/press_video/20100513/233925199.html.

[486]The accident at the "Raspadskaya"—12 Weeks Later, Finmarket, July 28, 2010, http://www.finmarket.ru/main/article/1596882.

[487]Sergey Slastunov, Methane coal mine safety in Russia—key issues of the coal industry, Mining informational and analytical bulletin, Moscow, 2011, № 12.

[488]Tatiana Zykova, Coal without the right to live. Rostehnadzor needs to regain the authority to halt unsafe mines without a trial, Rossiyskaya Gazeta, July 7, 2007.

[489]Dmitry Ivanov, Raspadskaya will not launch soon, Trud newspaper, May 26, 2010.

shutdown of mine production following a detected safety violation became quite a complex process for the regulator. As a result, the regulator could issue countless orders to mine managers, who would just ignore them with no serious impact on the profitability and functioning of the mines.[490]

During the decade before the accident, Raspadskaya mine was considered a leader in safe coal production in Russia. For example, in August 2002, the mine was selected for its exemplary safety record as a site for a field meeting of the State Council to discuss the problems of developing Russian coal mining; the meeting was attended by the President of Russia, ministers, and all the top executives of the Russian energy sector.[491] Nevertheless, after the accident, a very different picture of the working conditions at the mine emerged. In the 16 months before the accident, the regulator issued more than 1400 orders to eliminate safety violations at the mine, but the majority of them were ignored because the legislatively approved penalties were not enough of a deterrent. On four occasions, the regulator threatened to disqualify the mine director, but they could not act on their threats because there was no legal framework to back them up after the deregulation of state oversight.[492]

Deficient Remuneration System of Coalminers Promoting Information Distortion About the Real Level of Methane Concentration in the Mine

After the accident, workers at the mine revealed disgusting safety management practices that were informally introduced by the owners and the top management of the mine, and which could have led to the accident.[493] For example, the owners and the management set up a remuneration system based on the coal production performance of each shift, so that there was a direct dependence between the volume of coal produced and the earnings of each miner. A comprehensive system for degassing the coal roadways had been installed in the mine; but the performance-related pay system motivated coalminers to illegally block the methane detectors—which indicated dangerous levels of methane in the air when the new highly productive coal-plow machines were in operation. The methane detectors would have otherwise automatically halted coal production on the whole roadway until

[490]Tatiana Zykova, Coal without the right to live. Rostehnadzor needs to regain the authority to halt unsafe mines without a trial, Rossiyskaya Gazeta, July 7, 2007.

[491]Speech of Vladimir Putin at Russian State Council on the development of the coal industry, The Kremlin, Aug. 29, 2002.

[492]Alexander Terentyeva, Andrei Kotov, Investigators want to prosecute former director of "Raspadskaya" coal mine, Vedomosti, May 18, 2010.

[493]Revelation by miners about the practice of blocking methane detectors reported by journalists of "Week with Marianne Maksimovskaya", REN TV, May 15, 2010 and "Kvant" local TV channel of Keverovo region (May 10–16, 2010), http://www.youtube.com/watch?v=0IIWCvgnEWc.

the air had been degasified. The blocking of the methane detectors thus ensured continuous coal production and maintained the coalminers income.[494]

This led to a situation where the degassing system recorded completely inaccurate data about the concentration of methane within the mine. Therefore during any criminal investigation, it would be hard to prove the direct guilt of managers, because they had not directly ordered workers to continue coal production in methane-saturated air. Because the system would not have recorded a high concentration of methane, it would appear that there had been no such high concentration during the hours before any accident, and that the blowout must either have had an unavoidable cause—such as a sudden unpredictable methane emission—or been the fault of the miners who had blocked the methane sensors. Therefore, the miners themselves were unofficially allowed to make decisions about their own productivity and income… as well as safety. Because Kuzbass has a high rate of unemployment, thousands of young, strong but poorly educated workers were glad to get a coalmine job with a salary exceeding the average income in the region; the owner of the Raspadskaya mine, at a meeting with its personnel before the accident, stated that "*if somebody does not like the job, he could leave the company and the management of the coal mine could easily hire cheap Chinese coal miners*".[495] As a result, the blocking of methane detectors became a widespread practice in many Kuzbass coalmines.

In 2007, there were two methane blowout accidents on the nearby Ul'yanovskaya and Yubileynaya mines, in which 149 coalminers perished. In both accidents, it was revealed that there had been extensive blocking of the methane detectors, leading to the accumulation of methane to dangerous concentrations. The owners and management of both mines had put in place a payroll scheme that unofficially forced coal miners to violate safety rules in order to increase the profitability of coal production.[496] In the case of the Ul'yanovskaya mine, prosecutors found out in addition that management had corrupted representatives of the state safety regulator, who developed a strong loyalty to the managers so as to turn a blind eye to safety violations. In return, the coalminers even provided computers and office supplies to the regulator.[497] After the accident at the Raspadskaya mine, investigators found out that, at the time of the disaster, 40 % of the methane detectors in the mine were inoperative: the seals on 150 out of 400 detectors had been

---

[494]Revelation by miners about the practice of blocking methane detectors reported by journalists of "Week with Marianne Maksimovskaya", REN TV, May 15, 2010 and "Kvant" local TV channel of Keverovo region (May 10–16, 2010), http://www.youtube.com/watch?v=0IIWCvgnEWc.

[495]Revelation by miners about the practice of blocking methane detectors reported by journalists of "Week with Marianne Maksimovskaya", REN TV, May 15, 2010 and "Kvant" local TV channel of Keverovo region (May 10–16, 2010), http://www.youtube.com/watch?v=0IIWCvgnEWc.

[496]The cause of methane gas explosions in "Ul'yanovskaya" and "Yubileynaya" mines was pursuit of profit, Rossiyskaya Gazeta, June 6, 2007.

[497]Natalia Telegina, Victor Dyatlikovich, Why mines explode, Russian Reporter, May 20, 2010, № 19 (147).

tampered with.[498] This could also explain the massive fire, which continued for years after the accident burning out 300 km of coal roadways: over the years when methane detectors had been routinely blocked, neither management nor regulators nor coalminers had received any real data about the actual methane concentration in the air throughout the mine. After the accident, the owner of the coalmine admitted that he could not imagine that a disaster of such a tremendous scale could ever take place in reality.[499]

Ironically, the financial results of this economy based on biased incentives were devastating for the mine owners who had unofficially supported such dangerous practices.[500] After the accident, restoration costs were estimated at US$280 million, on top of the losses of the burned-out coal.[501] Six months after the accident, the market value of OJS "Raspadskaya", which had been the most lucrative coal company in Russia, had dropped by 59 % from US$6 billion to US$1.88 billion.[502] Total losses from the accident exceeded hundreds of times the profits resulting from implementing the remuneration system that had motivated miners to violate safety standards. The accident provoked social upheaval in the miners' hometown, where 3000 participants of a post-accident meeting and hundreds of other workers clashed with the police. This attracted the attention of government executives and the accident site was visited by Prime Minister Vladimir Putin. The accident led to new regulations on coal mining in Russia, bringing, amongst other changes, a new legal framework for the payment of miners.

Criminal charges of responsibility for the accident were brought against the mine's former director, his deputy, the managers of the coalmine plots and some mechanics at grassroots level. Unfortunately, for the majority of coal miners and residents of the Kuzbass region, the investigators could not find direct evidence that owners of the mine had given orders to miners to violate safety rules; some commentators attribute this to the close relations between Roman Abramovich, the Russian billionaire whose Evraz Holding company partly owns OJS "Raspadskaya", and high-ranking Russian politicians.[503]

---

[498] The former head of "Raspadskaya" was forced to resign four times, ITAR-TASS, May 18, 2010.

[499] Press-conference "How to prevent explosions in the mines? The opinions of scientists", RIA Novosti, May 13, 2010, http://ria.ru/press_video/20100513/233925199.html.

[500] Revelation by miners about the practice of blocking methane detectors reported by journalists of "Week with Marianne Maksimovskaya", REN TV, May 15, 2010 and Kvant' local TV channel of Keverovo region (May 10–16, 2010), http://www.youtube.com/watch?v=0IIWCvgnEWc.

[501] Anatoly Dzhumaylo, "Raspadskaya" produced losses, Kommersant, Sep. 21, 2012.

[502] Olga Alekseyeva, "Raspadskaya" is not for sale, Gazeta.RU, Oct. 6, 2010, http://www.gazeta.ru/business/2011/10/06/3792066.shtml.

[503] "Get to the truth. How did tame miners in Mezhdurechensk and who will be responsible for the tragedy at the mine", "Week with Marianna Maksimovskaya" Ren TV, May 22, 2010, http://www.youtube.com/watch?v=JnD9HSG3XPM.

Let us stress the remarkable similarity in all aspects between the Raspadskaya coal mine burnout disaster and the subprime financial crisis described in Sect. 2.2.3—in terms of incentives, the response of employees, the lack of culpability of management for designing an incentive structure that led directly to unethical or irresponsible behavior of employees, the emphasis on short-term profits over long-term benefits… and the fact that the ultimate disaster wiped out all previous gains.

Raspadskaya Coal Mine Burnout: Why Risks Were Concealed

- The owners and management focused on **short-term profitability instead of the long-term resilience of the coal mining business**. They created a sophisticated unofficial payroll scheme, which motivated coalminers to knowingly break safety rules. As a result, miners were potentially implicated in any possible methane blowout. This approach ensured that miners kept quiet about risky working practices.
- **Government oversight over Russian coalmining had been deregulated,** which allowed the management of the coalmine to violate safety rules with impunity.
- **Habituation/wishful thinking/overconfidence/self-suggestion/self-deception:** the owners and management of the mine totally underestimated the impact of a possible blowout of methane/coal-dust/air mixture during the intensive exploration of methane-saturated coal belts by powerful coal-plow machines, in parallel with the systematic desensitization of the methane detectors.

2.1.11 Fukushima-Daiichi Nuclear Disaster (Japan, 2011)

Nancy G. Leveson, Professor at MIT, on the widespread hindsight bias exhibited by experts during the analysis of causes of disasters:

After an incident: easy to see where people went wrong, what they should have done or avoided; easy to judge about missing a piece of information that turned out to be critical; easy to see what people should have seen or avoided [, but] almost impossible to go back and understand how world looked to somebody not having knowledge of outcome.

We need to mobilize "scientific imagination" in the process of decision

Hiroyuki Kameda (Lessons learned from the 2011 Great East Japan Earthquake, 2012)

Twenty-five years after Chernobyl, Japan repeated many of the mistakes of the Soviet nuclear industry during the Fukushima-Daiichi (Fukushima-1) nuclear disaster.

2.1.11.1 Summary of the Disaster

On March 11, 2011 at 2:46 p.m., a seaquake of magnitude 9.0–9.2 on the Richter scale occurred 70 km from the east coast of the Tohoku region in Japan. This was the largest earthquake ever recorded in Japan, and the United States Geological Survey considered that it was the fifth largest recorded worldwide since 1900.[504] The earthquake generated a large-scale tsunami, which reached the coastlines of Iwate, Miyagi and Fukushima prefectures approximately 50 min after the main shock, destroying hundreds of kilometers of coastline infrastructure and killing more than 18,800 people.[505]

There were five NPPs located in the disaster zone on the east coast of Japan. Several were hit by the tsunami but, at the Fukushima-Daiichi plant owned by Tokyo Electric Power Co. (TEPCO), the largest electric utility in Japan, it led to a severe nuclear disaster—level 7, the highest level on the International Nuclear Event Scale. The plant had 6 reactors (Units 1–6) and large pools with spent nuclear fuel, but only Units 1–3 were operating when the seaquake occurred: Units 5 and 6 were shut down for routine inspection, and Unit 4 was on reconstruction. The emergency shutdown (SCRAM) system on all operating reactors was activated successfully after the main shock. The maximum ground acceleration at the

[504]Christina Nyquist, The March 11 Tohoku Earthquake, One Year Later. What Have We Learned? March 9, 2012, U.S. Geological Survey.

[505]Prof. Dr. Wolfgang Kröger, Fukushima: Need for Reappraisal of Nuclear Risks? ETH Zürich, Keynote SRA-Europe 21st Annual Conference, Zurich, June 18–20, 2012.

Fukushima-Daiichi plant was 550 Gal (550 cm/s),[506] while the containment vessels were designed to retain functionality up to a seismic ground acceleration of 270 Gal and important buildings, structures, and equipment piping systems were designed to withstand 180 Gal.[507] Although the ground acceleration of the seaquake was beyond design limits, Unit 1 only had a leakage of coolant.[508] However, the plant lost all AC power sources because the earthquake had destroyed both external transmission lines and the Shin-Fukushima transformer station. DC power sources (diesel generators and batteries) generated electricity to cool the reactors for the next 51 min—until the tsunami reached the plant.[509] The maximum designed height of the protective seawall of the NPP was 5.7 m.[510] Vulnerable objects like seawater pumps were located beyond the seawall—4 m above sea level; diesel-generators and batteries were inside the reactor buildings—10 m above sea level.[511] But the tsunami waves generated by the Tohoku seaquake had built up to a height of 14–15.5 m by the time they hit the plant.[512] As a result, Fukushima-Daiichi NPP lost all sources of electricity to cool the reactors of Units 1, 2 and the spent fuel pool of Unit 4; Unit 3 had battery power for about 30 h; emergency diesel engines provided emergency power only to Units 5 and 6. Damage to the reactor core – and the resulting meltdown of nuclear fuel—began on Unit 1 3 h and 15 min after the tsunami struck, on Unit 3 after 43 h and on Unit 2 after 76 h.[513] There were 257 tons of nuclear fuel in the three operational reactors—Units 1 and 2 were fuelled by low-enriched uranium (LEU) and Unit 3 was fuelled by mixed oxide (MOX) fuel that contained plutonium—and 264 tons of spent nuclear fuel in the pool of Unit 4 at time of the disaster.[514,515]

[506] Prof. Dr. Wolfgang Kröger, Fukushima: Need for Reappraisal of Nuclear Risks? ETH Zürich, Keynote SRA-Europe 21st Annual Conference, Zurich, June 18–20, 2012.

[507] Fukushima Nuclear Accident Analysis Report, Tokyo Electric Power Company, Inc., June 20, 2012, pp. 8, 15.

[508] The official report of The Fukushima Nuclear Accident Independent Investigation Commission, The National Diet of Japan, Executive summary, July 5, 2012, pp. 17, 30.

[509] Ibid, p. 13.

[510] Fact Finding Expert Mission of the Fukushima Dai-Ichi NPP Accident Following the Great East Japan Earthquake and Tsunami, IAEA mission report, 24 May–2 June 2011, p. 11.

[511] The official report of The Fukushima Nuclear Accident Independent Investigation Commission, The National Diet of Japan, Executive summary, July 5, 2012, p. 14.

[512] Akira Izumo, Facts, Lessons Learned and Nuclear Power Policy of Japan after the Accident, Agency for Natural Resources and Energy, Ministry of Economy, Trade and Industry of Japan, January 24, 2012.

[513] The official report of The Fukushima Nuclear Accident Independent Investigation Commission, The National Diet of Japan, Executive summary, July 5, 2012, p. 13.

[514] Overview of facility of Fukushima Daiichi Nuclear Power Station, TEPCO, http://www.tepco.co.jp/en/nu/fukushima-np/outline_f1/index-e.html.

[515] The Status of Nuclear Fuel Stored at the Fukushima Daiichi and Fukushima Daini Nuclear Power Plants, Citizens' Nuclear Information Center (Japan), Jan. 31, 2013, http://www.cnic.jp/english/newsletter/nit154/nit154articles/03_nf.html.

The accident resulted in the release of approximately 900–940 PBq of radioactive substances into the atmosphere,[516,517] compared with the 5200 PBq estimated for the 1986 Chernobyl accident. Thus, the Japanese government reported to the International Atomic Energy Agency (IAEA) that the release was 1/6 of the emissions from the Chernobyl accident when converted to iodine. One hundred and fifty thousand residents were evacuated for a long time[518] because of radioactive contamination: 1800 km^2 of the Fukushima Prefecture have levels that would give a potential cumulative dose of 5 mSv/year or more.[519]

2.1.11.2 Environmental and Economic Consequences of the Disaster

It is expected that more than 40 years will be needed to remove the melted nuclear fuel on the plant and to clean vast areas contaminated by radiation: the Japan Center for Economic Research estimates that the cleanup from the accident may require 20 trillion yen[520] which is around US$200 billion or 4.2 % of the Japanese GDP. Since the disaster, the plant has been contaminating 400 tons of water daily to cool the melted reactors and spent nuclear pools. In the two years since the disaster, 280,000 tons of contaminated water have been stored in tanks at the plant.[521] In April 2011, TEPCO—in violation of the Convention on the Prevention of Marine Pollution by Dumping of Wastes and Other Matter but with the approval of the Japanese government—dumped 11,000 tons of low-radioactive water into the Pacific Ocean.[522] According to the opinion of Juan Carlos Lentijo, leader of the International Atomic Energy Agency mission team, "*it will be nearly impossible to ensure the time for decommissioning such a complex facility in less than 30–40 years as it is currently established in the roadmap*".[523]

After the accident, the Japanese government dramatically changed its position about the role of nuclear energy in the country's energy balance, aiming to reduce its input from 35 % in the 2010s to zero by 2035. Before the accident, the country

[516]The official report of The Fukushima Nuclear Accident Independent Investigation Commission, The National Diet of Japan, Executive summary, July 5, 2012, p. 39.

[517]Fukushima Accident, World Nuclear Association, updated 13 January 2014.

[518]The evacuation map in the following official government site (http://www.meti.go.jp/english/earthquake/nuclear/roadmap/pdf/141001MapOfAreas.pdf) shows the most dangerous area called Area3, which is defined as follows: "*Area3: Areas where it is expected that the residents have difficulties in returning for a long time*".

[519]The official report of The Fukushima Nuclear Accident Independent Investigation Commission, The National Diet of Japan, Executive summary, July 5, 2012, p. 38.

[520]The Fukushima Nuclear Accident and Crisis Management—Lessons for Japan-U.S. Alliance Cooperation, The Sasakawa Peace Foundation, Sep. 2012, p. 38.

[521]Mari Yamaguchi, IAEA: Japan nuke cleanup may take more than 40 years, Associated Press, April 22, 2013.

[522]Eben Harrell, Fukushima: Dumping into the Sea, Time, April 05, 2011.

[523]Mari Yamaguchi, IAEA: Japan nuke cleanup may take more than 40 years, Associated Press, April 22, 2013.

had been planning to get more than 50 % of its energy from nuclear energy by 2030. In 2011–2012, Japan faced a serious shortage of electricity for industrial and domestic needs—by mid-May 2011, only 17 out of the remaining 50 reactors in the country were operating due to intensive safety inspections. From May until July 2012, all Japanese nuclear reactors were suspended. This led to additional spending of US$40 billion on hydrocarbon fuel imports.[524] The shutdown of Japan's entire nuclear fleet has had profound economic consequences for the country due to the US$134 billion trade deficit in 2013 brought about by increased fossil fuel imports and lower productivity. Higher electricity prices and increased CO_2 emissions are also of concern. Together with on-going dire economic problems of the "two lost decades" following the bursts of the stock market and real-estate bubbles in Japan in 1990, this additional stress is catalyzing a reassessment of these political decisions. In June 2014, the three major business lobbies urged the Industry Minister to expedite restart of the nuclear reactors. "*The top priority in energy policy is a quick return to inexpensive and stable supplies of electricity*", they said.[525] There thus seems to be a rising political will to reinstall the nuclear industry as a major source of energy in Japan. For instance, in July 2014, Kyushu's Sendai nuclear power plant has been given draft approval to restart by Japan's Nuclear Regulation Authority (NRA), having met the greatly upgraded safety requirements published in July 2013. This is a major step towards actually returning to service, after Kyushu committed some US$3 billion on post-Fukushima upgrades for its nuclear plants. So far, ten more PWRs (pressurized water reactors) are queued for approval by NRA, the reconstituted safety regulator, plus seven BWRs (boiling water reactors), which required more major upgrading and also need formal approval from local government.[526]

After the accident, the National Diet, Japan's legislative body, established the Fukushima Nuclear Accident Independent Investigation Commission (NAIIC) with the legal authority to request—and take action to obtain—any necessary documents or evidence required. During the investigation, the commission performed interviews with 1167 people and organized 900 h of hearings. The commission concluded that the "*accident at the Fukushima Daiichi Nuclear Power Plant cannot be regarded as a natural disaster. It was a profoundly manmade disaster – that could and should have been foreseen and prevented. The accident was clearly 'manmade'. We believe that the root causes were the organizational and regulatory systems that supported faulty rationales for decisions and actions, rather than issues relating to the competency of any specific individual. We found an organization-driven mindset that prioritized benefits to the organization at the expense of the public*".[527]

---

[524]Nuclear Power in Japan, World Nuclear Association, updated Feb. 2014.

[525]World Nuclear Association, weekly digest 11 and 18 July 2014.

[526]Ibid.

[527]The official report of The Fukushima Nuclear Accident Independent Investigation Commission, The National Diet of Japan, Executive summary, July 5, 2012, p. 21.

2.1.11.3 Risk Concealment Before the Disaster

Common Interests of the Japanese Government and Private Corporations Towards the Development of the Civil Nuclear Industry in Japan

Japan began to develop civil nuclear energy in the mid 1960s. Nuclear energy has been a national strategic priority since the oil crisis in 1973 because of Japan's heavy dependence on imported fuel, which provided 84 % of its energy requirements in the 2010s.[528] Before the accident, nuclear energy was both promoted and regulated by the Nuclear and Industrial Safety Agency (NISA), working under the authority of the Ministry of Economy, Trade & Industry (METI). There was a deep seated conflict of interests: the primary goal of NISA was to protect society from radiation threat but, at the same time, NISA focused on the energy independence of Japan—which meant supporting large low-cost electricity production from a large number of nuclear plants, and maintaining a stable financial climate for private operators of nuclear plants to enable the further development of nuclear energy. For decades, a cozy relationship developed between operators, regulators and academics, which led to a situation where "*the regulators and the operators prioritized the interests of their organizations over the public's safety, and decided that Japanese nuclear power plant reactor operations 'will not be stopped'. Because the regulators and operators have consistently and loudly maintained that 'the safety of nuclear power is guaranteed', they had a mutual interest in averting the risk of existing reactors being shut down due to safety issues, or of lawsuits filed by anti-nuclear activists. They repeatedly avoided, compromised or postponed any course of action, and any regulation or finding that threatened the continued operation of nuclear reactors*".[529]

Unlearned Lessons from Three Mile Island and Chernobyl Accidents

The Chernobyl disaster had little influence on Japanese nuclear safety measures because of the national perception of Japan's unique technical culture, which was assumed to be better able to avoid, or else endure, such catastrophes. In 1986, the country was at the peak of a three-decade long "Economic Miracle" and the Japanese felt "*great pride in its global reputation for excellence in engineering and technology*".[530] As a result, "*the regulators also had a negative attitude toward the importation of new advances in knowledge and technology from overseas… At a time when Japan's self-confidence was soaring, a tightly knit elite with enormous financial resources had diminishing regard for anything 'not invented*

[528]Nuclear Power in Japan, World Nuclear Association, updated February 2014.

[529]The official report of The Fukushima Nuclear Accident Independent Investigation Commission, The National Diet of Japan, Executive summary, July 5, 2012, p. 43.

[530]The official report of The Fukushima Nuclear Accident Independent Investigation Commission, The National Diet of Japan, Executive summary, July 5, 2012, p. 9.

here".[531,532] They came to the conclusion that "*Japanese plants are safe, because we are Japanese*".[533] Leonid Bol'shov, head of the Russian Safety Institute of Atomic Energy Sciences, established after the Chernobyl accident, stated: "*we did not learn the lesson after the Three Mile Island accident and so we were faced with Chernobyl. After Chernobyl, we have learned the lessons, but it seems that the Japanese have not learned them and now they have been faced with Fukushima. In 1992, I went to Japan to various facilities, including nuclear power plants, where we were shown simulators for operators. We asked staff of the station: 'Do operators simulate severe accidents?' They replied to him: 'No, we have a good station'. That is why we need to learn from the mistakes of others. A point of view, that abroad everything is bad and in my country everything is good, is very dangerous*".[534] The Japanese never believed that a beyond-design accident would ever happen, and they never prepared for one. In addition, according to Akihisa Shiozaki, an attorney who was instrumental in putting together the first independent, non-governmental investigation of the Fukushima nuclear disaster, the government and the industry were reluctant to consider worst-case scenarios because of Japan's unique history: after World War II and the destruction of Nagasaki and Hiroshima by nuclear weapons, the Japanese population vehemently opposed all use of nuclear power in their country. Therefore, at the beginning of civil nuclear development in Japan, the government undertook a campaign to persuade people of "*the absolute safeness*" of nuclear power. "*Absolute safeness meaning that there was no risk that something could go wrong, no risk that a meltdown could happen. Well, that myth of absolute safeness developed over the years into a culture where it almost became a taboo to even talk about this... Discussing a worst-case scenario was feared because it might bring panic to the citizens. And therefore it was omitted from the regulatory discussions*". Eric Feldman, a law professor at the University of Pennsylvania, considers that there were significant political and economic forces backing nuclear power, and that as a result, "*talking about worst-case scenarios was avoided not simply because it would scare people, but because such fear would mean that local communities would oppose the building of reactors, and without local support the reactors would not be built*".[535]

Consequently, the government and the industry thought they did not need to implement serious safety improvements learnt from the experience of foreign nuclear accidents because Japanese stations were already designed for severe disasters like high-magnitude earthquakes. In 1991, this assumption led to a situation in which safety measures applied on Japanese nuclear stations became voluntary

---

[531] Ibid, p. 16.

[532] Ibid, p. 9.

[533] Asmolov: Fukushima Disaster was Predictable, The Mainichi Daily News, 26 April 2011.

[534] "It could have been prevented," Leonid Bol'shov, head of the Russian Safety Institute of Atomic Energy Sciences, about the differences between Fukushima and Chernobyl, Kommersant, 26 April 2011, № 73 (4614).

[535] Disasters, Rebuilding and Leadership—Tough Lessons from Japan and the U.S., The Wharton School, University of Pennsylvania, October 2013, p. 3.

and independent from the control of regulators: "*the accident management, including expedient and flexible measures that might be required under actual situations, shall be considered and implemented by the operators based on their 'technical competency' and 'expertise', but [it] shall not require authority to regulate the specific details of measures*".[536] After the accident, it was revealed that Japanese nuclear operators had ignored and/or delayed implementation of many IAEA recommendations and guidelines about safety measures generated by nuclear accidents elsewhere in the world.[537,538,539] Moreover, reluctance to reveal the failure of Japanese nuclear plants to conform to international standards, and fears of the possible restructuring of the nuclear community in Japan, led to the decision by the Japanese nuclear regulator to decline overseas scientific assistance to Fukushima-Daiichi NPP.[540] The NAIIC commission stated: "*this was a disaster 'Made in Japan'... Its fundamental causes are to be found in the ingrained conventions of Japanese culture: our reflexive obedience; our reluctance to question authority; our devotion to 'sticking with the program'; our groupism; and our insularity... This conceit [disregard for anything 'not invented here'] was reinforced by the collective mindset of Japanese bureaucracy, by which the first duty of any individual bureaucrat is to defend the interests of his organization. Carried to an extreme, this led bureaucrats to put organizational interests ahead of their paramount duty to protect public safety. Only by grasping this mindset can one understand how Japan's nuclear industry managed to avoid absorbing the critical lessons learned from Three Mile Island and Chernobyl; and how it became accepted practice to resist regulatory pressure and cover up small-scale accidents. It was this mindset that led to the disaster at the Fukushima Daiichi Nuclear Plant*".[541]

Concealment of Minor Incidents Was a Decade-Long Practice Within the Nuclear Industry

In 2002, the government of Japan launched an investigation into the widespread practice of falsifying routine safety inspection data at TEPCO NPPs because the

---

[536]The official report of The Fukushima Nuclear Accident Independent Investigation Commission, The National Diet of Japan, Executive summary, July 5, 2012, p. 28.

[537]The official report of The Fukushima Nuclear Accident Independent Investigation Commission, The National Diet of Japan, Executive summary, July 5, 2012, p. 75.

[538]The official report of The Fukushima Nuclear Accident Independent Investigation Commission, The National Diet of Japan, Chapter 1. Was the accident preventable? July 5, 2012, pp. 53–57.

[539]James M. Acton, Mark Hibbs, Why Fukushima Was Preventable, Carnegie Endowment for International Peace, March 2012, p. 2.

[540]The official report of The Fukushima Nuclear Accident Independent Investigation Commission, The National Diet of Japan, Executive summary, July 5, 2012, p. 80.

[541]Ibid, p. 80.

true data had been deleted. In the end, TEPCO confirmed 200 cases of data falsification between 1977 and 2002. Tsunehisa Katsumata, appointed as president of TEPCO after the falsification scandal, revealed "*serious cases of inappropriate conduct in which employees should have reported cracks in the shroud to the national government [and] failure to keep records of problems. The engineers involved were afraid that, if they notified the national government of the problem, they would have to shut down the plant for a longer period of time than planned. This fear resulted in a conservative mentality that led them to avoid reporting problems to the national government as long as they believed that safety was secured. Engineers, who were reluctant to report problems, therefore eventually came to believe that they would be allowed not to report faults if the faults did not pose an immediate threat to safety and, as a result, they went as far as to delete factual data and falsify inspection and repair records*".[542] The same practice occurred within other nuclear operators—for instance, in 2007, Hokuriku Electric Power confessed to hiding a nuclear incident on Shika NPP in 1999.[543] Nevertheless, according to research of James Acton and Mark Hibbs, "*the relationship between NISA and the Japanese government, on the one hand, and that between NISA and industry, on the other, was not fundamentally challenged*" after the falsification scandal.[544]

Tragic Underestimation of the Threat of High-Wave Tsunamis to TEPCO's Nuclear Power Plants

In 2003, after the restart of nuclear plants that were suspended in the falsification scandal (the suspension and the restart cost TEPCO about US$1.9 billion[545]), TEPCO "*implemented a company-wide program to reduce cost, including measures to curb maintenance expenditures*".[546] NISA helped operators to reduce costs on safety installations by allowing that "*actions should be taken autonomously by the operator*". Moreover, "*Since 2006, the regulators and TEPCO were aware of the risk that a total outage of electricity at the Fukushima Daiichi plant might occur if a tsunami were to reach the level of the site… NISA knew that TEPCO had not prepared any measures to lessen or eliminate the risk, but failed to provide specific instructions to remedy the situation... NISA informed the operators that they did not need to consider a possible station blackout because the*

[542]Mr Katsumata Speech BGM 2003, Reconstruction after Misconduct—the Pursuit of Excellence, TEPCO, http://www.tepco.co.jp/en/news/presen/pdf-1/0310-e.pdf.

[543]Steve Stecklow, Nuclear Safety Reports Called Into Question, The Wall Street Journal, Aug. 3, 2007.

[544]James M. Acton, Mark Hibbs, Why Fukushima Was Preventable, Carnegie Endowment for International Peace, March 2012, p. 24.

[545]Nuclear Power in Japan, World Nuclear Association, updated Feb. 2014.

[546]2003 Annual Report, TEPCO, pp. 2, 19, http://www.tepco.co.jp/en/corpinfo/ir/tool/annual/pdf/ar2003-e.pdf.

probability was small and other measures were in place. It then asked the operators to write a report that would give the appropriate rationale for why this consideration was unnecessary".[547]

An important reason why the regulators and TEPCO underestimated the risk of a high-wave tsunami was that the Japanese nuclear industry had focused so much on the possibility of earthquakes. They felt confident that they had made comprehensive calculations that guaranteed safety from beyond-design accidents. In the 1960s, when the Fukushima-Daiichi plant was designed by American companies General Electric (who designed the boiling water or BWR reactors) and EBASCO (who designed the plant), its foundations were at a height of 35 m above sea level on a bluff, but civil engineering staff of TEPCO lowered the bluff by 25 m in order to mitigate the threat posed by earthquakes and reduce the cost of running the seawater pumps.[548] The maximum expected height of a tsunami wave near Fukushima-Daiichi NPP was only 3.1 m above sea level, based on 13 earthquake tsunami statistics dating from 1611. Among them, the 1960 Chilean Earthquake tsunami, at 3.122 m, was the largest tsunami to have hit the Fukushima coastline since 1611.[549] Nevertheless, since 1498, there had been 12 tsunamis off the coast of Japan and the Russian Kuril Islands with maximum amplitudes of more than 10 m—half of which had maximum amplitudes over 20 m—generated by earthquakes with magnitudes between 7.4 and 9.2.[550] In particular, the Jogan Jishin earthquake in AD 869 occurred near the Fukushima Daiich Nuclear power plant, and created the largest tsunami in this region until that of 2011. The Active Fault and Earthquake Research Center (AFERC) developed a detailed study the Jogan Jishin earthquake tsunami.[551] Based on the study,[552] Mr. Okamura, a researcher at AFERC, warned in 2010 the nuclear and industrial safety subcommittee, the seismic and structural design subcommittee, and the working group for "earthquake, tsunami and geological features, the ground", that there was a possibility for a huge earthquake and tsunami near Fukushima. The meeting was held at the Ministry of Economy, Trade and Industry (METI) in June 24, 2010, and TEPCO officers were included as member of the working group. Mr. Okamura asked the officer of TEPCO (Mr. Nishimura) why the official report of the meeting did not

[547]The official report of The Fukushima Nuclear Accident Independent Investigation Commission, The National Diet of Japan, Executive summary, July 5, 2012, p. 16.

[548]The official report of The Fukushima Nuclear Accident Independent Investigation Commission, The National Diet of Japan, Chap. 1. Was the accident preventable? July 5, 2012, p. 23.

[549]Ibid, p. 23.

[550]James M. Acton, Mark Hibbs, Why Fukushima Was Preventable, Carnegie Endowment for International Peace, March 2012, p. 12.

[551]https://unit.aist.go.jp/actfault-eq/Tohoku/jogan_tsunami_e.html.

[552]Namegaya, Y., Satake K., and Yamaki, S. (2010). Numerical simulation of the AD 869 Jogan tsunami in Ishinomaki and Sendai plains and Ukedo river-mouth lowland, Annual Report on Active Fault and Paleoearthquake Researches, Geological Survey of Japan/AIST, No. 10, 1–21 (2010).

mention the Jogan Jishin earthquake and the associated risk of a huge tsunami. He called for a thorough investigation of the risk of unexpected tsunamis. In spite of Mr. Okamura's warning, TEPCO never prepared for the risk.[553]

The design of BWR reactors located on the ocean coastline of Japan came from American experience of reactors sited near rivers, which had never been intended to face sudden high-level waves or flash flooding. American engineers placed backup emergency diesel generators and DC batteries in turbine buildings around 4 m above sea level, and TEPCO agreed with this solution because nobody expected a tsunami wave of more than 3.1 m.[554] NISA never objected to this solution because the regulator had focused for decades on earthquake-resistant solutions rather than dealing with any possible tsunami threat. NISA also preferred to fund academic grants on earthquake safety, thereby marginalizing tsunami safety.[555] During the construction of the Fukushima-Daiichi NPP, Toshiba engineers wanted to improve on the General Electric design, but TEPCO blocked any major changes: "*TEPCO, conservative by nature, didn't allow the Japanese companies building the plant to make any alterations to GE's basic design… [TEPCO] told the Japanese makers to build the plants exactly in the same way as those of foreign makers… TEPCO was very bureaucratic*".[556] And once the Fukushima-Daiichi plant was operating, many engineers there were worried about the placement of the generators: "*If an earthquake hits and destroys some of the pipes above, water could come down and hit the generators. DC batteries were also located too close to the diesel generators. It's not at all good in terms of safety. Many of the middle-ranking engineers at the plant shared the same concern*".[557] In 2002, when the Japan Society of Civil Engineers issued a new tsunami assessment method for nuclear power plants, TEPCO raised the estimation of the maximum tsunami to which the Fukushima Daiichi Nuclear Power Plant could be exposed to 5.7 m and notified NISA. However, the company made only minor improvements, which did not affect the position of emergency generators and backup batteries.[558]

---

[553]The certified agenda of the METI meeting and extracts of the meeting (in Japanese) were communicated to us and translated by Professor Kaizoji and are available from the authors upon request.

[554]Reiji Yoshida, GE plan followed with inflexibility, The Japan Times, July 14, 2011.

[555]James M. Acton, Mark Hibbs, Why Fukushima Was Preventable, Carnegie Endowment for International Peace, Mar. 2012, p. 29.

[556]Reiji Yoshida, GE plan followed with inflexibility, The Japan Times, Jul 14, 2011.

[557]Ibid.

[558]The official report of The Fukushima Nuclear Accident Independent Investigation Commission, The National Diet of Japan, Chap. 1. Was the accident preventable? July 5, 2012, pp. 24–25.

Internal Risk Communication Failure

The risk of a potentially severe accident never appeared in TEPCO's list of risks. Any question about operating risks and nuclear safety was under the competence of the on-site plant department and would never have been raised at central risk management meetings.[559] Masatoshi Toyota, a former senior vice president of TEPCO and one of the executives who oversaw the construction of the Fukushima plant, stated: "*I didn't know until March 11 that the diesel generators were placed in the turbine buildings. If I had known, I would have definitely changed that*".[560] The Japanese "*reluctance to question authority*" and their slow bureaucratic system, geared only to passing on good news, led to a situation when executives had little understanding of the real condition of their plants and were fully satisfied with reassuring reports from the stations. Moreover, the TEPCO corporate system "*tolerated or encouraged the practice of covering up problems*"[561] so that "*utilization of risk information was insufficient, and the risk of [a station blackout] was not widely recognized by the management*".[562] The regulator required that "*nuclear reactor facilities shall be designed such that safe shutdown and proper cooling of the reactor after shutting down can be ensured in case of a short-term total AC power loss*".[563] However "*short-term*" blackout, for the majority of nuclear plants, meant just 30 min or less, because of the high-performance repair service of transmission lines in Japan after earthquakes. Nuclear executives "*fundamentally believed that, if we lost off-site power, we would be back up on the grid in no more than about half an hour*".[564] After the accident at the Fukushima-Daiichi plant, TEPCO Chairman Tsunehisa Katsumata said that the possibility of an unanticipated tsunami—resulting in a blackout lasting days rather than hours—had not been communicated internally to him when he was president of the company in 2008, because "*such [a] tsunami would not happen in reality*".[565]

This assumption was clearly demonstrated by the complete lack of response from TEPCO management to a chain of great natural disasters and scientific warnings in the preceding few years. Thus, the company's engineers took no account of an incident on the French Blayais NPP in December 1999, when the extratropical storm Martin brought a combination of high tide and strong winds, flooding the

---

[559]The official report of The Fukushima Nuclear Accident Independent Investigation Commission, The National Diet of Japan, Executive summary, July 5, 2012, p. 44.

[560]Reiji Yoshida, GE plan followed with inflexibility, The Japan Times, July 14, 2011.

[561]James M. Acton, Mark Hibbs, Why Fukushima Was Preventable, Carnegie Endowment for International Peace, Mar. 2012, p. 28.

[562]Ibid, p. 27.

[563]Ibid, p. 28.

[564]Ibid, p. 28.

[565]The official report of The Fukushima Nuclear Accident Independent Investigation Commission, The National Diet of Japan, Executive summary, July 5, 2012, p. 77.

plant and cutting its off-site power supply.[566] In 2006, a group of junior employees at TEPCO, inspired by one of the consequences of the Indian Ocean tsunami in December 2004—which flooded seawater pumps at the Madras NPP in India[567]—estimated "*that if … a 10 m tsunami occurred, there was a risk that the emergency seawater pump would cease to function and core damage could occur; and that if … a 14 m tsunami occurred, there was a risk that electrical equipment would cease functioning as the building flooded, making it impossible to use the emergency diesel generator, external AC power supply, or DC power supply, thereby causing the loss of all power sources*". The company shared this information with NISA.[568] The group also asked for 25 million dollars to implement appropriate measures, but TEPCO executives said that the study session had been conducted as training for junior employees, and that the company did not really expect such a large tsunami.[569]

In February 2008, TEPCO engineers, following new simulations taking recent events and the longer historical perspective into account, stated that a tsunami wave between 9.3 and 15.7 m in height could hit Fukushima-Daiichi NPP. However Sakae Muto, Deputy General Manager of TEPCO's Nuclear Power Plant Division in 2008, and others thought that no urgent action was required because such a tsunami was very unlikely.[570] Therefore, they did not convey the results of the simulations to the president of TEPCO. During NAIIC commission hearings, Tsunehisa Katsumata, who was president of TEPCO between October 2002 and June 2008, confirmed that he had never received any information about the threat of a tsunami leading to a total blackout at a TEPCO station during his presidency.[571] "*Information [about tsunami risks] was stopped at [TEPCO] headquarters*"[572] because "*the majority view in the company was that no major tsunami was likely*".[573] In fact, TEPCO bureaucrats reported the results of the simulations to NISA only three years after the simulations, on March 7, 2011, four days before the disaster.[574] It is remarkable that this risk was not revealed to NISA when the regulator was making the decision to issue a new ten-year license for the

---

566 James M. Acton, Mark Hibbs, Why Fukushima Was Preventable, Carnegie Endowment for International Peace, March 2012, pp. 22–23.

567 Ibid, p. 11.

568 The official report of The Fukushima Nuclear Accident Independent Investigation Commission, The National Diet of Japan, Chapter 1. Was the accident preventable? July 5, 2012, p. 26.

569 Earthquake Report № 443, Japan Atomic Industrial Forum, Inc, Jun 14, 2012.

570 The official report of The Fukushima Nuclear Accident Independent Investigation Commission, The National Diet of Japan, Chapter 1. Was the accident preventable? July 5, 2012, p. 29.

571 Stream record of NAIIC/Jikocho 12th Commission Meeting, May 14, 2012 No. 1 http://www.youtube.com/watch?v=BW4L-InZWg8 (1:25–1:33; 1:47–1:55; 2:00–2:04; 2:14–2:18).

572 Ibid.

573 TEPCO chairman blames politicians, colleagues for Fukushima response, The Asahi Shimbun, May 15, 2012.

574 James M. Acton, Mark Hibbs, Why Fukushima Was Preventable, Carnegie Endowment for International Peace, March 2012, pp. 1, 12.

forty-year-old Unit 1 in February 2011, less than a month before the disaster.[575] Unfortunately, it was only after the accident at the Fukushima-Daiichi plant that NISA ordered operators to reconstruct all seawalls around coastal NPPs with a minimum height of 15 m.[576]

External Risk Communication Failure

Ultimately, before the disaster, nobody from central government, the prefectural governments, or the local community was aware of the operating risks of the Fukushima-Daiichi NPP. Thus, Yuhei Sato, the governor of Fukushima Prefecture at the time of the accident, said that he did not know that the safety systems on the plant did not include anti-tsunami measures.[577] And Katsutaka Idogawa, mayor of the nearby city of Futaba, said, "*Ever since I was appointed as the mayor, I kept expressing our concern about the nuclear power plant to TEPCO and NISA. They kept telling us there is no need to worry, that the plant is absolutely safe*".[578] People from the towns hosting nuclear power plants testified that "*there was no communication about potential issues that are out of human control*". They always heard "*how safe the plants are*".[579]

2.1.11.4 Risk Concealment After the Disaster

Repetition of the Mistakes Done During Chernobyl Accident in the Japanese Crisis Response Actions to the Nuclear Disaster

Just as in the aftermath of the Chernobyl accident, the insufficient or falsely reassuring information transmitted from the operators at Fukushima-Daiichi NPP to TEPCO headquarters and the authorities delayed crucial appropriate reactions from officials to the most serious challenge for Japan since the Second World War. Fact is that, at least 8 h after the tsunami, TEPCO headquarters still believed that "*Reactor (Unit 1) was shut down and nuclear steam is [being] cooled by the isolation condenser*" and spread this information to NISA, the government and the public.[580] This message implied that the situation at the station was not too dangerous in the view of TEPCO executives, NISA and the government. The isolation

---

[575]Ibid, p. 16.

[576]Ibid, p. 18.

[577]The official report of The Fukushima Nuclear Accident Independent Investigation Commission, The National Diet of Japan, Executive summary, July 5, 2012, p. 81.

[578]Ibid, p. 73.

[579]Ibid, p. 76.

[580]Plant Status of Fukushima Daiichi Nuclear Power Station (as of 0AM March 12th), TEPCO, Press Release, Mar 12, 2011, http://www.tepco.co.jp/en/press/corp-com/release/11031203-e.html.

condenser in Unit 1 was a passive non-electronic backup cooling system, which is located under BWR reactors to cool the stream from the reactor core and restore condensed water to the reactor by gravity. After the earthquake, this system had started up automatically, but the operators of Unit 1 soon recognized that the condenser was cooling the core too quickly and shut it down manually in order to protect the reactor from damage.[581] After the tsunami, when operators lost all information from all gauges, they could no longer evaluate the condition of the condenser. Moreover, communication between the on-site emergency control center and each control room was limited to a single wired telephone line.[582] In addition, it was impossible to walk anywhere on the site of the plant in the first few hours, because of debris caused by the tsunami.[583] According to an investigation by Japanese public broadcasting corporation NHK, involving dozens of interviews with TEPCO staff[584] over the previous 40 years, the isolation condenser had never been tested during safety checks—so nobody on the plant knew that, during correct functioning of the condenser, it should emit a loud sound and large amounts of redundant steam escape through holes in the wall of the reactor building. Due to the absence of data from instruments, and their own lack of technical knowledge, operators and management of the plant assumed that the isolation condenser was working properly, even though staff at the plant heard no noise and observed little or no steam flowing from holes in the wall of the reactor building.[585] As a result, they conveyed incorrect information to their superiors. In reality, reactor core damage and the meltdown of nuclear fuel on Unit 1 set in just 3 h and 15 min after the tsunami hit. But it was only 21 h after the tsunami that TEPCO confirmed that the isolation condenser had not yet been used, remarking reassuringly that "*we are implementing a measure to reduce the pressure of the reactor containment vessels in order to fully secure safety. The reactor water level is decreasing, we will continue injecting water step by step*".[586] During the investigation, TEPCO officials declared that they had failed to respond promptly to the loss of cooling functions because they mistakenly believed that an emergency core cooling system was still functioning after the tsunami.[587] They probably believed that the cooling function could be restored within that time frame. So they issued a

[581]The official report of The Fukushima Nuclear Accident Independent Investigation Commission, The National Diet of Japan, Chapter 2. Escalation of the accident, July 5, 2012, p. 8.

[582]James M. Acton, Mark Hibbs, Why Fukushima Was Preventable, Carnegie Endowment for International Peace, March 2012, p. 16.

[583]NAIIC/Jikocho 6th Commission Meeting 2012/3/14, http://www.youtube.com/watch?v=4cEM6cvLm2s (0:23–0:28).

[584]Meltdown. Oversights in the Reactor Cooling System, NHK Documentary, Yoshihiro Nemoto, Jun 11, 2013.

[585]The official report of The Fukushima Nuclear Accident Independent Investigation Commission, The National Diet of Japan, Chapter 2. Escalation of the accident, July 5, 2012, pp. 78, 82–89.

[586]Plant Status of Fukushima Daiichi Nuclear Power Station (as of 1PM March 12th), TEPCO, Press Release, Mar 12, 2011, http://www.tepco.co.jp/en/press/corp-com/release/11031219-e.html.

[587]TEPCO officials unaware of cooling system shutdown, The Asahi Shimbun, December 19, 2011.

report to the Prime Minister's office (the Kantei) reassuring them: "*There will be no problem for eight hours even if no cooling [of the reactors] occurs*".[588] Moreover, the Chairman of the Nuclear Safety Commission of Japan visited the Kantei several hours after the disaster and said that "*[t]he situation is not one in which radiation is leaking to the outside atmosphere. While there are problems with the power source, the nuclear chain reaction has been completely stopped. The only thing left is to cool the reactors*".[589] In reality, with no cooling taking place, pressure inside the reactors of Unit 1 and Unit 3 reached twice the design limit between 8 and 11 h after the tsunami.[590] The main threat of high pressure inside BWR reactors is the possibility of the containment vessel exploding, releasing huge quantities of radioactive materials—dozens to hundreds of times more than were emitted at Chernobyl.[591] US protocols on handling accidents at similar reactors call for venting before pressure exceeds the design limit.[592] According to the NAIIC commission report, "*[t]he actual on-site situation of the vent in Unit 1 was not communicated to NISA or the Prime Minister's office [Kantei], which helped create an atmosphere of distrust between TEPCO's on-site management, the regulatory agencies and the Prime Minister's office*".[593] TEPCO was unwilling to vent, because they knew that venting radioactivity would cast doubt on the safety of the nuclear industry around the nation.[594] Exposure to radiation is a very sensitive theme for the Japanese after the nuclear destruction of Hiroshima and Nagasaki. In addition, the staff of the plant would not easily be able to vent the reactors: without electricity for the air compressors, the manual opening of vent valves was a very laborious and time-sensitive task, which would require preventive evacuation of nearby residents due to radiation emissions.[595] In the NAIIC hearing, Sakae Muto—executive vice president of TEPCO and general manager of the nuclear power plant division at the time of the disaster—confirmed that he was aware that TEPCO could not manage such a severe accident independently, but did not inform the Prime Minister of Japan or other authorities about TEPCO's

[588]WHAT WENT WRONG: Fukushima flashback a month after crisis started, The Asahi Shimbun, April 12, 2011.

[589]Ibid.

[590]The official report of The Fukushima Nuclear Accident Independent Investigation Commission, The National Diet of Japan, Chapter 2. Escalation of the accident, July 5, 2012, pp. 20, 40.

[591]NAIIC/Jikocho 16th Commission Mtg. 2012/5/28. www.youtube.com/watch?v=rjl2H4zY7DE (0:49–0:50).

[592]Phred Dvorak, Reactor Team Let Pressure Soar, The Wall Street Journal, April 23, 2011.

[593]The official report of The Fukushima Nuclear Accident Independent Investigation Commission, The National Diet of Japan, Executive summary, July 5, 2012, p. 33.

[594]Eric Talmadge, Mari Yamaguchi, How first 24 h shaped Japan's nuclear crisis, Associated Press, July 3, 2011.

[595]The official report of The Fukushima Nuclear Accident Independent Investigation Commission, The National Diet of Japan, Chapter 2. Escalation of the accident, July 5, 2012, p. 21.

needs.[596] Consequently, the Japanese government did not take emergency measures such as large-scale military helicopter assistance to transport diesel generators, air compressors, batteries, pumps, and so on to the site on the day of the disaster. Instead, the unhurried TEPCO tried to transport the needed staff by car, on roads disrupted by traffic jams! After the disaster, the Japanese military said that forces did not move in because they were not requested by TEPCO. Japan's Self-Defense Forces had personnel and equipment at the ready 15 miles away, but "*we have to wait for TEPCO to come to us and request help*".[597]

Absence of Decisiveness in Taking Urgent and Costly Solutions

There was a lack of fresh water sources at the site for emergency reactor cooling. There was plenty of salt water from the ocean—but using salt water would render the reactors permanently inoperable, resulting in billions of dollars of assets being lost. Akira Omoto, a former TEPCO executive and a member of the Japan Atomic Energy Commission, confessed that "*[TEPCO] hesitated because it tried to protect its assets*".[598] Operators and management at the plant did not have the credentials to make this decision, and available TEPCO managers were reluctant to take responsibility for it because the President of TEPCO was out of Tokyo on a business trip and the Chairman was in China. According to the NAIIC commission "*The top three management members (the president, chairman, and vice president) were unavailable when the accident broke out. [The chairman] only found out that the president had been away after the accident happened. A lack of a sense of impeding crisis was obvious from the fact that [the chairman] made no contact with the president after the president's return from abroad until his return to the head office... At the time of the accident, neither the Chairman nor the President of TEPCO were present or accessible, an inconceivable situation for an operator of nuclear power plants. The Chairman and the President also had different understandings of the emergency response structure, a fact that very likely contributed to the delay in TEPCO's response to the accident*".[599] But 24 h after the tsunami, a hydrogen explosion occurred on Unit 1, and the Prime Minister of Japan Naoto Kan intervened. The NAIIC commission described his influence: "*[Because] the situation deteriorated and the planned government accident response systems failed to function, control of the emergency response was taken by the Kantei, with Prime Minister Kan at the center of an ad hoc group of*

[596]Stream record of NAIIC/Jikocho 6th Commission Meeting, March 14, 2012, http://www.youtube.com/watch?v=4cEM6cvLm2s (1:18–1:20).

[597]Norihiko Shirouzu, Phred Dvorak, Yuka Hayashi, Andrew Morse, Bid to 'Protect Assets' Slowed Reactor Fight, The Wall Street Journal, March 19, 2011.

[598]Ibid.

[599]The official report of The Fukushima Nuclear Accident Independent Investigation Commission, The National Diet of Japan, Executive summary, July 5, 2012, pp. 77, 33.

politicians, advisors and the chairman of NISA. This group included people who were neither experts nor had an adequate understanding of the on-site situation".[600] TEPCO finally began to pump seawater into Unit 1 reactor 29 h after the tsunami; and it was not until the third day after the tsunami that they started to use seawater for the other reactors.[601] At the time, Naoto Kan complained that TEPCO "*has almost no sense of urgency whatsoever*" and on the sixth day after the disaster, when the situation had reached the worst-case scenario—meltdown of three reactors and a fire in the spent nuclear fuel pool of Unit 4—the government shifted full responsibility for dealing with the situation from NISA and TEPCO to the military.[602]

Struggle Between Political Camps as a Major Obstacle to the Adequate Risk Information Transmission in Crisis Situation

TEPCO and NISA communicated only the most limited information to the government from the beginning of the disaster. The main reason for this was the complicated political environment inside Japan. Prime Minister Naoto Kan represented the Democratic Party of Japan (DPJ), which in 2009 defeated the Liberal Democratic Party (LDP); center-right conservatives had been in power from 1955 until 2009. The LDP was the main lobbying force for nuclear energy in Japan, had a close relationship with American Republicans and represented the interests of the Japanese elite and bureaucracy. Naoto Kan and the DPJ, on the other hand, had a center-left political position and aimed to "*overthrow the ancient régime locked in old thinking and vested interests, solve the problems at hand, and create a new, flexible, affluent society. We shall build a society governed with transparent, just, and fair rules*".[603] DPJ politicians distrusted the United States[604] and criticized the fact that the Japanese nuclear industry originated from the American nuclear industry. After the disaster, Naoto Kan tried to close down all the nuclear stations in Japan. It was only after elections in December 2012, when the LDP won back a majority, that the nuclear strategy of Japan was turned round again: the new Prime Minister Shinzo Abe declared the restart of existing NPPs and started a program to

---

[600]Ibid, p. 34.

[601]The official report of The Fukushima Nuclear Accident Independent Investigation Commission, The National Diet of Japan, Chapter 2. Escalation of the accident, July 5, 2012, pp. 21–22.

[602]Frustrated with TEPCO, Kan turns to SDF in nuclear crisis, The Mainichi Daily News, March 17, 2011.

[603]Our Basic Philosophy—Building a Free and Secure Society, The Democratic Party of Japan, April 1998, http://www.dpj.or.jp/english/about_us/philosophy.html.

[604]The Fukushima Nuclear Accident and Crisis Management—Lessons for Japan-U.S. Alliance Cooperation, The Sasakawa Peace Foundation, September 2012, p. 58.

build new ones.[605] Obviously, in March of 2011, NISA and TEPCO executives—who had had a cozy relationship for decades and shared the old history of falsification and risk information concealment within the industry—represented the camp opposite to that of Naoto Kan. In the crisis, this led to mutual distrust. Thus, the NAIIC commission found out that "*TEPCO had been reporting to NISA, as was the standard protocol, that it was in the process of venting [on Unit 1]. But there is no confirmation that the venting decision was conveyed to senior members of METI, or to the Kantei. This failure of NISA's function and the scarcity of information at TEPCO headquarters resulted in the Kantei losing faith in TEPCO*".[606] METI executive Banri Kaieda described the situation vividly: "*from immediately after the breakout of the accident, communicating and sharing information among the accident site, the Kantei, and TEPCO headquarters was like the telephone game 'whispering down the lane'*".[607] During the NAIIC testimony, Naoto Kan stated that when he questioned them about the reasons for the delay with the venting process on the first day of the disaster, TEPCO answered: "*we do not know*".[608] Moreover, he remarked that TEPCO did not have the same technical background as General Electric, which designed the BWR reactors on Fukushima-Daiichi NPP, and questioned whether TEPCO staff fully understood their structure and technical pattern.[609] Due to the absence of valid information from TEPCO executives and NISA, the Kantei "*participated in discussions of technical matters regarding the nuclear reactors. Prime Minister Kan asked for the mobile phone number of the head of the plant at Fukushima, leaving the top management of TEPCO out of the loop*".[610] This mistrust reached its peak 15 h after the tsunami when Naoto Kan, dissatisfied with the lack of information about conditions at the plant and the venting process on Unit 1, flew by helicopter to Fukushima-Daiichi NPP in order "*to understand the situation, as he could not obtain any meaningful information from the members of NISA, the NSC, or the technical advisor from TEPCO*".[611] TEPCO also failed to inform him immediately when Unit 1 exploded 24 h after the tsunami—so the leader of Japan was informed about the explosion by TV news! In the ensuing telephone call with TEPCO executives, Prime Minister Kan exclaimed: "*What the hell is going on?*", because nothing had been said to him for about an hour.[612] Kenichi Shimomura, former deputy director

---

[605] Hiroko Tabuchi, Japan's New Leader Endorses Nuclear Plants, The New York Times, December 30, 2012.

[606] The official report of The Fukushima Nuclear Accident Independent Investigation Commission, The National Diet of Japan, Executive summary, July 5, 2012, p. 18.

[607] Ibid, pp. 78–79.

[608] Stream record of NAIIC/Jikocho 16th Commission Meeting, May 28, 2012, www.youtube.com/watch?v=rjl2H4zY7DE (0:26–0:29).

[609] Ibid (2:23–2:26).

[610] The official report of The Fukushima Nuclear Accident Independent Investigation Commission, The National Diet of Japan, Executive summary, July 5, 2012, p. 74.

[611] Ibid, p. 80.

[612] Japan PM to nuclear power firm: "What the hell's going on?", Reuters, Mar 15, 2011.

general for public relations and chief spokesman of the Kantei, described the impossible situation in which Naoto Kan found himself: "*Being a leader without information... information about the nuclear crisis was a luxury we did not have in the prime minister's office at the time*".[613] Although TEPCO and the regulators "*had agreed on how to deal with the vent and the injection of seawater, the Kantei was unaware of this, and intervened, resulting in further disorder and confusion*".[614] In its turn, TEPCO and NISA were also unaware of the details of discussions held among Kantei members, since Prime Minister Kan surrounded himself with numerous experts of his own, independent from them.[615] As a result, managerial decisions about the plant and the evacuation were discordant. In addition, both sides held press conferences and briefings separately. Ultimately, the Japanese public and the international community concluded that the authorities and the industry could not communicate, manage and take responsibility for such a challenging event, and this caused panic both inside Japan and worldwide.

Distortion of Information About the Condition of the Plant Led to Inadequate Governmental Crisis Response Measures

The lack of reliable information about the situation on the plant led to a chaotic organization by the Kantei of the evacuation of residents in areas neighboring the plant. There were further examples of risk information concealment. The Associated Press found and reported one case: "*Yukio Edano, the chief Cabinet spokesman, is the face of Japan's government. At 7:45 p.m. [around four hours after the tsunami], his job was to make an unprecedented statement to the nation – but made it sound routine and reassuring. 'We have declared a nuclear emergency,' he said from behind a podium in the press conference room at the prime minister's office. 'Let me repeat that there is no radiation leak, nor will there be a leak'. He was wrong. Recently released TEPCO documents revealed that radiation was detected at the plant perimeter at 5:30 p.m. [two hours after the tsunami], but the utility apparently didn't fax those readings to the government until shortly after 9 p.m.*".[616]

The NAIIC commission revealed that "*regarding the disclosure of an increase of reactor vessel pressure at Unit 2, TEPCO issued a press release about seawater injection at 23:00 on March 14, but made no disclosure about an increase in*

---

[613]Disasters, Rebuilding and Leadership—Tough Lessons from Japan and the U.S., The Wharton School, University of Pennsylvania, October 2013, p. 4.

[614]The official report of The Fukushima Nuclear Accident Independent Investigation Commission, The National Diet of Japan, Executive summary, July 5, 2012, p. 34.

[615]The Fukushima Nuclear Accident and Crisis Management—Lessons for Japan-U.S. Alliance Cooperation, The Sasakawa Peace Foundation, September 2012, p. 100.

[616]Eric Talmadge, Mari Yamaguchi, How first 24 h shaped Japan's nuclear crisis, Associated Press, July 3, 2011.

radiation dosage at the entrance of the plant that occurred between 19:00 and 21:00 on the same day. TEPCO also downplayed the severity of the situation in their disclosure regarding the critical problems in the suppression chamber of Unit 2; moreover, there was a significant delay between the time when TEPCO informed the Kantei and when it disclosed the information publicly. TEPCO noted that they did not inform the public of an increase in reactor vessel pressure at Unit 3, as of 8:00 on March 14, because NISA had banned the release. In fact, the Kantei had merely instructed TEPCO to inform them of the contents of releases when they were made. In obeying NISA's order to halt the release of this crucial information, TEPCO effectively prioritized its own interests and those of NISA over the greater good of the public and their right to be informed".[617] As a result, "*the central government was not only slow in informing municipal governments about the nuclear power plant accident, but also failed to convey the severity of the accident. Due to the breakdown in communication from the central government in the post-accident time period, neither the Fukushima prefectural government nor the central government were aware of each other's actions: for example, the Fukushima prefectural government unilaterally ordered that residents within a two-kilometer radius of the plant be evacuated, based on prior emergency prevention training. This was followed 30 minutes later by the central government ordering the evacuation of residents within a three-kilometer radius. Similarly, the speed of information in the evacuation areas varied significantly depending on the distance from the plant. Residents were informed of the accident a few hours after it occurred, but they did not receive any information about the situation or the accident, or information that would be useful for their evacuation. Many residents had to flee with only the barest necessities and were forced to move multiple times or to areas with high radiation levels. There was great confusion over the evacuation, caused by prolonged shelter-in-place orders and voluntary evacuation orders. Some residents were evacuated to high dosage areas because radiation monitoring information was not provided*".[618]

The government also did not use data from the SPEEDI system (System for Prediction of Environment Emergency Dose Information), which had been designed in the 1980s to make forecasts of radiation dispersal, because "*the information was incomplete*": the earthquake and tsunami had injured some sensors near the plant. "*Without knowing the strength of the releases, there was no way we could take responsibility if evacuations were ordered*," said Keiji Miyamoto, who manages SPEEDI.[619] However, some experts consider that "*officials there did not want to take responsibility for costly evacuations if their estimates were later called into question*".[620] On the sixth day after the disaster happened, the U.S.

---

[617]The official report of The Fukushima Nuclear Accident Independent Investigation Commission, The National Diet of Japan, Executive summary, July 5, 2012, p. 80.

[618]Ibid, p. 19.

[619]Norimitsu Onishi and Martin Fackler, Japan Held Nuclear Data, Leaving Evacuees in Peril, New York Times, August 8, 2011.

[620]Ibid.

State Department advised American citizens to evacuate from the area within a 50-mile (80-km) radius of the Fukushima Daiichi NPP,[621] but, at the same time, the Japanese government started to evacuate citizens from within a 20-km radius. TEPCO officially confirmed the meltdown of all three reactors only after 2 months when all rods inside the reactors were melted: the "*delay in confirming the meltdowns at Fukushima suggested the utility feared touching off a panic by disclosing the severity of the accident earlier*".[622] Taken together, these events have provoked strong public anxiety and distrust of the Japanese nuclear industry and the government. During his testimony, Yuhei Sato (Governor of Fukushima Prefecture at the time of the accident) told the commission that "*the government failed to provide the necessary information at the time of the accident: "I still cannot trust the government*"".[623] This likely contributes to the fact that the majority of Japanese support the phasing out of nuclear power.[624]

Fukushima-Daiichi Nuclear Disaster: Why Risks Were Concealed

- The distinctive position of the nuclear industry within the Japanese economy and the **misplaced loyalty of regulators concerning shortcomings in the design and operation of Japanese NPPs**, which allowed plant operators to neglect basic safety rules and conceal the occurrence of many safety violations from regulators and the public with impunity.
- The **national arrogance of both executives and regulators in the Japanese nuclear industry**, who refused to learn from the experience of other countries that had faced nuclear accidents, or to implement IAEA's recommendations and advanced safety requirements. The Japanese preferred to rely on their supposed technical superiority over the rest of the world. They assumed that falsifying data about minor equipment faults would never lead to catastrophic results and that the Japanese attitude toward work would always compensate for minor imperfections in reactor design during natural disasters.
- **Habituation/wishful thinking/overconfidence/self-suggestion/self-deception** among representatives of the Japanese nuclear industry concerning the low probability of a severe nuclear accident caused by a tsunami.

[621]The Fukushima Nuclear Accident and Crisis Management—Lessons for Japan-U.S. Alliance Cooperation, The Sasakawa Peace Foundation, September 2012, p. 48.

[622]Shinichi Saoshiro, Tepco confirms meltdowns at 2 more Fukushima reactors, Reuters, May 24, 2011.

[623]The official report of The Fukushima Nuclear Accident Independent Investigation Commission, The National Diet of Japan, Executive summary, July 5, 2012, p. 77.

[624]Martin Adams, Powering ahead. Perspectives on Japan's energy future, The Economist Intelligence Unit Limited, 2012, p. 4.

- TEPCO's focus on the **short-term profitability of operations and on ongoing cost reduction** provoked reluctance among executives to reveal the risks of NPPs—whether to IAEA specialists, representatives of local authorities or emergency services, investors or local residents—because this would entail additional expenses on advanced safety measures.
- The **specific national risk perception and organizational culture**: Japanese corporate mentality is based on unconditional submission of employees to their supervisors and does not approve of employees asking embarrassing questions. This makes the working environment uncomfortable for whistleblowers.
- The **absence of a prompt risk assessment system**, and the long chain of communication between field staff and senior management, made urgent decision-making difficult during the disaster: field staff had no authority for even minor on-site decisions during the development of the critical situation.
- The **political struggle** between the Democratic Party and the Liberal Democratic Party, which generated massive distortion of information about the real condition of the plant after the disaster. Both parties used the accident in their own political interests.
- Misleading comments from the Kantei, NISA and TEPCO about the accident during the first days after the disaster to the Japanese people and the international community were caused by the following factors: **lack of information and misjudgment about the real scale of the disaster** in the first days; the **absence of objective estimates of possible consequences of the disaster**; **fear of massive panic** in Japan and in nearby countries because nuclear accidents and radiation are the most dangerous threats in public perception; **reluctance to confess that regulation of the Japanese nuclear industry had been defective**, and that concealing the imperfections and risks of Japanese NPPs had been common practice for decades.

2.1.12 Other Cases of Risk Information Concealment

2.1.12.1 Minamata Mercury Poisoning (Japan, 1932–1968)

In 1956, in Minamata city in Japan, a strange epilepsy-like neurological disease was discovered among locals, as well as in their cats and dogs. They called the disease "*dancing cat fever*". Initially, scientists thought that it was an infectious disease but, when they tested marine creatures on the coast nearby, they discovered extremely high levels of mercury contamination, which was determined as being caused by industrial wastewater discharge from the adjacent Chisso Corporation chemical factory. The factory used mercury sulfate as a catalyst in the production of acetaldehyde, and had been discharging the compound into Minamata Bay for 25 years. And seafood from the bay had been the main diet of local residents and their domestic animals for decades.

Chisso Corporation knew about the potential damage to the health of locals and to the environment, but was reluctant to construct expensive wastewater treatment facilities. Moreover, the company continued to discharge mercury-contaminated waste after the cause of the disease had been confirmed. It lobbied to cut back pollution control regulation, and obstructed detailed investigation by scientists and the media. Ultimately, 2265 victims have been officially certified—1784 of whom died from the poisoning—and over 10,000 people have received financial compensation from the company, which paid out a total of more than US$170 million.[625]

During the Fukushima-Daiichi nuclear disaster described above, many commentators compared the neglectful behavior of TEPCO with the actions of Chisso Corporation during the Minamata crisis. They concluded that, in the intervening 50 years, the Japanese industry had not changed in its willingness to risk the health, and even the lives, of local residents through its activities.[626,627]

Minamata Mercury Poisoning: Why Risks Were Concealed

- Chisso Corporation prioritized **short-term profitability over the long-term resilience** of the chemical factory, or the protection of public health and the environment.

2.1.12.2 Asbestos Crisis (Worldwide, 1970s)

Asbestos became a very popular construction material at the end of the 19th and the beginning of the 20th centuries because of its resistance to fire, heat, electrical and chemical damage, and its sound absorption, average tensile strength, and affordability. The first evidence that asbestos fibers cause lung cancer and mesothelioma was discovered among asbestos miners, and had been scientifically proven by the 1930s to 1940s. Nowadays, the World Health Organization estimates that about 125 million people around the world are annually exposed to asbestos in the workplace, and about 100,000 workers die each year from an asbestos-related disease.[628]

---

[625]Minamata disease, Wikipedia, http://en.wikipedia.org/wiki/Minamata_disease.

[626]Christine L. Marran, Contamination: From Minamata to Fukushima, Asia-Pacific Journal, May 2011.

[627]Timothy S. George, Fukushima in Light of Minamata, Asia-Pacific Journal, Mar. 12, 2012.

[628]Mesothelioma & Asbestos Worldwide, http://www.asbestos.com/mesothelioma/related-issues.php.

In the United States, it took more than three decades for the government to impose strict regulations concerning the working conditions of employees dealing with asbestos. Regulations were finally developed as the consequence of a lawsuit during which specific documents were provided proving that industry officials knew of the dangers of asbestos and tried to conceal them from workers to avoid the costs of improving the safety conditions of workplaces. During an exemplary lawsuit, it was stated that "*[t]he manufacturers put a lethal risk of harm in (the plaintiff's) work environment, then allowed him unwittingly to confront the risk with tragic results, on a daily basis*".[629] The asbestos industry had also been hiding health risks from customers, because of the fear of losing whole markets. After the risks were revealed, dozens of American firms had to file for bankruptcy due to asbestos liabilities—and with 600,000 claims from individuals so far, the total cost of asbestos compensation is estimated to be more than US$200 billion.[630] Nevertheless, China and India still consume large amounts of asbestos imported from Russia, Canada and Kazakhstan.[631]

Asbestos Crisis: Why Risks Were Concealed

- The **priority of short-term profitability**, and the industry's reluctance to confess the harmfulness of asbestos, thereby destroying the market and generating millions of lawsuits seeking compensation for health damage.

2.1.12.3 Savar Building Collapse (Bangladesh, 2013)

On 24 April 2013, in Savar in the Greater Dhaka Area of Bangladesh, 1127 workers at garment factories died when the Rana Plaza building collapsed on them. There are more than 5000 competing garment factories in Bangladesh, which provide cheap labor for the tailoring of many world-famous brands. The average monthly salary of a sewing machine operator is only US$38, but the garment industry produces garments for up to US$20 billion and provides Bangladesh with 77 % of its exports.[632,633] The Rana Plaza was originally designed as a six-story building for shops and offices, but the owner of the plaza illegally constructed three additional floors using low-quality materials—and sited five garment factories there, deploying heavy machinery, which generated excessive vibrations.

[629]Bill Burke, Shipyards, a Crucible for Tragedy. Part 1: How the war created a monster, The Virginian-Pilot, May 6, 2001.

[630]Michelle J. White, Understanding the Asbestos Crisis, University of California, May 2003.

[631]Mesothelioma & Asbestos Worldwide. http://www.asbestos.com/mesothelioma/related-issues.php.

[632]Victor Luckerson, Bangladesh Factory Collapse: Is There Blood on Your Shirt? TIME, May 2, 2013.

[633]Farid Ahmed, Death toll from Bangladesh building collapse climbs above 700. CNN, May 7, 2013.

The day before the collapse, local authorities discovered cracks in the building and issued an order to evacuate the whole building. The personnel on the lower floors with shops and a bank were not permitted to their workplaces until inspectors had confirmed the safety of the building; but managers of the garment factories insisted that their staff should go to work, otherwise they would all lose their monthly salary.[634] Moreover, they misled the sewers by telling them that the building had been inspected and declared safe.[635] The motives of the managers were simple: if operations were shut down, they would be fined by their customers—world-famous high street clothing brands—for delays with shipping, and could lose contracts in a highly competitive market. Two years earlier in 2011, Walmart and GAP had refused to sign a new industry agreement to pay Bangladeshi factories a higher price, so the garment industry could not afford safety upgrades on their sewing factories.[636]

Savar Building Collapse: Why Risks Were Concealed

- **Short-term profitability** in a highly competitive market took priority over the safety of personnel.
- The owners of the garment factories were **afraid of losing customers** in case of a prolonged time-out of the factory.

2.2 Financial Sector

Financial genius is a short memory in a rising trend

John Kenneth Galbraith

K Street has replaced the Invisible Hand of perfect competition with the Visible Fist of money and corruption

Dr. Woody Brock in American Gridlock

Over a protracted period of good times, capitalist economies tend to move from a financial structure dominated by hedge finance units to a structure in which there is a large weight to units engaged in speculative and Ponzi finance... The greater the weight of speculative and Ponzi finance, the smaller the overall margins of safety in the economy and the greater the fragility of the financial structure

Hyman Minsky, 1992

---

634 Arun Devnath, Mehul Srivastava, 'Suddenly the Floor Wasn't There,' Factory Survivor Says. Bloomberg, 25 April 2013.

635 Syed Zain Al-Mahmood, Rebecca Smithers, Matalan supplier among manufacturers in Bangladesh building collapse. The Guardian, 24 Apr. 2013.

636 Arun Devnath, Mehul Srivastava, 'Suddenly the Floor Wasn't There,' Factory Survivor Says. Bloomberg, 25 Apr. 2013.

As Jeremy Grantham said when asked what investors would learn from this crisis:

"*In the short term, a lot. In the medium term, a little. In the long term, nothing at all. That is the historical precedent*". Or as John Kenneth Galbraith put it, markets are characterized by "*Extreme brevity of financial memory … There can be few fields of human endeavor in which history counts for so little as in the world of finance*"

2.2.1 Barings Bank Collapse (Singapore-UK, 1995)

In February 1995, Barings PLC—the oldest and the most reputable bank in Britain—collapsed from the unauthorized trading of Nick Leeson, a Singapore-based trader at the bank, who single-handedly lost about US$1.4 billion (£827 million).

2.2.1.1 Risk Concealment Before the Disaster

During its 233-year history, Barings Bank financed the British side of the Napoleonic Wars, the Louisiana Purchase and the Erie Canal and was the personal bank of Queen Elizabeth II. Francis Baring, one of the founders of the bank, described it as "*unquestionably the first merchant in Europe – first in knowledge and talent, and first in character and opulence*".[637] In 1986, Margaret Thatcher's Conservative government suddenly deregulated the British financial sector and allowed traditional commercial banks to provide investment bank services (for instance, securities brokerage and securities underwriting). They wanted to

[637]The Annual Register, Or, A View of the History, Politics, and Literature for the Year, London, J. Dodsley, 1825, p. 400.

increase the competitive advantage of British banks on the international markets and ensure the status of London as one of the world's financial centers. Following the deregulation, Barings Bank secured more than 50 % of its total profit from securities.[638]

In 1989, Nick Leeson came to Barings in London from Morgan Stanley, to take a back-office position focusing on the control of security transactions. Soon, he was sent to Indonesia for another back-office project, working with stock certificates and bearer bonds. In 1992, he was appointed to the position of general manager and head trader at Barings Securities (Singapore) Limited (BSS). There, he had clearance for "*transacting futures and options orders for clients or for other firms within the Barings organization, and riskless arbitraging of price differences between Nikkei futures traded on the SIMEX [Singapore Exchange] and Japan's Osaka exchange*".[639] Given his lack of trading experience, many of his deals were not making profit, and he began to hide his losses via an already existing secret "client" account, number 88888. By February 1995, he had generated losses of over US$1.4 billion (£827 million) and brought Barings to bankruptcy. Incredibly, neither external auditors, nor supervisors, nor regulators had detected Leeson's true position prior to the collapse.[640] Leeson admitted that the main reason why he had concealed his own losses and falsified his profits was "*fear of failure*"—because the surrounding environment, both on the Singapore stock exchange and within Barings bank, extolled success and profits and despised failure and losses: if his true losses were revealed, it would highlight his "*incompetence, negligence and failure*".[641] He also expected that the consequences of his deception would be dramatic, but not catastrophic for the entire Barings business.

From 1992 to 1995, Barings headquarters and internal control did not spot the hidden losses, for several reasons. Firstly, Leeson, as both general manager and head trader of BSS, combined the functions of trader and back-office manager. These roles were usually divided between different people, from different departments of an investment bank and with different professional missions: earning money through deals as a trader, keeping risks within acceptable limits and keeping deals legal as a manager. In this period, BSS "*was operated almost entirely by Leeson alone*".[642]

[638]Documentary "Going Rogue", Journeyman Pictures, Dec. 2011.

[639]Glyn A. Holton, Barings Debacle, Risk Encyclopedia, May 2013.

[640]Report to the Board of Banking Supervision Inquiry into the Circumstances of the Collapse of Barings, Bank of England, 18 July 1995, Conclusion chapter, subsection "Outline".

[641]Documentary "Going Rogue", Journeyman Pictures, Dec. 2011.

[642]Report to the Board of Banking Supervision Inquiry into the Circumstances of the Collapse of Barings, Bank of England, 18 July 1995, Conclusion chapter, subsection: "Why was the True Position not Noticed Earlier?".

Secondly, Leeson had more than 5 years' experience as a back-office clerk in Morgan Stanley and Barings Bank; he had a very clear understanding of the shortcomings of the internal control system of a huge financial institution. Thus, his secret "88888" account escaped detection by several external and internal auditors—who accepted a forged guarantee letter, and sent reports to top managers back in London assuring them that Leeson's transactions were credible. Two other major rogue traders—Jerome Kerviel from Société Générale, who lost US$6.9 billion in unauthorized trading, and Kweku Adoboli from UBS who lost US$2.3 billion—also worked in back-office departments before their trading careers.[643]

Thirdly, both the senior management and those responsible for control functions at Barings had a merchant banking background: they knew little about derivatives[644] and associated them with tremendous risks. Later, Leeson characterized Barings managers as "*idiots*"[645] who did not understand the basics of the futures trading business: "*How little did the management of Barings know about what was going on? They had no clue. In 1994 [they] came from London, New York, and Tokyo to receive an award from SIMEX for the 'Highest Customer Volume'*".[646] Because they knew so little about derivatives, the management team was blinded by falsified profits from Singapore, which had a direct influence on their annual bonuses; they believed that Leeson was making fully matched trades at no real risk to Barings.[647] Thus, they continued to send money to Singapore to cover Leeson's losses, comforting themselves that Leeson would bring them millions: "*[Barings was] driven to make profits, profits, and more profits...*".[648] "*[I]t was their greed that lay at the root of the whole problem. They did not want to know about the internal structure of the firm*".[649] For instance, in 1993, Leeson's "transactions" appeared to be making 10 % of all Barings' profits.

Finally, Barings had an exclusive relationship with the Bank of England: according to Lord Hollick, the British central bank had an "*informal regulatory regime*" concerning Barings.[650] This fact allowed Barings to violate restrictions on

---

643Documentary "Going Rogue", Journeyman Pictures, Dec. 2011.

644A derivative itself is merely a contract between two or more parties. A price derivative is related to the changing price of an underlying asset. Derivatives are synthetic securities—in general, they are not used to purchase the underlying asset, but to hedge from the risks of the changing price of the underlying asset, and generate income from the difference in prices on the underlying asset. The most common underlying assets include stocks, bonds, commodities, currencies, interest rates and market indexes.

645Documentary "25 Million Pounds", Director Adam Curtis, 1999.

646Gareth Hutchens, Barings wake up call unheeded: Leeson, Sydney Morning Herald, Oct. 20, 2012.

647Report to the Board of Banking Supervision Inquiry into the Circumstances of the Collapse of Barings, Bank of England, 18 July 1995, Conclusion chapter, subsection: "Why was the True Position not Noticed Earlier?".

648How Leeson broke the bank. BBC, June 22, 1999.

649Lords Hansard entry for 21 Jul 1995 (150721-14). http://www.parliament.the-stationery-office.co.uk/pa/ld199495/ldhansrd/vo950721/text/50721-14.htm.

650Ibid.

regulatory capital or capital adequacy: "*[Barings bank's] capital base was only $250 million, [but] at the end of 1994 I had $500 million in Singapore, so twice the capital base of the bank. I think it was 10 times the legal limit that [a bank] could lend to a subsidiary, which the Bank of England had allowed to happen*".[651] In 1993, the bank's chairman Peter Baring commented to Brian Quinn, Director of the Bank of England: "*The recovery in profitability has been amazing following the reorganization, leaving Barings to conclude that it was not actually terribly difficult to make money in the securities markets*".[652] Obviously, the regulator was pleased that its efforts towards deregulation seemed to be leading to greater profitability in the British banking sector…

After the fraud was revealed, Leeson was sentenced to $6^1/_2$ years in a Singaporean prison, but was released in 1999 for good conduct and due to colon cancer. Since this time, he has become a risk perception and risk control systems consultant and conference speaker. He also explained his story in the book "*Rogue Trader*", which has since been adapted as a Hollywood blockbuster.

Barings Bank Collapse: Why Risks Were Concealed

- By authorizing the use of unfamiliar and risky financial instruments, Barings managers gave priority to **short-term profitability over the long-term financial stability of the oldest bank in the UK**.
- The **climate of wishful thinking** at the bank made it uncomfortable for people to spread warnings, or make a sober assessment of suspicious operations or phenomenal earnings.
- **Habituation/false reassurance/self-suggestion/self-deception** among executives at the Bank of England and Barings Bank concerning the low probability of massive losses from **deregulation and innovative financial instruments** (derivatives). The tendency among decision-makers not to see the **whole picture about risks**.
- The widely accepted "**success at any price**" **organizational culture** within the investment banking industry, and the fear of being blamed as incompetent, forced Nick Leeson to start to hide his own losses, leading to a fatal spiral.
- **Ignorance about derivatives and their associated risks** among executives of the bank and representatives of the internal control department, which allowed Leeson to falsify data with impunity for 3 years.

[651]Gareth Hutchens, Barings wakeup call unheeded: Leeson, Sydney Morning Herald, Oct. 20, 2012.

[652]Shelagh Heffernan, Modern Banking, John Wiley & Sons, 2005, p. 381.

2.2.2 *Enron's Bankruptcy (USA, 2001)*

It is difficult to get a man to understand something when his salary depends on his not understanding it

Upton Sinclair

2.2.2.1 Summary

In December 2001, the American company Enron went bankrupt, losing US$63.4 billion in assets. At the time, this was the largest bankruptcy in US history, though this record was subsequently beaten shortly thereafter by WorldCom in 2002 in a similarly accounting scandal—with lost assets of US$107 billion—and then again in 2008 by Lehman Brothers, who lost more than US$600 billion in assets.[653]

Enron had a predecessor, Metallgesellschaft AG, whose aim it copied. Metallgesellschaft was formerly one of Germany's largest industrial conglomerates based in Frankfurt, with 20,000 employees and revenues in excess of 10 billion US dollars in 1993. It had over 250 subsidiaries specializing in mining, specialty chemicals (Chemetall), commodity trading, financial services, and engineering (Lurgi).[654] The unsuccessful maturity transformation in Metallgesellschaft's hedging long term contracts with short term futures[655,656] was echoed by banks and derivative dealers in the financial disasters discussed below. It is interesting to note also that, upon Metallgesellschaft's collapse in 1993, virtually all its staff were hired by Enron in the early days.[657]

Enron focused on wholesale merchant and commodity market businesses, management of retail customer energy services, operation of gas transmission systems, and management of energy-related assets and broadband services through approximately 3500 domestic and foreign subsidiaries and affiliates. For years before the bankruptcy, Enron executives routinely practiced fraud in the firm financial records in order to increase Enron's perceived revenue, so that the company's value in the market—and their own income—would continue to grow in the

---

[653]These loss estimations for these three bankruptcies assume that the asset valuations before the bankruptcy were true. Since extraordinary inflation of asset valuations has been one of the causes for the bankruptcies, these considerable loss figures must be taken with a grain of salt.

[654]Megallgesellschaft AG, http://en.wikipedia.org/wiki/Metallgesellschaft.

[655]Brennan, M.J. and N.I. Crewe, Hedging long maturity commodity commitments with short dated futures contracts. In: M A H Dempster & S R Pliska, eds. Mathematics of Derivative Securities. Cambridge: University Press, pp. 165–189 (1997).

[656]Medova, E.A. and A. Sembos, Price protection strategies for an oil company. In: S W Wallace and W T Ziemba, eds. Applications of Stochastic Programming, MPS-SIAM Series in Optimization. Philadelphia: SIAM, pp. 575–608 (2005).

[657]Professor M.A.H. Dempster (University of Cambridge, UK), personal communication (Dec. 22, 2014).

short-term. In 2000, Enron's falsified revenue exceeded US$101 billion and the company was ranked the seventh largest company in the USA.[658] Enron's falsified revenue came from trading operations, such as the trading of energy derivatives. As a result of the bankruptcy, 59,000 shareholders—including several pension funds and university endowments—lost a total of more than US$60 billion, up to 25,000 Enron employees were deprived of a total of US$2 billion in pension funds and stock options, and 20,000 creditors got back only 14 to 25 cents on every dollar lent to Enron.[659] In the wake of Enron's collapse, Arthur Andersen accountancy firm failed in 2002 following the irreparable damage to its reputation: Arthur Andersen was Enron's external and internal auditor, and one of the "Big Five" accounting firms in the world along with Price Waterhouse Coopers, Deloitte & Touche, Ernst & Young and KPMG.

Later, accounting frauds were found at WorldCom, Tyco, HealthSouth, and other companies, in investigations triggered by the undermined trust of millions of American and foreign investors in the credibility of financial reporting and audit processes in the United States. Moreover, another accounting scandal at the Italian dairy producer Parmalat revealed a €14.3 billion hole in the company's account sheets in 2003; an investigation found that senior managers at Bank of America, Citigroup, Deloitte & Touche and Grant Thornton were also involved in the fraud.[660] To restore public confidence, the US Congress hurriedly enacted on July 30, 2002, the Sarbanes-Oxley Act, also known as the 'Public Company Accounting Reform and Investor Protection Act' (in the Senate) and the 'Corporate and Auditing Accountability and Responsibility Act' (in the House), regulating the disclosure of information by U.S. public company boards and management, public accounting firms and investment banks. Under the new legislation, directors or accountants found to have knowingly concealed financial risks from investors and regulators in a firm's accounting would face decades in prison. Unfortunately, these changes could not prevent inflating earnings and routine concealment of debts—through off-balance-sheet partnerships or repo agreements—during the subsequent real estate bubble and the collapse of Lehman Brothers in 2008. Ernst & Young, Lehman Brothers' auditors, knew about debt concealment in the company (repo agreement 105 temporarily removed as much as $50 billion in bad assets from their balance sheet in 2008), but did not inform the Board of Directors or the regulators.[661] "*So the mere passage of a statute does not appear to*

[658]Report of investigation of Enron Corporation and related entities regarding federal tax and compensation issues, and policy recommendations. Volume I: Report, U.S. Joint Committee on Taxation, Feb. 2003, p. 5.

[659]Dick Carozza, Interview with Sherron Watkins. Constant Warning, Fraud Magazine, January/February 2007.

[660]The Parmalat scandal, Special Report, World Finance, June 24, 2011.

[661]Grant McCool, Ernst & Young accused of hiding Lehman troubles, Reuters, Dec. 21, 2010.

serve as a remedy for bad human behavior" as John Nugent observed.[662] According to some researchers "*Enron is too complex a story to avail of one single explanation for its rise and fall*".[663]

2.2.2.2 Risk Concealment Before the Disaster

Origins and the Rise to Stardom

Enron was created in 1985 by the merger of two natural-gas pipeline companies, Houston Natural Gas and InterNorth Inc., in order to develop the first nationwide natural gas pipeline system. The new company aimed "*to become the premier natural gas pipeline company in North America*"[664] after federal deregulation of natural gas transportation in the United States. Ken Lay, chairman and CEO of Enron from 1985 to 2002, was an economics Ph.D. who started his career in Nixon's Administration as undersecretary of energy, and a supporter of deregulation of the US energy sector.[665] In the early 1980s, Lay was an executive at Florida Gas Transmission. After the debt-financed merger of Houston Natural Gas and

[662] Stephen Nzuve, The Impact of the Enactment of the Sarbanes Oxley Act in the United States, 2002 on the Improvement of Corporate Finance and Good Governance Behavior, University of Nairobi, Sep. 5, 2012.

[663] Dennis Tourish, Naheed Tourish, Charismatic leadership and corporate cultism at Enron: The elimination of dissent, the promotion of conformity and organizational collapse, Leadership, Nov. 2005.

[664] Dick Carozza, Interview with Sherron Watkins. Constant Warning, Fraud Magazine, January/February 2007.

[665] Documentary "Enron: The Smartest Guys in the Room", Director Alex Gibney, 2005.

InterNorth Inc., Enron was left with a monster debt of US$4.3 billion. Moreover, because of deregulation, Enron lost exclusive rights on its pipelines in the highly competitive natural gas market, and natural gas prices went down: gas producers were able to sell sources directly to end users, and required pipelines to run their volumes for a simple transport tariff.[666] Searching for a new strategy, which could generate profits and cash flow in the long-term, was a real challenge for the Enron management team.[667]

In 1987, Lay started to hear about unauthorized—and sometimes fabricated—oil commodities trading at Enron Oil Trading (EOT), one subdivision of the former InterNorth Inc. Traders were running double books and using undisclosed accounts to "*[move] excess profits from one [accounting] quarter to the next through entities operating outside the Enron books*"[668,669] Later, internal investigation found out that traders made these actions in Enron's interests. The unit seemed profitable: in 1985, the company as a whole showed US$79 million losses, recovering to make US$557 million profit in 1986, while EOT earned US$10 million in 1985 and US$28 million in 1986 without need of capital investments in infrastructure.[670] This case demonstrated to Enron's management the possibility of earning money without investment in low-profit real energy infrastructure by using a flexible accounting approach. Ultimately, the management sent a fax to Enron Oil with the following message: "*please keep making us millions...*".[671] In 1987, Enron Oil Trading generated losses of up to US$1 billion, but ultimately Enron managed to reduce them to US$142 million; however, an EOT executive was found guilty of fraud and sentenced to 1 year in prison.[672]

In spite of these losses, Lay realized that commodity derivatives were a new prospective market for Enron, which could gain billions with proper management. When in 1989 Jeffrey Skilling, then a management consultant at the Houston office of McKinsey & Company, offered to establish "Gas Bank" service for Enron, insuring long-term fixed prices on natural gas for buyers and suppliers by means of futures and options, Lay offered Skilling an executive position at Enron Finance Corporation to enable him to implement this "*trading partners*" strategy.[673] It was the first step in Enron's transition from an old-fashioned regional energy company to an innovative nationwide energy-trading corporation. During

[666]Stephen V. Arbogast, Resisting Corporate Corruption, M & M Scrivener Press, 2008, p. 10.

[667]C. William Thomas, The Rise and Fall of Enron. When a company looks too good to be true, it usually is, Journal of Accountancy, Apr. 2002.

[668]Stephen V. Arbogast, Resisting Corporate Corruption, M & M Scrivener Press, 2008, p. 43.

[669]Matti Rantanen, Systems Intelligence in Leadership and Everyday Life Reasons, Systems Analysis Laboratory, 2007, p. 175.

[670]Stephen V. Arbogast, Resisting Corporate Corruption, M & M Scrivener Press, 2008, pp. 12, 19.

[671]Bethany McLean, Peter Elkind, The Smartest Guys in the Room: The Amazing Rise and Scandalous Fall of Enron, Portfolio Trade, 2003, p. 20.

[672]Documentary "Enron: The Smartest Guys in the Room", Director Alex Gibney, 2005.

[673]Gary M. Cunningham, Jean E. Harris, Enron And Arthur Andersen: The case of the crooked E and the fallen A, Global Perspectives on Accounting Education, Volume 3, 2006, p. 31.

the following few years, Enron successfully became the largest energy trader in the United States with more than 25 % of the country's gas and electricity transactions, Houston became the Wall Street of energy trading,[674] and Enron launched significant energy trading on the British deregulated wholesalers market and initiated operations in more than 30 countries. The company even changed its mission statement, announcing its intention "*to become the world's leading energy company*"[675] and later issued a new one: "*we make markets*".[676] As an interesting fact, John LeBoutillier, former Professor of Harvard Business School where Jeffrey Skilling graduated near the top of his class, remembered that Skilling stated at that time: "*I'd keep making and selling the product [in spite of the fact that it was discovered that his fictional company produces a potentially lethal product]. My job as a businessman is to be a profit center and to maximize return to the shareholders. It's the government's job to step in if a product is dangerous*".[677]

In 1997, Jeffrey Skilling was promoted to President and Chief Operating Officer of Enron, his mission to bring about the full-scale transformation of the company into a global energy-trading corporation. Under his leadership, Enron's annual revenues rose from about US$9 billion in 1995 to US$100 billion by 2000.[678] In 1999, Enron launched Enron Online, an internet trading site for electronic commodities: Enron was counterparty to every transaction conducted on the platform, with credit risk management to ensure a safe trading environment. By the following year, Enron Online had traded a total of US$335 billion.[679] By 2000, trading operations produced about 99 % of the company's income, 88 % of income before tax and 80 % of identifiable assets.[680] In 1996, Fortune named Enron the most innovative company in America, and they continued to award this title to Enron for the next six years until the company went bankrupt.[681] A tribute to Enron from the magazine's April 2000 edition starts in this way: "*Imagine a country-club dinner dance, with a bunch of old fogies and their wives shuffling around halfheartedly to the not-so-stirring sounds of Guy Lombardo and his All-Tuxedo Orchestra. Suddenly, young Elvis comes crashing through the skylight,*

[674] Dick Carozza, Interview with Sherron Watkins. Constant Warning, Fraud Magazine, January/February 2007.

[675] Ibid.

[676] Peter C. Fusaro, Ross M. Miller, What Went Wrong at Enron: Everyone's Guide to the Largest Bankruptcy in U.S. History, John Wiley & Sons, Inc, 2002, p. 70.

[677] Ibid, p. 28.

[678] Yuhao Li, The Case Analysis of the Scandal of Enron, International Journal of Business & Management, Oct. 2010, 5(10), 37.

[679] C. William Thomas, The Rise and Fall of Enron. When a company looks too good to be true, it usually is, Journal of Accountancy, Apr. 2002.

[680] Clinton Free, Mitchell Stein, Norman Macintosh, Management Controls: The Organizational Fraud Triangle of Leadership, Culture and Control in Enron, The Organization, July/August 2007.

[681] Dick Carozza, Interview with Sherron Watkins. Constant Warning, Fraud Magazine, January/February 2007.

complete with gold-lamé suit, shiny guitar, and gyrating hips. Half the waltzers faint; most of the others get angry or pouty. And a very few decide they like what they hear, tap their feet ... start grabbing new partners, and suddenly are rocking to a very different tune. In the staid world of regulated utilities and energy companies, Enron Corp. is that gate-crashing Elvis. Once a medium-sized player in the stupefyingly soporific gas-pipeline business, Enron in the past decade has become far and away the most vigorous agent of change in its industry".[682]

Political Context and Network

Active lobbying for deregulation of the energy and financial sectors played an important role in Enron's growth. Lay had a cozy relationship with the Bush family as a devoted friend and major contributor to the gubernatorial and presidential election campaigns of George H. W. Bush, George W. Bush and other Republicans.[683,684,685,686,687,688] This familiarity helped Lay and Enron to benefit from the easing of government control in several spheres.

Firstly, George H. W. Bush was Vice President of the USA during the eight-year presidency of Ronald Reagan, who in his turn was an apologist for deregulation in many spheres, including finance, transport and energy. Reagan declared his position in the following way: "*Government is not the solution to our problem; government is the problem*"[689]; "*We who live in free market societies believe that growth, prosperity and, ultimately, human fulfillment are created from the bottom up, not the government down... [We] believe in the magic of the marketplace*".[690] From 1989 to 1993, George H. W. Bush continued Reagan's deregulation strategy as President of the USA in his own right. In the dozen years of Republican power, new principles were established for the federal deregulation of the American wholesale and retail electricity markets. In the 1990s, this energy deregulation continued on a state level. For example, Pete Wilson, the Republican Governor of California from 1991 to 1999, implemented state electricity deregulation in 1996: the state sold their own power plants and bought electricity from a single

[682]Brian O'Reilly, The Power Merchant, Fortune Magazine, April 17, 2000.

[683]Documentary "Enron: The Smartest Guys in the Room", Director Alex Gibney, 2005.

[684]Enron's Kenneth Lay Is a Bush Family Friend, The Los Angeles Times, December 02, 2001.

[685]John Nichols, Ken Lay—Guilty. George Bush—Guilty. The Nation, May 25, 2006.

[686]Bush-Lay letters suggest close relationship CNN, February 17, 2002.

[687]Josh Gerstein, Enron's Close Ties to Bush, ABC News, December, 10, 2001.

[688]Arianna Huffington, Ken Lay on Trial: Why are the Media Forgetting the Bush/Cheney Connection? The Huffington Post, April 26, 2006.

[689]Inaugural address of Ronald Reagan, January 20, 1981.

[690]Remarks at the Annual Meeting of the Boards of Governors of the World Bank Group and International Monetary Fund, September 29, 1981.

wholesale pool, the California Power Exchange.[691] However in 2000–2001, an electricity crisis erupted in California. Enron energy traders manipulated the supply of electricity—creating an artificial power shortage and causing blackouts by shutting down Californian power plants—to jack up state wholesale prices by 1000 %; the price of natural gas (in Enron's pipelines) jumped by the same amount.[692] Enron earned billions on Californian energy contracts and overpriced natural gas. After the 2000 presidential elections, George W. Bush, on Lay's recommendation, appointed Pat Wood (former chairman of the Public Utility Commission of Texas) as Chairman of the Federal Energy Regulatory Commission, which regulates the transmission and wholesale sale of electricity, natural gas and oil in interstate commerce, and so on. Spencer Abraham, Bush's Secretary of Energy, had previously received campaign contributions from Enron as Republican senator for Michigan.[693] In spring 2001, when Gray Davis, the Governor of California, asked George W. Bush's Republican administration for a federal response to the state's electricity crisis, Bush refused any federal government intervention or price controls. He maintained that California legislators had left too many regulatory restrictions in place in the electricity market, and that the federal government had nothing to do with energy companies manipulating the market; and he personally did not see Enron's role in the California crisis.[694,695,696] The passive attitude of the Bush administration was likely motivated by the wider political context, in particular, given that California had voted for the Democratic candidate Al Gore in the recent presidential elections, and Democrat Gray Davis had presidential ambitions for the 2004 election cycle.[697] Moreover, Davis had earlier signed the nation's first state law requiring car makers to limit auto emissions—damaging the interests of oil companies and car manufacturers, both heavyweight supporters of the Republican Party through campaign contributions. This crisis helped Republicans reverse the gubernatorial election result by the electoral recall of the incumbent, for only the second time in American history: ultimately, Davis was succeeded by Republican Arnold Schwarzenegger in November 2003.[698] In an analogous political context, in 2005, the federal govern-

---

[691] Paul H. Dembinski, Carole Lager, Andrew Cornford and Jean-Michel Bonvin, Enron and World Finance. A Case Study in Ethics. p. 26.

[692] Greg Palast, California Reamin', The Guardian, May 22, 2001.

[693] Robert O'Harrow Jr., Lucy Shackelford, Enron Case. Political Players. Washington Post. http://www.washingtonpost.com/wp-srv/business/enron/4a.html.

[694] Don Van Natta, Enron's Many Strands: The Administration; Bush's California Energy Stance Faulted, New York Times, May 8, 2002.

[695] Jason Leopold, Lay and Skilling in the Dock. Enron and the Bush Administration, Counterpunch, Feb. 1, 2006.

[696] Frank Pellegrini. Bush's Enron Problem, Time Magazine, January 10, 2002.

[697] Clancy Sigal, Notes From Los Angeles; A Gray Future for California Voters, The New York Times, Oct. 17, 2002.

[698] Wikipedia's profile of Joseph Graham "Gray" Davis, Jr. http://en.wikipedia.org/wiki/Gray_Davis.

ment responded similarly passively to the Hurricane Katrina crisis, when Democrat Kathleen Blanco was Governor of Louisiana. And in 2008, Republican Bobby Jindal duly won the gubernatorial election in Louisiana.

Secondly, US accounting practice is based on state regulation, and both the Texas-registered Enron and the Houston office of Arthur Andersen were under the jurisdiction of the Texas State Board of Public Accountancy (TSBPA).[699] Mike Conaway was appointed as TSBPA chairman until 2004 during George W. Bush's term as Governor of Texas (1995–2000). In the 1980s, Conaway was chief financial officer of Arbusto/Bush Exploration.[700] The worst falsifications of Enron's accounting reports occurred while Conaway was at the TSBPA.

Weak Regulatory Climate

Thirdly, Enron benefited from the weakening regulatory oversight over energy futures trading. In 1989, early in George H.W. Bush's presidency, Enron started trading natural gas commodities and commodity derivative financial contracts. From this time, along with the investment banks, Enron lobbied for the removal of regulatory restriction on over-the-counter (OTC) derivatives—and particularly energy derivatives—from the Commodity Futures Trading Commission. In 1989, the Securities and Exchange Commission (SEC) "*began requiring that managers make specific disclosures of financial contingencies and off-balance-sheet arrangements when a particular 'trend, demand, commitment, event or uncertainty' was 'reasonably likely'. [However], if management determined that the contingency was not reasonably likely to occur, no disclosure was required*".[701] And on January 30, 1992, SEC accepted the mark-to-market accounting method for the energy contracts of Enron Gas Services group, which later allowed Enron to voluntarily calculate its revenue by the market value of derivative trading and to create the illusion of being "*larger*" than General Electric, Citigroup, or IBM.[702] Active lobbying continued after George H. W. Bush lost his second presidential race in 1992—for which Lay was co-chairman of Bush's re-election committee. Derivative traders also found support from Alan Greenspan, Chairman of the U.S. Federal Reserve during four US presidencies (Reagan, Bush, Clinton and Bush), and from Robert Rubin and Lawrence Summers, Secretaries of the US Treasury during Clinton's terms—who were all ardent apologists for deregulation in the

---

[699]Gary M. Cunningham, Jean E. Harris, Enron And Arthur Andersen: The case of the crooked E and the fallen A, Global Perspectives on Accounting Education, Volume 3, 2006, p. 33.

[700]Mike Conaway's profile of The Washington Post, 2004. http://www.washingtonpost.com/wp-srv/elections/2004/candidates/24150/.

[701]Paul H. Dembinski, Carole Lager, Andrew Cornford and Jean-Michel Bonvin, Enron and World Finance. A Case Study in Ethics. p. 69.

[702]B. G. Dharan, W. R. Bufkins, Red Flags in Enron's Reporting of Revenues & Key Financial Measures, July 23, 2008, p. 3.

financial sector.[703] This deregulation would ultimately be a significant catalysis in the dynamics ending with the global financial and economic crisis in 2008–2009. In 2001, Harvey who has represented each of the "Big Five" accounting firms,[704] including Arthur Andersen, was appointed as Chairman of the SEC in George W. Bush's administration.[705] Over decades of lobbying, the SEC budget was consciously reduced[706]; meanwhile the complexity of derivatives trading was rising.

Ultimately, the informal relationship between Enron's senior managers and American political decision makers and regulators resulted in the weakening of government oversight of the energy and financial sectors. This allowed Enron's management to implement several sophisticated methods of accounting, ensuring the regular growth of Enron's revenue and allowing debts to be hidden off the balance sheet in special purpose entities—which together resulted in the permanent growth of Enron's capitalization and multi-million-dollar earnings for executives. Worthy of mention, according to findings of Peter Fusaro and Ross Miller: "*Kenneth Lay, was not only on good terms with George W. Bush, he was a strong supporter of Al Gore's environmental program. It seemed that Enron would get in bed with any politician who could wield influence on its behalf*'.[707]

Stellar Performance

Thus, the tremendous annual growth of Enron's revenues, from about US$13 billion in 1996 to US$138.7 billion for the first 9 months of 2001, was achieved by the aforementioned mark-to-market accounting method and the "*merchant model*". These methods were based on reporting, for deals on the Enron Online platform and elsewhere, "*the entire value of each trade on which it was a counterparty as its revenue, rather than reporting as revenues only its trading or brokerage fees*".[708] A similar accounting approach was also applied in other companies like Dynegy, Reliant Energy and El Paso, even though investment banks used the more conservative "*agent model*" based on brokerage fees alone.[709] According to the calculations of Bala Dharan and William Bufkins, Enron's revenues were "*increased as much as fifty times, compared to what they would have been under*

---

[703]Rick Schmitt, Profit and Loss, Stanford Magazine, March/April 2009.

[704]Biography of Harvey L. Pitt, The American Academy in Berlin GmbH, http://www.americanacademy.de/de/home/person/harvey-l-pitt.

[705]Robert O'Harrow Jr., Lucy Shackelford, Enron Case. Political Players. Washington Post. http://www.washingtonpost.com/wp-srv/business/enron/4a.html.

[706]Arthur Gudikunst, ENRON: A Study of FAILURES, Who, How and Why, Bryant College Working Paper Series, Faculty Newsletter, Sep., 2002, p. 10.

[707]Peter C. Fusaro, Ross M. Miller, What Went Wrong at Enron: Everyone's Guide to the Largest Bankruptcy in U.S. History, John Wiley & Sons, Inc, 2002, p. xiii.

[708]B. G. Dharan, W. R. Bufkins, Red Flags in Enron's Reporting of Revenues & Key Financial Measures, July 23, 2008, p. 7.

[709]Ibid, p. 7.

more traditional accounting... We estimate that an adjustment for both MTM [mark-to-market] accounting and merchant accounting would have pushed down Enron's reported revenues to US$6.3 [billion] in 2000 instead of the reported US$100.8 [billion]... This allowed Enron to report expected benefits from future transactions into current period income ... An article in the Texas edition of the Wall Street Journal on September 20, 2000... referred to the soaring stock prices of Enron... and continued: 'What many investors may not realize is that much of these companies' recent profits constitute unrealized, non-cash gains. Frequently, these profits depend on assumptions and estimates about future market factors, the details of which the companies do not provide, and which time may prove wrong'... Enron used revenues – not profits [reported profits were microscopic relative to revenues] – as its primary financial objective, performance driver and measure of success. Enron's use of distorted, 'hyper-inflated' revenues was... important to it in creating the impression of innovation, high growth and spectacular business performance".[710]

And, with consulting support from the Houston office of Arthur Andersen, Enron financial executives used special purpose entities (or special purpose vehicles or financial vehicle corporations) to hide debts and losses from Enron's balance sheets in order to keep Enron's credit rating on investment grade and keep down the cost of capital borrowing. When Jeffrey Skilling became Enron's chief operating officer in 1997, CFO Andrew Fastow developed a network of 3000 special purpose entities, which were unconsolidated on Enron's balance sheet. He did this "*to accomplish favorable financial statement results, not to achieve bona fide economic objectives or to transfer risk. Some transactions were designed so ... Enron could have kept assets and liabilities (especially debt) off its balance sheet... They allowed Enron to conceal from the market [between 1997 to 2001] very large losses resulting from Enron's merchant investments by creating an appearance that those investments were hedged – that is, that a third party [partnership with companies Chewco, LJM1, LJM2, Condor, Raptor I–IV and other] was obligated to pay Enron the amount of those losses – when in fact that third party was simply an entity in which only Enron had a substantial economic stake... These transactions resulted in Enron reporting earnings from the third quarter of 2000 through the third quarter of 2001 that were almost $1 billion higher than should have been reported... In virtually all of the transactions, Enron's accounting treatment was determined with extensive participation and structuring advice from Andersen, whose management reported to the Board of Enron. Enron's records show that Andersen billed Enron $5.7 million for advice in connection with the LJM and Chewco transactions alone, above and beyond its regular audit fees*".[711] In the 1990s, there was the longest bull market in the

[710]Ibid, pp. 1, 10, 13.

[711]W. C. Powers, Jr., R. S. Troubh, H. S. Winokur, Jr., Report of investigation by the special investigative committee of the Board of directors of Enron Corp. , Austin, TX, February 1, 2002, pp. 4–5.

history of the USA,[712] but in the early 2000s the market was under the impact of the dot-com bubble burst and the 9/11 attack. "*If the market reversed, mark-to-market accounting required the recognition of losses, possibly enormous losses... Enron hid, delayed or ignored the loss. Andersen apparently did not question any of the values assigned to the contracts or object to tactics to hide, delay or ignore losses. Some of Enron's most abusive special purpose entities were created to avoid reporting mark-to-market losses*".[713] Due to accounting falsifications, from 1996 to 2000, Enron's declared that its market value grew by more than $4^1/_2$ times, reaching over US$60 billion—70 times earnings and six times book value.[714] In this five-year period, Enron paid five executives more than US$500 million via options, bonus payments and salaries.[715]

Because of the cozy relations between Enron executives and American politicians, regulators failed to prevent the worst-case scenario from happening in the Enron case. However, during the five years leading up to the bankruptcy, other audiences failed to recognize—or helped to cover—the concealment of risk by Enron executives: the company's auditors, its board of directors, other employees, investment banks and the media.

Arthur Andersen

In the 1990s, Arthur Andersen actively developed an accounting consulting practice as a supplement to their main auditing practice. The units within Arthur Andersen competed with each other, failed to communicate about the problems of their clients and sought only permanent growth of revenue regardless of the source of that revenue, the quality of clients or the legality of their recommendations.[716] Enron was Arthur Andersen's second largest client worldwide—the largest was WorldCom, which filed for bankruptcy in 2002—and the largest client in the accountants' Houston office.[717] The Houston office of Arthur Andersen provided both auditing and the new consulting service to Enron. Andersen consultants helped to implement aggressive accounting and the use of special purpose entities (more than 70 % of the fees that Andersen received from Enron came from consulting) while at the same time, the Andersen audit unit earned US$1 million a

---

[712]C. William Thomas, The Rise and Fall of Enron. When a company looks too good to be true, it usually is, Journal of Accountancy, April 2002.

[713]Gary M. Cunningham, Jean E. Harris, Enron And Arthur Andersen: The case of the crooked E and the fallen A, Global Perspectives on Accounting Education, Volume 3, 2006, pp. 40–41.

[714]Paul M. Healy and Krishna G. Palepu. The Fall of Enron, Journal of Economic Perspectives. Spring 2003, 17(2), 3.

[715]Dan Ackman, Pay Madness At Enron, Forbes Magazine, March 22, 2002.

[716]Jennifer Sawayda, Arthur Andersen: An Accounting Confidence Crisis, Daniels Fund Ethics Initiative, University of New Mexico, pp. 2, 6.

[717]Gary M. Cunningham, Jean E. Harris, Enron And Arthur Andersen: The case of the crooked E and the fallen A, Global Perspectives on Accounting Education, Volume 3, 2006, p. 31.

week for internal and external auditing (less than 30 % of Enron's payments to Andersen).[718] Enron had no fraud examiners and no internal audit department.[719] Enron outsourced their own "internal audit" to Arthur Andersen and many of Enron's internal accountants and controllers were former Andersen executives.[720] Joseph Berardino, Andersen's chairman, testified that "*in the previous year (2000), Andersen had received $52 million in fees from Enron, of which only $25 million could be directly attributed to the audit. Of those fees, $13 million were clearly for consulting work and the remaining $14 million is arguably related to the audit because it is work that can 'only be done by auditors'*".[721] The bonuses of staff at the Houston office of Arthur Andersen depended on Enron's stable growth, and many Andersen employees, "*[l]ured by promises of undreamt-of-wealth... aspired to work for Enron and were therefore very reluctant to 'rock the boat' with the company*".[722] Ultimately, this led to a situation where auditors approved falsified accounting reports in order to earn more.

Carl Bass, among other Andersen auditors, expressed concern over Enron's practice of mark-to-market accounting and use of special purpose entities—but immediately after Bass' complaint David Duncan, Andersen senior executive at the Houston office, removed Bass from the Enron account.[723] Obviously, if Bass had revealed his finding to the Texas State Board of Public Accountancy—which as we have noted was under the control of a friend of George W. Bush and Ken Lay—he could have lost his job and put a cross on his carrier as an auditor in Texas state without any assurance that the falsification case would even be properly investigated by the TSBPA. In the end, despite being later recognized as an accounting hero, Bass lost his auditor's license along with other former auditors of Enron.[724]

Moreover, Andersen headquarters had a weak system of internal control over its regional units, and Andersen executives were delighted by the continuous growth of the Houston office's revenue, so they avoided asking what would be considered awkward questions about the details of consulting and audit practice. After the falsification was revealed, Andersen's Houston office immediately destroyed thousands of e-mails and papers relating to their auditing and consulting for Enron

---

[718]Ibid, p. 43.

[719]Dick Carozza, Interview with Sherron Watkins. Constant Warning, Fraud Magazine, January/February 2007.

[720]C. William Thomas, The Rise and Fall of Enron. When a company looks too good to be true, it usually is, Journal of Accountancy, April 2002.

[721]Peter C. Fusaro, Ross M. Miller, What Went Wrong at Enron: Everyone's Guide to the Largest Bankruptcy in U.S. History, John Wiley & Sons, 2002, p. 127–128.

[722]Paul H. Dembinski, Carole Lager, Andrew Cornford and Jean-Michel Bonvin, Enron and World Finance. A Case Study in Ethics. pp. 196–197.

[723]Jennifer Sawayda, Arthur Andersen: An Accounting Confidence Crisis, Daniels Fund Ethics Initiative, University of New Mexico, p. 6.

[724]Ibid, p. 8.

from 1997 to 2001[725]; consequently, Andersen was found guilty only of obstruction of justice for shredding these documents and company files, and they were fined just US$0.5 million.[726] Nevertheless, investors were left in doubt about Andersen's accounting reports for other clients over the previous decade. As a result, the company failed to restart its business after the Enron case, as auditing is based mainly on trust and on the reputation of the auditor.

According to some researchers, "*Enron's board of directors simply did not understand what was going on; they trusted that Jeffrey Skilling's and Andrew Fastow's labyrinthine special purpose entities made sound financial sense; after all, both Skilling and Fastow had graduated from top MBA programs. Thus, neither the auditors nor the Board of Directors performed effectively their function of monitoring the activities of insiders for the benefit of outsiders…*[727] *The Auditing Committee of the Board of Directors continued to rely on its public auditing firm, Arthur Andersen, who continued to write favorable opinion letters that ENRON's accounting was 'adequate to provide reasonable assurance as to the reliability of financial statements*'".[728] Others suggest that Enron's board of directors kept silent for financial reasons: "*Each director received nearly $350,000 per year for serving on Enron's board. That amount was double the high end of normal large public company director fees. The board routinely bragged about Enron's management team. One may ask how much of their 'Enron can do no wrong' attitude was impacted by the fees they received?*".[729]

Enron's Culture

Under the leadership of Jeffrey Skilling, who got the nickname of Darth Vader for his ruthless behavior,[730] a "*cut-throat*" corporate culture, unusual for an energy company, was developed at Enron. This culture would have been more appropriate for an investment bank. Because the mark-to-market accounting approach allowed the recording of profits from long-term deals in the current year, it put enormous pressure on traders to keep providing gigantic new deals, which continued Enron's revenue and market capitalization growth. According to Peter Fusaro and Ross

---

[725] Ibid, pp. 5, 8.

[726] Elizabeth K. Ainslie, Indicting Corporations Revisited: Lessons of the Arthur Andersen Prosecution, American Criminal Law Review, Vol. 43:107, p. 107.

[727] Paul H. Dembinski, Carole Lager, Andrew Cornford and Jean-Michel Bonvin, Enron and World Finance. A Case Study in Ethics. pp. 196–197.

[728] Arthur Gudikunst, ENRON: A Study of FAILURES, Who, How and Why, Bryant College Working Paper Series, Faculty Newsletter, Sep. 2002, p. 3.

[729] Dick Carozza, Interview with Sherron Watkins. Constant Warning, Fraud Magazine, January/February 2007.

[730] Peter C. Fusaro, Ross M. Miller, What Went Wrong at Enron: Everyone's Guide to the Largest Bankruptcy in U.S. History, John Wiley & Sons, 2002.

Miller: "*Enron's corporate culture was essentially focused on two things: The first was profits and the second was how to make even greater profits. The firm didn't strive to create long-lasting business relationships and had little desire to be involved in anything that smacked of the low margins associated with retail-oriented business*".[731] At Enron, it was not the quality of a deal but the size of the deal and the maintenance of a constant 'deal flow'[732] that mattered: "*Good deal versus bad deal? Didn't matter. If you could give it a positive Net Present Value it got done*".[733] As soon as a deal was done, the trader immediately received compensation and forgot about the future of the deal. Therefore, the entire staff of Enron was focused on short-term output.[734] Enron preferred to hire "*the best and the brightest*" young MBA school graduates, who at first were too inexperienced to understand the real flaws of Enron's corporate system—but were very smart, aggressive, and hungry to make short-term money. Enron paid extremely large rewards to traders who met their earnings targets: some common traders were able to earn up to US$15 million a year.[735]

The corporate message was simple: "*If you were smart enough and tough enough to work at Enron, you deserved to live like last year's Oscar winner*".[736] On bonus day, luxury car dealers set up showrooms around the Enron headquarters building. In 2000, the base salary at Enron exceeded the peer group average by 51 %, bonus payments by 382 % and stock options by 484 %.[737] Because employees' pension funds were invested in Enron stock and significant compensation was in the form of stock options, employees were motivated to increase Enron's capitalization by any means. In exchange for such large compensation, Enron claimed enormous productivity, total loyalty to superiors and faith in Enron's unique path: the staff were nicknamed "*Enronians*" and "*believers*".[738]

Moreover, Jeffrey Skilling created the most rigid system of selection and ranking of personnel in corporate America. The system was called the Performance

---

[731]Peter C. Fusaro, Ross M. Miller, What Went Wrong at Enron: Everyone's Guide to the Largest Bankruptcy in U.S. History, John Wiley & Sons, 2002, p. 47.

[732]Clinton Free, Mitchell Stein, Norman Macintosh, Management Controls: The Organizational Fraud Triangle of Leadership, Culture and Control in Enron. The Organization, July/Aug. 2007.

[733]Paul H. Dembinski, Carole Lager, Andrew Cornford and Jean-Michel Bonvin, Enron and World Finance. A Case Study in Ethics. p. 196.

[734]Clinton Free, Mitchell Stein, Norman Macintosh, Management Controls: The Organizational Fraud Triangle of Leadership, Culture and Control in Enron. The Organization, July/August 2007.

[735]Ibid.

[736]Brian Cruver, Anatomy of Greed: Telling the Unshredded Truth from Inside Enron, Basic Books, 2003, p. 191.

[737]Dan Ackman, Pay Madness At Enron, Forbes Magazine, Mar. 22, 2002.

[738]Dennis Tourish, Naheed Tourish, Charismatic leadership and corporate cultism at Enron: The elimination of dissent, the promotion of conformity and organizational collapse, Leadership, Nov. 2005.

Review Committee (PRC), and it focused on assessing the amount, profitability and permanency of the deals an employee had brought into Enron during the previous six months. Every six months, 15 % of staff found themselves on the bottom rank of the PRC rating; if they remained there in the next review, they were fired.[739,740] Such a system strengthened competition and atomization between traders, which prevented anyone in the company being honest with anyone else about the risks they were taking: "*Clearly, the switch from affirmation to punishment within Enron meant that employees regularly received mixed messages. On the one hand, they were the cleverest and best in the world – a form of positive reinforcement, or love bombing, that it would be hard to better. On the other, they could be branded as 'losers', and fired at any time. Consistent with general cultic norms, the overall effect was disorientation, an erosion of one's confidence in one's own perceptions and, most crucially, a further compliance with the group's leaders that strengthened conformist behavior in general… It is clear that Enron management regarded kindness as a show of weakness. The same rigors that Enron faced in the marketplace were brought into the company in a way that destroyed morale and internal cohesion. In the process of trying to quickly and efficiently separate from the company those employees who were not carrying their weight, Enron created an environment where employees were afraid to express their opinions or to question unethical and potentially illegal business practices. Because the rank-and-yank system was both arbitrary and subjective, it was easily used by managers to reward blind loyalty and quash brewing dissent... [There was a] prevailing culture [of] 'the undiscussability of the undiscussable also undiscussable'... [A] former senior manager's summary of the internal culture: 'There was an unwritten rule... a rule of 'no bad news.' If I came to them with bad news, it would only hurt my career*'".[741] "*Paranoia flourished and trading contracts began to contain highly restrictive confidentiality clauses. Secrecy became the order of the day for many of the company's trading contracts, as well as its disclosures*".[742] "*Enron Gas Services was developing a reputation as a predatory place where people would sell each other out to survive*".[743]

---

[739]Paul H. Dembinski, Carole Lager, Andrew Cornford and Jean-Michel Bonvin, Enron and World Finance. A Case Study in Ethics. p. 196.

[740]C. William Thomas, The Rise and Fall of Enron. When a company looks too good to be true, it usually is. Journal of Accountancy, April 2002.

[741]Dennis Tourish, Naheed Tourish, Charismatic leadership and corporate cultism at Enron: The elimination of dissent, the promotion of conformity and organizational collapse, Leadership, November 2005.

[742]C. William Thomas, The Rise and Fall of Enron. When a company looks too good to be true, it usually is, Journal of Accountancy, April 2002.

[743]Paul H. Dembinski, Carole Lager, Andrew Cornford and Jean-Michel Bonvin, Enron and World Finance. A Case Study in Ethics. p. 28.

This internal climate of concealment about risk soon extended to communication with external audiences: Mark Koenig, Enron's former head of investor relations, testified the following "*I wish I knew why I did it. I did it to keep my job, to keep the value that I had in the company, to keep working for the company. I didn't have a good reason*".[744] Sherron Watkins—Enron vice president, subordinate of Andrew Fastow, former auditor at Arthur Andersen and ultimately famous in the USA as an internal whistle-blower—sent an anonymous memo to Lay about the possibility of a wave of accounting scandals after the unexpected resignation of Jeffrey Skilling as CEO in August 2001; later she sent a signed letter to Lay, and visited him personally.[745] Watkins began her letter with these selfish words: "*Has Enron become a risky place to work? For those of us who didn't get rich over the last few years, can we afford to stay?*".[746] She honesty outlined to Lay possible risks from accounting fraud, but the motivation for her whistleblowing attempts was not concern about investors (many of them were pension funds and education organizations; she mentioned them once in the letter), but apparently about the personal wealth of employees—their options and pension funds—and her career prospects: "*I am incredibly nervous that we will implode in a wave of scandals. My eight years of Enron work history will be worth nothing on my resume, the business world will consider the past successes as nothing but an elaborate accounting hoax*".[747] Nevertheless, she recommended to cover-up the problems ("*clean up quietly if possible*").[748] Her letter was found only during the investigation, when the falsifications were revealed, but she had kept silent about Enron's frauds for years before the bankruptcy. Apparently, it was impossible for a conscientious person to survive unscathed in Enron's "*cut-throat*" environment for 8 long years…

Enron and the Investment Banks

Investment banks made serious money from underwriting merger deals, while broker fees brought insignificant profits in comparison. They got large investment-banking fees from Enron transactions, they were investors in Enron's off-balance-sheet special purpose entities and had credit exposure to Enron.[749] Therefore,

---

[744]Clinton Free, Mitchell Stein, Norman Macintosh, Management Controls: The Organizational Fraud Triangle of Leadership, Culture and Control in Enron. The Organization, July/August 2007.

[745]Gary M. Cunningham, Jean E. Harris, Enron And Arthur Andersen: The case of the crooked E and the fallen A, Global Perspectives on Accounting Education, Volume 3, 2006, p. 34.

[746]Text of Watkins' Letter to Lay After Departure of Chief Executive, The New York Times, January 16, 2002.

[747]Ibid.

[748]Peter C. Fusaro, Ross M. Miller, What Went Wrong at Enron: Everyone's Guide to the Largest Bankruptcy in U.S. History, John Wiley & Sons, Inc, 2002, p. 127.

[749]Paul H. Dembinski, Carole Lager, Andrew Cornford and Jean-Michel Bonvin, Enron and World Finance. A Case Study in Ethics. pp. 31–23.

investment bank analysts were urged to publish positive analytic reports about Enron, and sell-side bank traders promoted Enron's stock among their clients—though with an average annual growth of over 65 %, Enron stocks did not require a lot of effort to promote.[750] If the occasional dissident—like John Olson, an analyst at Merrill Lynch—made a "*sell*" recommendation on Enron stocks or published "*neutral*" reports, he or she would be fired, since all the investment banks had a close relationship with Enron's executives. Later, in gratitude for the dismissal of John Olson, Merrill Lynch received at least US$45 million in fees from Enron deals.[751] Enron's impressive projections to become "*the world's leading company*" attracted investors from all around the world. The company declared: "*We believe wholesale gas and power in North America, Europe and Japan will grow from a US$660 billion market to a US$1.7 trillion market over the next several years. Retail energy services in the United States and Europe have the potential to grow from US$180 billion to $765 billion in the not-so-distant future. Broadband's prospective global growth is huge – it should increase from just US$17 billion today to $1.4 trillion within five years. Taken together, these markets present [a several] trillion [dollar] opportunity for Enron... Our stock price is going to go to $120 per share*".[752,753] The registered maximum for an Enron share was US$90 in August 2000; as things turned out, the value of the share in late November 2001 was less than US$1. A Goldman Sachs analytic report extolled Enron: "*Enron has built unique and, in our view, extraordinary franchises in several business units in very large markets*".[754] In early 2001, according to Thomson First Call, 13 of Enron's 18 analysts recommended to buy Enron's stocks.[755] Moreover, 10 out of 15 analysts who followed Enron were still rating the stock as a "*buy*" or a "*strong buy*" as late as November 8, 2001, when Enron finally confessed to accounting falsifications.[756]

---

[750]B. G. Dharan, W. R. Bufkins, Red Flags in Enron's Reporting of Revenues & Key Financial Measures, July 23, 2008, pp. 2,4.

[751]Richard A. Oppel Jr., Merrill Replaced Research Analyst Who Upset Enron, New York Times, July 30, 2002.

[752]Dennis Tourish, Naheed Tourish, Charismatic leadership and corporate cultism at Enron: The elimination of dissent, the promotion of conformity and organizational collapse, Leadership, November 2005.

[753]Text of Sherron Watkins' Testimony at House Hearing on Enron. The New York Times, Feb. 14, 2002.

[754]Bethany McLean, Is Enron Overpriced? Fortune, March 5, 2001.

[755]Ibid.

[756]The Watchdogs didn't Bark: Enron and the Wall Street Analysts, Hearing before the Committee on Governmental Affairs United States Senate, One Hundred Seventh Congress, Second Session, U.S. Government Printing Office, Washington, February 27, 2002.

The Academics Are Impressed

Impressed by the company's fantastic growth in the market, Harvard University prepared a case study about Enron's success for MBA students; Business Week, Forbes, Fortune and other business magazines and newspapers were dazzled by the "Enron Miracle" and published articles portraying the company in a favorable light.[757] For example, Fortune evaluated Enron stocks among its "*10 stocks to last the decade... that should put your retirement account in good stead and protect you from those recurring nightmares about stocks that got away*"[758]; Skilling was named "*The #1 CEO in the USA*" for embracing innovative theories of business and receiving enormous income from these innovations.[759] Ultimately, this was a tale of "*individual and collective greed born in an atmosphere of market euphoria and corporate arrogance. Hardly anyone ... wanted to believe the company was too good to be true... Many kept on buying the stock, the corporate mantra and the dream*".[760] It was only on March 5, 2001 that Fortune magazine published the first serious investigation into the accounting practices of Enron—an article by Bethany McLean entitled "*Is Enron Overpriced?*"—which provoked great attention from investors to the problems.[761]

The Demise

At the peak of Enron's stock price, Ken Lay and Jeffrey Skilling secretly began to sell their stock options. At the same time, they assured the employees that the stock would probably rise.[762] On October 16, 2001, after Skilling's resignation in August and following the activity of Sherron Watkins, Lay was forced to announce the first quarterly losses for more than four years, of US$618 million; net assets declined to US$1.2 billion.[763] Sherron Watkins assessed Skilling's resignation as a trigger that would lead to unearth the cover-ups in Enron's "success story": "*I believe that the probability of discovery significantly increased with Skilling's*

757Dick Carozza, Interview with Sherron Watkins. Constant Warning, Fraud Magazine, January/February 2007.

758David Ryneck, 10 Stocks to Last the Decade, Fortune, August 14, 2000.

759Clinton Free, Mitchell Stein, Norman Macintosh, Management Controls: The Organizational Fraud Triangle of Leadership, Culture and Control in Enron. The Organization, July/August 2007.

760C. William Thomas, The Rise and Fall of Enron. When a company looks too good to be true, it usually is, Journal of Accountancy, April 2002.

761Bethany McLean, Is Enron Overpriced? Fortune, Mar. 5, 2001.

762Yuhao Li, The Case Analysis of the Scandal of Enron, International Journal of Business & Management, October 2010, Vol. 5 Issue 10, p. 37.

763Clemens von Frentz, Enron—Chronicle of a record bankruptcy, Manager Magazine, June 25, 2003.

shocking departure. Too many people are looking for a smoking gun".[764] Soon, Enron revealed that the company had overstated its earnings during the last four years by US$586 million, and hidden US$3.1 billion in debts.[765] After these statements, it became clear to everybody that Enron had been falsifying its accounts for years. From the first investigations of possible accounting fraud, the Enron fall began…

Lay probably counted on the help of George W. Bush, his close friend, but due to tremendous media pressure after the fraud revelations, Bush tried to disown his relationship with Lay, and Enron's massive contributions to his election campaigns. Several Enron executives pleaded guilty, and got decades-long jail sentences with multi-million-dollar fines. Ken Lay died in July 2006 before serving his sentence. Jeffrey Skilling and Andrew Fastow were sentenced to 24 and 10 years in prison respectively.

Enron's Bankruptcy: Why Risks Were Concealed

- **Close and corrupting relationships between Enron executives and representatives of the US political elite led to deregulation changes** that allowed Enron to build a flawed business model. The risks of such a model could be hidden with impunity because of the absence of a strict regulatory framework, and extensive informal relationships between Enron executives, regulators and politicians. Employees of Enron and Arthur Andersen were afraid to reveal risks to the public because they feared they would not find support from regulators, who seemed to have a cozy relationship with Enron's management team.
- The business model was geared to constantly raising the earnings of Enron executives by maintaining the permanent growth of the company's market value. This growth could be achieved by a continual increase of Enron's short-term revenue figures and low debts. Therefore, **Enron's executives corrupted their auditors and several investment banks with lucrative years-long contracts for reaching the required figures.**
- **Wishful thinking of the board of directors, and among investors, employees and the media**—they preferred to believe only in what they wanted to believe, and ignored facts and early warnings. The unwillingness of the majority of investors to go deep into Enron's complex financial operations while the company was steadily expanding in the market.
- **Unfathomable complexity** of the financial engineering through which Enron generated its false financial results was key. This was a precursor to

[764]Peter C. Fusaro, Ross M. Miller, What Went Wrong at Enron: Everyone's Guide to the Largest Bankruptcy in U.S. History, John Wiley & Sons, Inc, 2002, p. 126.

[765]Corporate Fraud: Stop History from Repeating Itself, Kroll investigative service, 2011.

the absolute impossibility of penetrating the CDO-squared structure of the mid-2000s. It was not just an unwillingness, it was an inability.

- The **reluctance of Enron executives to confess any shortcomings of the created business model** in the early stages of Enron's ascent, because doing so could lead to accusations of incompetence and the collapse of capitalization. The **fear of criminal prosecution after the majority of the falsifications** had occurred caused Enron's management to continue distorting information about the real situation within the company until bankruptcy.
- **A "success at any price" and "no bad news" culture**, the **secrecy** of deals at Enron, the **absence of internal control within the company and its frequent labor turnover**: all these processes were consciously implemented by executives to provide a **fragmentary picture of risks among employees**.

2.2.3 Subprime Mortgage Crisis (USA, 2007–2008)

The experience of France in the Belle Époque proves, if proof were needed, that no hypocrisy is too great when economical and financial elites are obliged to defend their interests.

Thomas Piketty

Thirty per cent of OTC derivatives are bought and seventy per cent are sold.

Michael A.H. Dempster

2.2.3.1 The LTCM 1998 Precursor in Addition to the Enron Precursor

The subprime mortgage crisis that started in 2007 in the USA had two notable precursors: (i) as reported earlier, Enron was a precursor with respect to the complexity of its financial engineering constructions; (ii) the hedge fund Long-Term Capital Management (LTCM), which collapsed of in 1998,[766] was a precursor with respect to excess leverage (its positions were quite straightforward and very easy to understand, unlike Enron's structures) and the potential for a single firm to have systemic catastrophic impact on the global financial system.[767]

With 1997 Memorial Nobel prize winners Myron S. Scholes and Robert C. Merton among its principals, LTCM developed arbitrage positions betting on the convergence of what was deemed mispriced spread between bonds and between equity pairs and also traded options. In its first few years, LTCM achieved remarkable returns but had to escalate its leverage to enormous proportions, as its capital base grew and investment opportunities decreased. Thus, at the beginning of 1998, the firm had equity of US$4.72 billion and had borrowed over US$ 124.5 billion with assets of around US$129 billion, for a debt to equity ratio of over 25 to 1.[768] The debt of LTCM was developed with counterparties being most of the important banks on Wall Street. Starting with the 1997 East Asian financial crisis followed by the 1998 Russian government bond default, the convergence arbitrages that LTCM had bet upon actually diverged, leading to huge losses. As LTCM's capital was composed of funds from the same financial professionals with whom it traded, its difficulties led Wall Street to fear that LTCM liquidation of its securities to cover its debt would further push price down in a positive feedback loop, and could cause a chain reaction with catastrophic losses throughout the financial system. On September 1998, the Federal Reserve supervised a bail out of LTCM involving 14 financial institutions for a US$3.6 billion recapitalization allowing to avoid further liquidation in order to prevent the vicious cycle that was feared to possibly collapse the entire world financial system.

Unfortunately, the Federal Reserve, the US treasury and regulators did not learn anything or choose to ignore the lessons offered by the LTCM debacle, allowing essentially the same leverage dynamics to develop industry wide with catastrophic consequences that are still echoing.

---

[766]See Donald Mackenzie, An Engine, Not a Camera: How Financial Models Shape Markets, The MIT Press, Aug. 29, 2008 for a detailed account.

[767]It is interesting to note that, in the combined 1200 pages of Rubin and Greenspan biographies, only about 1.5 pages are dedicated to the deepest banking crisis they oversaw while at the Treasury and the Fed. Should we conclude they were too scared still from the near death experience of the banking system in 1998 to discuss it in public? But their behavior in the 2007–8 crisis proved instead they had not even understood the event!

[768]Long-Term Capital Management's profile. http://en.wikipedia.org/wiki/Long-Term_Capital_Management.

2.2.3.2 Brief Summary of the Crisis

During the 2000s, an American real estate bubble was forming,[769] which burst during 2007–2008. More than eight million American households lost their homes due to foreclosure. More than US$17 trillion of household wealth was wiped out within 21 months after the burst. The American subprime mortgage crisis triggered a global financial and economic crisis in 2008–2009,[770] which caused the most severe recession in over 50 years. Total stock market losses exceeded US$30 trillion worldwide.[771] In order to prevent a total collapse of the world financial system, governments imperiled trillions of taxpayers' money on bailouts of private financial institutions, which were "*too big to fail*". This global salvage operation disrupted the stability of government finance not only in the USA, but also in many European countries. The US federal deficit (the amount by which federal spending exceeds federal income in a given fiscal year) grew from US$161 billion in 2007 to $1.4 trillion in 2009[772]; and total public debt (the total amount owed by the federal government, including debts from intra-governmental holdings) increased by US$3.5 trillion, from $8.8 trillion in the middle of 2007 to $12.3 trillion at the end of 2009.[773]

After the crisis, the Financial Crisis Inquiry Commission (FCIC) was created by the United States government to "*examine the causes of the current financial and economic crisis in the United States*". During hundreds of witness hearings under oath, numerous cases of risk information concealment were revealed, which had led to an inadequate perception of mortgage-related risks among US officials and financial industry executives. The commission stated: "*The crisis was the result of human action and inaction, not of Mother Nature... The captains of finance and the public stewards of our financial system ignored warnings and failed to question, understand, and manage evolving risks within a system essential to the well-being of the American public. Despite the expressed view of many on Wall Street and in Washington that the crisis could not have been foreseen or avoided, there were warning signs. The tragedy was that they were ignored or discounted... Little meaningful action was taken to quell the threats in a timely manner*".[774] Ultimately, the deregulation of the financial sector over decades, the highly fragmented state of US financial regulation and the desire of both government and financial institutions to ensure permanent growth of

[769]W.-X. Zhou, D. Sornette, Is There a Real-Estate Bubble in the US? Physica A: Statistical Mechanics and its Applications, 2006, 361, pp. 297–308.

[770]Markus K. Brunnermeier, Deciphering the Liquidity and Credit Crunch 2007–2008, Journal of Economic Perspectives, 2009, 23(1), 77–100.

[771]Justin Yifu Lin, Policy Responses to the Global Economic Crisis, Development Outreach, World Bank Institute, Volume 11, Issue 3, December 2009, pp. 29–33.

[772]Historical Budget Data—August 2013, Revenues, Outlays, Deficits, Surpluses, and Debt Held by the Public since 1973, Congressional Budget Office, August 12, 2013.

[773]The Daily History of the Debt Results: historical returns from 07/16/2007 through 12/31/2009, U.S. Department of the Treasury, Bureau of the Fiscal Service.

[774]The Financial Crisis Inquiry Report, The Financial Crisis Inquiry Commission, Washington, D.C., January 2011, p. xvii.

income and of the economy as a whole, in an illusionary belief in a perpetual money machine[775]—all of these together led to this crisis. There are strong arguments and compelling evidence that the financial meltdown was predictable and thus avoidable, and resulted from an exaggerated implementation of the free-market ideology and shareholder-value capitalism, with strong asymmetric information and misaligned incentives between shareholders and managers (the so-called agency problem), competition of the corporate-elites and, arguably, a shocking failure of leadership.[776]

From a macro-economic view point, the crisis had its roots in non-sustainable global unbalances, in particular in the exploding China trade surplus with respect to the US, and the associated growing US debt bought by the Chinese, Japanese and Germans: "*If foreigners hold the debt, the interest rate stays stable. Mercantilist only works as long as they are willing to take the losses with the inflation that is coming along. Avoids social consequences of supernormal growth rates for them; subsidizes us by buying our debt. Good deal for us: they give us goods and we give them paper. Herb Stein would say: unsustainable trends have to end*".[777] Even the New York Fed admitted that, because of this trade, rates were held artificially low and drove the search for yield leading up to the crisis.[778] This sounds all too familiar regarding the five years of Quantitative Easing (QE) that have followed and created many short lived bubbles and other unsustainable unbalances![779] The crisis has forced us to pay closer attention to the deregulated derivatives market, whose notional size amounts for more than US$600 trillion[780]—10 times more than the annual global world GDP. Nobody really knows what kind of threats derivatives could bring to the world financial system…

2.2.3.3 Risk Concealment Before the Disaster

Decades-Long Destruction of the Legislation that Followed the Great Depression

In 1933, after the Great Depression, the Banking Act, usually referred to as the Glass–Steagall Act, was passed. In four provisions of the Act, securities activity by

---

[775]D. Sornette and P. Cauwels, 1980–2008: The Illusion of the Perpetual Money Machine and what it bodes for the future, Risks 2, 103–131, 2014.

[776]Michael Lounsbury and Paul M. Hirsch, Editors, Markets on Trial: The Economic Sociology of the U.S. Financial Crisis, Emerald Group Publishing Limited (July 14, 2010).

[777]Meltzer on the Fed, Money, and Gold, EconTalk Episode with Allan Meltzer hosted by Russ Roberts, Library of Economics and Liberty (May 19, 2008) (http://www.econtalk.org/archives/2008/05/meltzer_on_the.html).

[778]Dempster, M.A.H. (University of Cambridge, UK), personal communication (Dec. 22, 2014).

[779]Sornette, D. and P. Cauwels, Financial bubbles: mechanisms and diagnostics, Review of Behavioral Economics (in press 2015) (http://ssrn.com/abstract=2423790).

[780]Sources: U.S. National statistics; International Monetary Fund; OECD, Bain Macro Trends Group Analysis, 2012; Top 10 Challenges for Investment Banks 2011, Accenture, 2010, Chapter "Challenge 2: Dealing with OTC Derivatives Reform".

commercial banks and affiliations between commercial banks and securities firms were restricted to avoid conflicts of interests. The creation of a single financial institution combining an investment bank, a commercial bank and an insurance company was prohibited. In 1934, the United States Securities and Exchange Commission (SEC) was established to regulate secondary trading of securities, by regulating stock exchanges and enforcing sanctions against criminal acts of fraud.[781] In the 1970s, American economists suggested that deregulation of the economy could increase competition within industries, reduce the price of goods and services through the interdependency of supply and demand, and enhance the economic growth of the United States. In 1971, during the Nixon presidency, the first step towards deregulation was made in the transportation industry with the deregulation of rail and truck transport. In 1978, during the Carter administration, the Airline Deregulation Act was passed.

In 1981, U.S. president Reagan promoted a new economic policy based on the reduction of government spending, federal income tax and capital gains tax, and on minimizing government intervention in the economy with the goal to stimulate jobs creation and productivity gains. To implement the "Reaganomics" strategy, Donald Regan—former chairman and CEO of the investment bank Merrill Lynch—was appointed to the position of Secretary of the Treasury, the American equivalent of finance minister. In 1982, the Garn–St. Germain Depository Institutions Act was passed in order to help savings and loan associations compete with mutual funds, which were offering more lucrative interest rates during the high inflation of the 1970s–1980s.[782] In a short speech describing his motivations for the Garn–St. Germain Depository Institutions Act, Reagan "*told an audience of S & L executives, bankers members of Congress, staffers, and journalists that the bill – which bore the names of Republican Senator Jake Garn of Utah and Democratic Congressman Fernand St Germain of Rhode Island – would cut S & Ls loose from the girdle of old – fashioned regulation. One of Reagan's campaign platforms was deregulation, to get government off the backs of businesses to help the struggling economy create new jobs. When Reagan took office in 1981, mortgage rates were in nosebleed territory: 14 percent. (And this was for home buyers with good credit.) A year later rates would be even higher – 16 percent. ... Reagan signed the Garn – St Germain bill, he said the legislation would create more housing, more jobs, and growth for the economy. 'All in all', 'he proclaimed', 'I think we've hit the jackpot*'".[783] The act broadened the range of legally permitted loans

[781]Matthew Sherman, A Short History of Financial Deregulation in the United States, Center for Economic and Policy Research, July 2009, pp. 3–4.

[782]Ibid, p. 7.

[783]Paul Muolo and Mathew Padilla, Chain of Blame. How Wall Street Caused the Mortgage and Credit Crisis, John Wiley & Sons, Inc., p. 52; But this book does not mention any information about the large contributions of the financial industry to Reagan's campaign and the appointment of its representatives in key positions in US regulatory bodies.

and investments, allowing banks to provide variable-rate mortgage loans. Mutual funds, established by investment banks, were serious competitors of commercial banks: the assets of mutual funds surged from US$3 billion in 1977 to more than US$740 billion in 1995, and US$1.8 trillion by 2000.[784] These funds were not regulated: according to the testimony of Paul Volcker, former chairman of the Federal Reserve, "*There was no regulation. It was kind of a free ride*".[785] These funds had complete freedom of investment activity and did not participate in the deposit insurance system. Clients' protection against losses was based only on the investment bank's reputation for protecting money market funds.[786] The rise of mutual funds urged the commercial banks to "*put a lot of pressure on [government] institutions to get higher-rate performing assets*".[787]

The Tax Reform Act of 1986 also promoted the house bubble and mortgage refinancing frenzy: "*And there was yet one more advantage to being a consumer finance company, especially one that was making loans (second liens) secured by a house. Congress passed the Tax Reform Act of 1986 – signed into law by President Reagan – which eliminated the ability of consumers to deduct interest payments on credit cards, auto loans, and all types of personal loans. Worried about a growing budget deficit, the politicians were hoping that by eliminating the tax deduction, this newfound money would feed the federal coffers. Consumers could no longer deduct the interest payments on their cars, credit cards, or personal loans, they might stop spending, which ultimately might hurt the consumer finance industry. Instead, it shifted borrowing – to some degree – away from personal loans to an asset class where Americans could still deduct the interest payments: the home*".[788]

In 1986–1987, the Federal Reserve allowed American banks to make up to 5 % of gross revenues from investment banking business, and to underwrite commercial paper (unsecured promissory notes issued by banks or corporations), municipal bonds, and mortgage-backed securities.[789] In August 1986, Alan Greenspan, a leading apologist for deregulation and the free market, was appointed as chairman of the Federal Reserve; he was to remain chairman for the following 18½ years until January 2006. In the same year, commercial banks obtained the right to get up to 10 % of their revenue from debt and equity securities; in 1996, this limit was raised to 25 %.[790] Banks also got permission to deal with derivatives: debt

[784]The Financial Crisis Inquiry Report, The Financial Crisis Inquiry Commission, Washington, D.C., Jan. 2011, p. 30.

[785]Ibid, p. 33.

[786]Ibid, p. 33.

[787]Ibid, p. 34.

[788]Paul Muolo and Mathew Padilla, Chain of Blame. How Wall Street Caused the Mortgage and Credit Crisis, John Wiley & Sons, Inc., pp. 36–37.

[789]Matthew Sherman, A Short History of Financial Deregulation in the United States, Center for Economic and Policy Research, July 2009, p. 9.

[790]Ibid, p. 9.

securities (allowed from 1983), interest and currency exchange rates (from 1988), stock indices (from 1988), precious metals such as gold and silver (from 1991), and equity stocks (from 1994).[791] During the 1980s and early 1990s, commercial banks began providing higher-risk loans with higher interest payments. They offered loans to oil and gas producers, financed leveraged buyouts of corporations, and funded residential and commercial real estate developers for international expansion.[792] As a result, during the savings and loan crisis in the late 1980s and early 1990s, 1034 savings and loan associations failed with US$160 billion losses.[793] By contrast, 584 banks had failed between 1934 and 1980 when there was a rigid legal framework.[794]

Later, Greenspan described the arguments for deregulation: "*Those of us who support market capitalism in its more competitive forms might argue that unfettered markets create a degree of wealth that fosters a more civilized existence. I have always found that insight compelling*".[795] "*The market-stabilizing private regulatory forces should gradually displace many cumbersome, increasingly ineffective government structures*".[796] The decline of government involvement in the economy had an ideological and geopolitical basis, coinciding as it did with the collapse of the Soviet Union, a unique example of total government control over political, social and economic activities. In addition, the financial lobby sought theoretical credibility for further deregulation from the academic world, and began to engage prominent professors and researchers to study the possible advantages of deregulated markets. They offered millions of dollars in funding and grants, tens of thousands of dollars in speaking fees and generous salaries for involvement on the boards of financial institutions.[797] In fact, the academic founders of modern finance theory did not need to be induced into their theoretical positions. They deeply believed in the creative logic of their work.[798] Unsurprisingly, all these elements combined to ensure the dominance of a free-market theory, supported by apparently solid scientific studies, which argued for the necessity to decrease the role of government in the economy. This research helped financial lobbyists to find

[791]The Financial Crisis Inquiry Report, The Financial Crisis Inquiry Commission, Washington, D.C., Jan. 2011, p. 35.

[792]Ibid, p. 35.

[793]Timothy Curry, Lynn Shibut, The Cost of the Savings and Loan Crisis: Truth and Consequences, FDIC Banking Review, Dec. 2000, pp. 26, 29.

[794]Failures and Assistance Transactions, Number of Institutions US and Other Areas (1934–1980), Federal Deposit Insurance Corporation. http://www2.fdic.gov/hsob/HSOBSummaryRpt.asp?BegYear=1934&EndYear=1980&State=1&Header=1.

[795]Remarks by Chairman Alan Greenspan, Before the Council on Foreign Relations, Washington, D.C., Nov. 19, 2002.

[796]The Financial Crisis Inquiry Report, The Financial Crisis Inquiry Commission, Washington, D.C., Jan. 2011, p. 28.

[797]Documentary film "Inside Job", Director Charles Ferguson, 2010.

[798]See Donald Mackenzie, An Engine, Not a Camera: How Financial Models Shape Markets, The MIT Press, Aug. 29, 2008.

a legal justification for deregulation, and convince politicians to disassemble the legal framework that had been in place since the Great Depression: from 1999 to 2008, the financial sector spent US$2.7 billion on reported federal lobbying. In addition, the sector contributed more than US$1 billion to political campaigns during this period.[799]

"We Had a 21st-Century Financial System with 19th-Century Safeguards"

In 1998, during the Clinton administration, Citibank announced a merger with Travelers Insurance Group—which owned Salomon Brothers investment bank—to establish the largest financial institution in the world, Citigroup Inc. It is remarkable that the deal was declared in violation of the Glass–Steagall Act of 1933, but the Federal Reserve made an exception for this merge. At the time of the deal, the Secretary of the Treasury was former Goldman Sachs executive Robert Rubin, who worked at Citigroup Inc. after the merger as a board member, chairman of the executive committee and chairman of the board of directors (1999–2009). Citigroup Inc. paid him up to US$126 million.[800] In 1999, after lobbying from the financial sector, Congress passed the Gramm-Leach-Bliley Act, which lifted all restrictions against the combination of banking, securities and insurance operations within a single financial institution. This paved the way for further mergers.[801] Ultimately, by 2005, the ten largest US commercial banks held 55 % of the industry's assets—twice the proportion held by the top ten in 1990.[802] Lawrence Summers, Rubin's successor as Secretary of the Treasury and a former academic economist and Harvard professor, said on the passing of the Gramm-Leach-Bliley Act: "*Today, Congress voted to update the rules that have governed financial services since the Great Depression and replace them with a system for the 21st century. This historic legislation will better enable American companies to compete in the new economy*".[803] After the crisis, the FCIC commission stated that, in fact, "*we had a 21st-century financial system with 19th-century safeguards*".[804]

---

[799]The Financial Crisis Inquiry Report, The Financial Crisis Inquiry Commission, Washington, D.C., Jan. 2011, p. xviii.

[800]William D. Cohan, Rethinking Bob Rubin From Goldman Sachs Star to Crisis Scapegoat, Bloomberg, Sep. 20, 2012.

[801]Matthew Sherman, A Short History of Financial Deregulation in the United States, Center for Economic and Policy Research, July 2009, p. 10.

[802]The Financial Crisis Inquiry Report, The Financial Crisis Inquiry Commission, Washington, D.C., Jan. 2011, p. xvii.

[803]Stephen Labaton, Congress Passes Wide-Ranging Bill Easing Bank Laws, The New York Times, Nov. 5, 1999.

[804]The Financial Crisis Inquiry Report, The Financial Crisis Inquiry Commission, Washington, D.C., Jan. 2011, p. xx.

Deregulation led to a situation where the banking, securities and insurance operations of the new merged financial institutions were still overseen by separate regulators: there was no single government regulator looking at all of their commercial activities. So neither government nor the executives of financial institutions had the whole picture of the risks involved in a complex combination of businesses with different interests—especially in the widening distribution of derivatives. For instance, the CEO of Citigroup told the FCIC commission that US$40 billion invested in highly rated mortgage securities would "*not in any way have excited my attention*", and the co-head of Citigroup's investment bank said he spent "*a small fraction of 1 % of his time on those securities*". The commission declared that "*too big to fail meant too big to manage. We conclude a combination of excessive borrowing, risky investments, and lack of transparency put the financial system on a collision course with crisis*".[805]

The complexity of understanding the principles of the creation and calculation of derivatives, together with continued lobbying from financial institutions, resulted in the absence of serious government regulation over innovative financial instruments. We should also stress the prevalence of finance theories that rationalized and legitimized both public deregulation and extraordinary private risk-seeking behavior, while generating such complexity in the financial engineering and the underlying cash flows beneath the financial structures that literally no one could have penetrated in the cases of Enron and of the global financial crisis.

When the Commodity Futures Trading Commission expressed their intention to discuss the possible regulation of over-the-counter (OTC) derivatives, their attempts to do so were suspended by Alan Greenspan, Robert Rubin and Lawrence Summers.[806] Greenspan testified that there was no need for government oversight, because "*regulation of derivatives transactions that are privately negotiated by professionals is unnecessary*".[807] In the 20 years from early 1990 to 2009, the unregulated global derivatives market—of which 90 % consisted of OTC derivatives—grew from US$10 trillion to US$605 trillion[808]; the world GDP in 2010 was approximately US$65 trillion.[809] After the mortgage crisis in autumn 2008, Greenspan admitted that "*Those of us who have looked to the self-interest of lending institutions to protect shareholders' equity (myself especially) are in a state of shocked disbelief*".[810] The FCIC commission considered that "*the enactment of legislation in 2000 to ban the regulation by both the federal and state governments*

[805]Ibid, p. xix.

[806]Rick Schmitt, Prophet and Loss, Stanford Magazine, March/April 2009.

[807]Testimony of Chairman Alan Greenspan, The regulation of OTC derivatives, Before the Committee on Banking and Financial Services, U.S. House of Representatives, July 24, 1998.

[808]Top 10 Challenges for Investment Banks 2011, Accenture, 2010, Chapter "Challenge 2: Dealing with OTC Derivatives Reform".

[809]In search of growth, The Economist online, May 25, 2011.

[810]Kara Scannell, Sudeep Reddy, Greenspan Admits Errors to Hostile House Panel, The Wall Street Journal, October 24, 2008.

of OTC derivatives was a key turning point in the march toward the financial crisis".[811]

Creation of Housing Bubble

In order to stimulate economic growth, the administrations of Bill Clinton and George W. Bush set aggressive goals to increase home ownership, which could generate activity in the construction industry and create millions of new jobs. Deregulated financial products and a Federal Reserve interest rate of 1.75 %, the lowest in the previous 40 years, supplied accessible credit for potential borrowers. From 1999 to 2007, the average house price nationwide increased by 67 %; in 110 metropolitan areas, the price doubled. The floor area of an average new home grew by 15 % in the decade from 1997 to 2007.[812] In 2005, more than 10 % of house sales were made for financial reasons by investors, speculators, or people buying second homes. Houses became a commodity—an asset—and could be mortgaged to get cash for putting children through college, medical bills, or sabbaticals to launch new businesses. As a result, home refinancing rose from US$460 billion in 2000 to US$2.8 trillion in 2003, despite stagnant wages.[813]

Before the deregulation of the early 1980s, lenders selected borrowers carefully, because they needed, for their own sake, to ensure that a borrower could pay a 30-year fixed-rate mortgage. The stability of financial institutions depended on the reliability of their debtors. Even in the 1990s, only the highest quality clients who could comply with tough requirements—known as "prime" borrowers—were eligible. For example, one requirement was that first-time home buyers should be able to make a 20 % down payment. However, deregulation and active encouragement from the government allowed lenders to lower the acceptable standard for borrowers, and provide credit for people with no credit history or proof of income—and the "subprime" market was born. It was made possible by the creation of a securitization pipeline: lenders packaged loans into residential mortgage–backed securities, and these securities were repackaged again into collateralized debt obligations (CDOs) by investment banks like Goldman Sachs, Merrill Lynch, Bear Stearns or Lehman Brothers. In their turn, CDOs were promoted among more conservative American investors (retirement systems, hospitals, endowment funds and the like) and global investors (pension funds and sovereign funds) as a "*super-senior*" and "*s*`securities from a range of different quality debtors[814,815]

[811]The Financial Crisis Inquiry Report, The Financial Crisis Inquiry Commission, Washington, D.C., January 2011, p. xxiv.

[812]Ibid, p. 5.

[813]Ibid, p. 5.

[814]Paul Muolo and Mathew Padilla, Chain of Blame. How Wall Street Caused the Mortgage and Credit Crisis, John Wiley & Sons, pp. 185, 219.

[815]The Financial Crisis Inquiry Report, The Financial Crisis Inquiry Commission, Washington, D.C., Jan. 2011, pp. 117, 119, 170, 278, 339, 393.

Economist James Grant described the "*mysterious alchemical processes [by which] Wall Street transforms BBB-minus-rated mortgages into AAA-rated tranches of mortgage securities*".[816] Companies like American International Group (AIG)—the largest insurance company in the world—insured the banks against potential default by credit default swaps (CDSs). By 2007, AIG had issued CDSs on $379 billion of underlying value.[817]

A strong positive feedback mechanism developed between the home price dynamics and the loan origination process: "*Why would the defendants overvalue the homes? Answer: because the higher the house value, the larger the loan Ameriquest [it was one of the largest sub-prime mortgage lender in the United States until shut down in 2007] could fund. The larger the loan, the higher the commission the friends could earn. On $9 million in retail loans that Ameriquest had extended on the 64 homes, the seven friends earned $172,400, which works out to almost two points (2 percent) per loan. The loan officers, most of whom were in their 20 s, found the borrowers by going through the company's 'turn – down files' where LOs stored the names of customers who had previously been rejected for loans. But the borrowers didn't receive kickbacks from the LOs – they were just happy to get a mortgage*".[818] This fed on the absence of information concerning payments of subprime borrowers: "*Mike McMahon, the stock analyst, saw the problem coming. 'They were way too optimistic on the life of the loans,' he said. 'Everyone was guessing with limited historical data.' Historical data? As far as securitization went, subprime mortgages had no history. 'Everyone was guessing,' said McMahon. 'These weren't Fannie, Freddie, and FHA loans where there's 40 years of past data to look at on how they'd perform*'".[819]

Corruptive Assessments of American Rating Agencies

In their book Chain of Blame, Paul Muolo and Mathew Padilla summarize the process as follows: "*Almost every mortgage they put into a bond was a loan made to a borrower who either had bad credit or was considered a stated – income risk. Stated – income mortgages worked like this: The borrowers stated their income*

[816]The Financial Crisis Inquiry Report, The Financial Crisis Inquiry Commission, Washington, D.C., Jan. 2011, p. 194.

[817]Matthew Richardson, Why the Volcker Rule Is a Useful Tool for Managing Systemic Risk, NYU Stern School of Business, 2012, p. 8.

[818]Paul Muolo and Mathew Padilla, Chain of Blame. How Wall Street Caused the Mortgage and Credit Crisis, John Wiley & Sons, p. 88.

[819]Ibid, p. 44.

and the lenders believed them. It was a wildly popular product and for obvious reasons: Borrowers got what they wanted even though they had to pay a slightly higher interest rate for it. Wall Street loved any type of loan that was paying a higher rate than the conventional or "A" paper rate of good credit quality mortgages sold to Fannie Mae and Freddie Mac, two congressionally chartered mortgage giants whose mission in life was to buy such loans. A higher – yielding mortgage meant that a Wall Street firm like Bear Stearns could create a higher – yielding bond to sell to an investor. Every time a bond salesman at Bear (or any other firm) sells a bond, he takes a fraction of the deal for himself. On a $50 million bond, the commission might be an eighth of a point, which works out to $62,500. Bond commissions are not openly publicized and can vary greatly depending on what type of bond is being sold. But one equation rings true – the higher the yield on the bond, the higher the bond sale commission. Subprime mortgages were the highest – yielding loans around that were backed by something tangible: a house".[820]

The CDOs' AAA rating was assigned by such respected rating agencies as Moody's, Standard & Poor's and Fitch. Investment banks "*paid handsome fees to the rating agencies to obtain the desired ratings*"[821] – between US$0.5 million to 0.85 million for every mortgage-related security. In the 1990s, the quality of the obligations was not in doubt but, with the subsequent growth of lending, it became harder to carefully track the quality of borrowers. The rating agencies knew exactly what they were doing. One S&P employee wrote: "*Rating agencies continue to create an even bigger monster – the C.D.O. market. Let's hope we are all wealthy and retired by the time this house of cards falters*". Another wrote in an instant message: "*We rate every deal. It could be structured by cows and we would rate it*".[822] After the crisis, Moody's executives testified: "*We had almost no ability to do meaningful research… The threat of losing business to a competitor* [Standard & Poor's or Fitch], *even if not realized, absolutely tilted the balance away from an independent arbiter of risk towards a captive facilitator of risk transfer… Bankers were pushing more aggressively, so that it became from a quiet little group to more of a machine… Subprime [residential mortgage–backed securities] and their offshoots offer little transparency around composition and characteristics of the loan collateral… Loan-by-loan data, the highest level of detail, is generally not available to investors*".[823] Moody's standard disclaimer – stating that "*The ratings… are, and must be construed solely as, statements of opinion and not statements of fact or recommendations to purchase, sell, or hold any securities*", gave the rating agency protection against any lawsuits from misled

[820]Ibid., pp. 7.

[821]The Financial Crisis Inquiry Report, The Financial Crisis Inquiry Commission, Washington, D.C., January 2011, p. 44.

[822]Jake Zamansky, The Chickens Come Home to Roost for Standard & Poor's, Forbes Magazine, February 5, 2013.

[823]The Financial Crisis Inquiry Report, The Financial Crisis Inquiry Commission, Washington, D.C., January 2011, p. 119.

investors. From 2000 to 2007, Moody's rated nearly 45,000 mortgage-related securities as AAA. In 2006 alone, when earnings on mortgage ratings reached US$887 million or 44 % of overall corporate revenue, Moody's was putting its AAA stamp of approval on 30 mortgage-related securities every working day. In 2007–2008 during the crash, 83 % of the AAA mortgage securities of 2006 were ultimately downgraded.[824]

Following the onset of the financial crisis in 2008, banks and other financial firms have collectively paid more than $40 billion as punishment for crisis-era misdeeds to the US government as off January 2015. But rating agencies have been also in the firing line and the US Justice Department has launched a lawsuit against S&P, a unit of McGraw Hill Financial, accusing S&P of giving top ratings to poor quality mortgage-backed securities between 2004 and 2007 and of knowingly misleading investors with inflated ratings of residential mortgage-backed securities and collateralized debt obligations (CDOs). According to the lawsuit, S&P gave the deceptive ratings so it could collect fees from the financial firms that sold the securities. The Justice Department and more than a dozen state attorneys general argue that "*S&P's relationships with the banks that designed the mortgage deals 'improperly influenced' the ratings criteria. It also accused S&P of falsely claiming that its ratings 'were objective, independent, uninfluenced by any conflicts of interest*'".[825] For fears of the negative impact towards shareholders, reputation damage and to avoid the embarrassment of paying the same or even more after a trial, ratings company Standard & Poor's has struck a $1.37 billion settlement with the U.S. Justice Department over mortgage ratings that S&P issued leading up to the 2008 financial crisis, a penalty large enough to wipe out the rating agency's entire operating profit for a year[826,827] According to the CNN announcement of February 3, 2015,[828] the settlement also resolves lawsuits with attorneys general in 19 states and the District of Columbia and S&P will pay separately $125 million to California's public pension fund to resolve claims that it was misled in three separate transactions. In January 2015, S&P already paid $58 million to the Securities and Exchange Commission and $19 million to settle similar charges with the attorneys general in New York and Massachusetts. Again, according to the CNN announcement of February 3,[829] "*S&P said the latest settlement 'contains no findings of violations of law by the company'. However, federal*

[824]Ibid, p. xxv.

[825]Ben Protess, S&P nears settlement with Justice Dept. over inflated ratings, The New York Times, January 12, 2015.

[826]Tom Huddleston, Jr., Fortune, JANUARY 28, 2015, http://fortune.com/2015/01/28/standard-poors-doj-settlement/.

[827]Evan Perez and Ben Rooney, S&P to pay $1.4 billion to settle U.S. charges, CNN Money (New York), February 3, 2015 (http://money.cnn.com/2015/02/03/investing/sp-mortgage-settlement/index.html).

[828]Ibid.

[829]Ibid.

prosecutors said the company has acknowledged the 'improper conduct that led to this settlement'".

Development of CDO Monsters

Once a mortgage securities package was sold, the lender had no need to monitor the financial situation of debtors, because all payments from borrowers were transmitted to the owners of mortgage securities. The minimum down payment was soon reduced to 3 % and, after Bush's "*Zero Down Payment Initiative*", it dropped to just US$500. These lucrative terms attracted millions of subprime (near-prime, non-prime, and second-chance lending) borrowers, who had to pay a higher rate than prime borrowers: in California, the average subprime borrower was paying US$600 per month more than a prime borrower on their mortgage payments because they had received a subprime loan.[830] The share of subprime mortgages in the US mortgage market increased from 7.4 % in 2002 to 23.5 % in 2007.[831] Subprime mortgages meant greater profitability for financial institutions. Investment banks wanted constant growth in the number of new CDOs (collateralized debt obligations), and encouraged lenders to issue new credit for everyone. In a permanently growing real estate market, such a model worked well: borrowers warmed up the property market by taking the new more accessible credit, and lenders did not have to worry about the creditability of borrowers because they were transferring risks through CDOs and other mortgage securities to investors, who in turn insured risks through CDSs. As this was developing in full force, Greenspan declared that the financial system had achieved unprecedented resilience.[832] AAA ratings for mortgage securities maintained the illusion of a high quality of assets, which seemed to be among the safest in the world. One FCIC commission witness compared this financial creativity with "*cheap sangria, when a lot of cheap ingredients [are] repackaged to sell at a premium. It might taste good for a while, but then you get headaches later and you have no idea what's really inside*".[833] One of the inventors of securitization testified that "*If you look at how many people are playing, from the real estate agent all the way through to the guy who is issuing the security and the underwriter and the underwriting group and blah, blah, blah, then nobody in this entire chain is responsible to anybody*".[834] As the commission put it, "*They all believed they could off-load their risks on a moment's notice to the next person in line. They were wrong. When*

[830]The Financial Crisis Inquiry Report, The Financial Crisis Inquiry Commission, Washington, D.C., January 2011, p. 11.

[831]Ibid p. 70.

[832]Ibid, p. 83.

[833]Ibid, p. 6.

[834]Ibid, p. 89.

borrowers stopped making mortgage payments, the losses—amplified by derivatives—rushed through the pipeline".[835]

For the financial sector, the real estate boom became a major source of profit, as a result of the higher interest rates paid by borrowers and the chain of bond placements by Wall Street: "*A consumer (usually subprime) buying a home or refinancing and trying to keep closing costs low would agree to pay a higher interest rate on the mortgage in return for paying no points (or fewer points) at the closing table. The higher yield on the loan made that mortgage more valuable to the wholesaler (Countrywide, Wells Fargo, Washington Mutual), because the wholesaler could sell it to Wall Street at a better price than a lower – yielding loan would garner. In the world of Wall Street, the higher the interest rate on a loan, the more valuable it became. Why? Answer: because the loan would be pooled into a bond, and bond investors loved higher – yielding assets. A higher yield of even just 1 percent more on a billion – dollar bond would translate into millions extra in income for the bondholder*".[836]

From 1978 to 2008, the amount of debt held by the financial sector increased from US$3 trillion to US$36 trillion; more than 33 % of all corporate income in the United States was generated by financial institutions in 2003, while in 1980 the proportion had been 15 %.[837] Before the 1980s, the majority of investment banks were private companies; a loyal employee would receive a bonus on retirement after a successful career lasting decades. After the 1980s, when investment banks became public companies and staff began to trade with shareholders' money, the compensation model completely changed: tremendously high annual bonuses urged executives and managers to focus on short-term financial results, increasing current capitalization and short-term profitability while ignoring the possible consequences of risky practice in the long-term. In 2007, Wall Street paid roughly US$33 billion in year-end bonuses to New York workers.[838] None of the executives wanted to overturn the defective mortgage market by revealing the shortcomings of the business model they had created. After the crisis Jamie Dimon, CEO of JP Morgan, testified that "*I blame the management teams 100 % and ... no one else*".[839] Federal, state and local government also benefited from the real estate boom through permanent economic growth, massive foreign investment in the US stock market, declining unemployment, rising revenues from individual and property taxes, etc. Any problem in the property market could lead to a cascade effect in many American industries, bankrupting millions of Americans, destroying profitability of the financial sector and bringing severe political consequences. So,

[835]Ibid, p. xxiv.

[836]Paul Muolo and Mathew Padilla, Chain of Blame. How Wall Street Caused the Mortgage and Credit Crisis, John Wiley & Sons, Inc., p. 68.

[837]The Financial Crisis Inquiry Report, The Financial Crisis Inquiry Commission, Washington, D.C., January 2011, p. xvii.

[838]Ibid, p. 63.

[839]Ibid, p. 18.

decision makers in both the financial sector and US government were reluctant to ask questions or embark on a detailed investigation of common business practice within the securitization pipeline. This unwillingness encouraged the institutions involved in the pipeline to conceal information about the real situation within their businesses.

For example, lenders hired thousands of young people, with no mortgage experience, to sell credit products "*to, in some cases, frankly unsophisticated and unsuspecting borrowers*".[840] Lenders promoted low monthly payments in the first few months after taking a loan and hid bigger fees in subsequent payments, which were seldom disclosed to borrowers. A study by two Federal Reserve economists confirmed that at least 38 % of borrowers with adjustable-rate mortgages (ARMs) did not understand the calculation of their interest rates.[841] In 2006, during the Federal Reserve's Home Ownership and Equity Protection Act hearings, consumers testified that ARM loans were sold to people speaking "*primarily non-English languages [migrants from Latin American and Asia], only to be pressured to sign English-only documents with significantly worse terms*".[842] The FCIC commission found out that lenders had made loans knowing that borrowers could not afford them, and that the percentage of borrowers who defaulted on their mortgages within the first few months had grown steadily. Lenders had been "*forcing them to pay into—pay loans that they could never pay off. Prevalent among these clients are seniors, people of color, people with disabilities, and limited English speakers and seniors who are African American and Latino*".[843]

In 2003, Washington Mutual ordered an internal study "*to explore what Washington Mutual could do to increase sales of Option ARMs, our most profitable mortgage loan*".[844] It revealed that Washington Mutual brokers "*felt these loans were 'bad' for customers … a lot of (loan) consultants don't believe in it … and don't think [it's] good for the customer*".[845] Nevertheless, the company motivated brokers to focus precisely on selling ARMs: after 2004, more than 50 % of all Washington Mutual mortgages were adjustable-rate, and the volume of ARMs sold by Washington Mutual rose from US$30 billion to US$68 billion in 2004. Unfortunately, other lenders came to the same conclusion—and nationwide ARM sales rose from US$65 billion in 2003 to US$255 billion in 2006. During the subsequent crash, it was ARMs that generated the majority of defaults by borrowers as well as the greatest losses for mortgage securities holders.

During the boom, executives of Countrywide—a company that was financing up to 20 % of all mortgages in the United States, around 25 million homebuyers—recognized that many of the loans they were originating could result in

[840]Ibid, p. 8.

[841]Ibid, p. 90.

[842]Ibid, p. 90.

[843]Ibid, p. 109.

[844]Ibid, p. 107.

[845]Ibid, p. 107.

"*catastrophic consequences*" and "*financial and reputational catastrophe*" for the firm. Angelo Mozilo, the co-founder and CEO of Countrywide, wrote in an internal e-mail: "*In all my years in the business, I have never seen a more toxic [product]*".[846] Nonetheless, Countrywide and the investment banks continued to sell these securities to investors, and insurance companies continued to insure them against default (in October 2010, Angelo Mozilo attained an agreement with SEC, in a settlement of the allegations against him that he misled Countrywide's investors. He was sanctioned to pay a record US$67.5 million in fines[847]).

The FCIC commission found that critical information was withheld from investors by other lenders too: Countrywide's portfolio consisted of 59 % non traditional loans, but Wells Fargo had 58 %, Washington Mutual 31 %, CitiFinancial 26.5 %, and Bank of America 18 %.[848] In some cases, lenders distorted information about the earnings and workplaces of applicants to ensure fast confirmation of loans: some debtors were categorized as "antiques dealers" or "light construction" workers.[849] After the bankruptcy of New Century Financial Corporation in 2007, it was revealed that 40 % of its mortgages were loans with little or no documentation. Consequently, mortgage fraud grew 20-fold between 1996 and 2005 and doubled again between 2005 and 2009.[850] More generally, this process can be described as a kind of fraud pandemic: "*In 2004 and 2005, home lenders originated $1.4 trillion in subprime loans – almost all of it winding up in ABSs, with the riskier bonds going into CDOs. Reporters from National Mortgage News and the Orange County Register began to investigate the outsourcing firms, interviewing not only the executives at those companies but also their rank – and – file workers who were hired – on a contract basis – to sit in hotel conference rooms, armed with a laptop, with orders to review one loan an hour. Mortgages were given a rating of a one, two, or three. One meant pass, two meant so – so, and three meant fail. 'You weren't supposed to fail loans unless they were horrendous,' one contract underwriter told the reporters. He also confessed that they were told by their supervisors at Clayton never to use a certain word – 'fraud'. Because competition was so stiff those years and because Merrill, Bear, J.P. Morgan, and other Wall Street firms were so hungry for product (which they could put into ABSs and CDOs), the goal, the underwriters said, was to pass as many loans as possible. Loan fraud is a fuzzy term that can mean many things, but in practice it boils down to two basic swindles: Either a borrower is lying about his or her income or the house is not worth what someone says it is. By late 2006, agents from the Federal Bureau of Investigation were describing loan fraud as pandemic in the*

[846]Ibid, pp. xxii, 20.

[847]Former Countrywide CEO Angelo Mozilo to Pay SEC's Largest-Ever Financial Penalty Against a Public Company's Senior Executive, SEC, Washington, D.C., Oct. 15, 2010, http://www.sec.gov/news/press/2010/2010-197.htm.

[848]Ibid, p. 20.

[849]Ibid, p. 12.

[850]Ibid, p. xxii.

United States, singling out stated – income loans (that is, so-called liar loans) being funded through mortgage brokers as a chief problem".[851] In spite of the warnings of the FBI about mortgage fraud, government regulators of the financial sector paid little attention to this inappropriate practice.

During the commission hearing, a criminologist observed that "*Lax or practically non-existent government oversight created what criminologists have labeled 'crime-facilitative environments', where crime could thrive*". "*The FBI did have severe limits*" because they were authorized to tackle the threat of terrorism; nevertheless "*[they] got virtually no assistance from the regulators, the banking regulators and the thrift regulators*".[852] One former Bear Stearns executive testified that a Federal Reserve representative, on hearing that the housing securitization market was on a shaky foundation, said: "*We don't see what you're talking about because incomes are still growing and jobs are still growing*". Regulators "*relied extensively on banks' own internal risk management systems*" and expected that "*markets will always self-correct*".[853]

Wall Street CEOs Reject Early Warnings

In June 2006, Richard Bowen, chief business underwriter of Citi, discovered that up to "*60 % of the loans that [were bought] and packaged into obligations were defective. If the borrowers were to default on their loans, the investors could force Citi to buy them back. He tried to alert top managers at the firm by 'email, weekly reports, committee presentations, and discussions'; but though they expressed concern, it 'never translated into any action'. He finally took his warnings to the highest level he could reach – Robert Rubin, the chairman of the Executive Committee of the Board of Directors and a former US treasury secretary. He sent Rubin and the others a memo with the words 'URGENT—READ IMMEDIATELY' in the subject line. Sharing his concerns, he stressed to top managers that Citi faced billions of dollars in losses if investors were to demand that Citi repurchase the defective loans. Rubin told the Commission in a public hearing in April 2010 that 'I do recollect this and that either I or somebody else, and I truly do not remember who, but either I or somebody else sent it to the appropriate people, and I do know factually that that was acted on promptly and actions were taken in response to it'. According to Citigroup, the bank undertook an investigation and the system of underwriting reviews was revised... There was no disclosure made to the investors with regard to the quality of the files they were purchasing... Bowen told the Commission that after he alerted management by sending emails, he went from*

[851] Paul Muolo and Mathew Padilla, Chain of Blame. How Wall Street Caused the Mortgage and Credit Crisis, John Wiley & Sons, Inc., p. 197.

[852] The Financial Crisis Inquiry Report, The Financial Crisis Inquiry Commission, Washington, D.C., January 2011, pp. 161, 164.

[853] Ibid, pp. 19, 170, 171.

supervising 220 people to supervising only 2, his bonus was reduced, and he was downgraded in his performance review".[854]

However, such practice was common not only in Citi, but also among other players of the securitization pipeline. Thus, Lehman Brothers CEO Richard Fuld was quickly eliminating internal critics who realized early that Lehman was heading for serious trouble. Any warnings from talented researchers and managing directors were ignored. There was a lack of communication and common understanding between the board of directors and senior management.[855] By December 2006, Goldman Sachs executives recognized "*the major risk in the mortgage business*", and they secretly decided – despite their own rule that "*clients' interests always come first*" – to sell all mortgage securities to their own clients. The prevailing attitude is only too clear from these comments: "*Distribute junk that nobody was dumb enough to take first time around*"; "*[They] structured like mad and traveled the world, and worked their tails off to make some lemonade from some big old lemons*"; "*How much of that sh—deal did you sell?*".[856,857] "*... If any other banking agencies in Washington were alarmed by the boom that occurred in subprime lending – $2.4 trillion in A – to D mortgages originated from the beginning of 2004 to the end of 2007, or 20 percent of all loans funded in the United States (a record) – they hardly voiced much concern, at least publicly. Perhaps because Wall Street was busy securitizing almost all of the loans being originated, they figured: If it's good enough for the Street it must be okay...*".[858]

The commission found out that "*the firm targeted less-sophisticated customers in its efforts to reduce subprime*".[859] In July 2007, Goldman Sachs failed to disclose to investors vital information about the low quality of one CDO, known as ABACUS 2007-AC1,[860] which months later lost investors almost all of their $150 million investment.[861] In July 2010, the SEC found that "*Goldman also acknowledged that its marketing materials for the subprime product contained incomplete information*" and sued Goldman Sachs for a US$550 million fine – the largest

---

[854]Ibid, p. 19.

[855]Robyn Alman, Richard Cudmore, Natalie McVeigh, Lehman Brothers: An Exercise in Risk Mismanagement, New England College of Business, 2009.

[856]The Financial Crisis Inquiry Report, The Financial Crisis Inquiry Commission, Washington, D.C., January 2011, p. 235.

[857]Elizabeth MacDonald, Goldman Sachs Accused of Misleading Congress, Clients, FOX Business, April 14, 2011.

[858]Paul Muolo and Mathew Padilla, Chain of Blame. How Wall Street Caused the Mortgage and Credit Crisis, John Wiley & Sons, p. 292.

[859]The Financial Crisis Inquiry Report, The Financial Crisis Inquiry Commission, Washington, D.C., Jan. 2011, p. 235.

[860]Press Release: Goldman Sachs to Pay Record $550 Million to Settle SEC Charges Related to Subprime Mortgage CDO, Firm Acknowledges CDO Marketing Materials Were Incomplete and Should Have Revealed Paulson's Role, U.S. Securities and Exchange Commission, Washington, D.C., July 15, 2010.

[861]Dan Wilchins, Karen Brettell, Richard Chang, Factbox: How Goldman's ABACUS deal worked, Reuters, April 16, 2010.

penalty ever paid by a Wall Street firm until that time – for withholding risk information. After the trial, one finance expert declared that this case was "*the most cynical use of credit information that I have ever seen... [It is like] buying fire insurance on someone else's house and then committing arson*".[862] In 2013, JP Morgan Chase was fined US$13 billion by the US government for overstating the quality of the mortgages the bank had been selling to investors before the subprime mortgage crisis.[863] And in 2014, the Bank of America also agreed to pay out a very large penalty – US$16.65 billion – to settle mortgage bond claims (by August 2014, large American banks paid a cumulative penalty of almost on US$127 billion).[864] These practices amounted to what John C. Bogle, the founder and previous CEO of Vanguard, has qualified as "*the general loss of the fiduciary principle*".[865]

Nobody Understood the Whole Picture of Risks

The FCIC commission declared: "*The mortgage pipeline also introduced leverage at every step. High leverage, inadequate capital, and short-term funding made many financial institutions extraordinarily vulnerable to the downturn in the market in 2007*".[866] Over-the-counter derivatives enabled derivatives traders at five major investment banks (Bear Stearns, Goldman Sachs, Lehman Brothers, Merrill Lynch, and Morgan Stanley) to operate with leverage ratios on their capital as high as 40 to 1. In other words, for every US$40 in assets, there was only US$1 in capital to cover losses; less than a 3 % drop in asset values could bankrupt any major investment bank.[867] Brokers at investment banks traded for and against the housing boom through credit default swaps. They often used clients' assets to raise cash for their own activities without informing clients. Warren Buffett, the chairman and CEO of Berkshire Hathaway Inc., testified that derivatives were "*very dangerous stuff*", difficult to understand for market participants, regulators, auditors and investors. He added that he didn't think he could manage a complex derivatives book.[868] However, it was revealed in November 5, 2011, that Buffett had sold equity derivatives (put options) to undisclosed buyers for $4.9 billion.

[862]The Financial Crisis Inquiry Report, The Financial Crisis Inquiry Commission, Washington, D.C., Jan. 2011, p. 236.

[863]Aruna Viswanatha, David Henry, Karen Freifeld, JPMorgan says 'mea culpa' in $13 billion settlement with U.S., Reuters, Nov. 19, 2013.

[864]Christina Rexrode, Andrew Grossman, Record Bank of America Settlement Latest in Government Crusade, The Wall Street Journal, Aug. 21, 2014.

[865]Bogle, J.C. The Fiduciary Principle (No Man Can Serve Two Masters). Journal of Portfolio Management 36, 15–25, 2009.

[866]The Financial Crisis Inquiry Report, The Financial Crisis Inquiry Commission, Washington, D.C., Jan. 2011, p. 134.

[867]Ibid, p. xix.

[868]Ibid, p. 49.

Liabilities on the so-called equity-index puts widened when four stock indexes fell from the levels they were at when Buffett made the contracts near the market peaks in 2006 and 2007. If the indexes are at zero when the agreements expire, the losses would be about $34 billion. Bloomberg communicated that Buffett's foray in equity derivatives had put pressure on Berkshire, with profits dropping 24 %.[869]

The leverage level was often hidden in derivatives positions, in off-balance-sheet entities or REPO transactions to prevent rumors about the real financial situation of investment banks on the market.[870] Serious doubts about the financial state of any firm could restrict access to the interbank lending market and bankrupt the firm. So it was not only external observers, but also investment bank executives, who failed to understand the real influence of OTC derivatives on their business. Even senior managers at the financial institutions lacked a sense of "the whole picture" of the risks of derivatives; and yet they continued to reassure investors, competitors, partners and the authorities of the financial stability of their organizations. For instance in April 2008 – just after the failure of Bear Stearns – Richard Fuld, CEO of Lehman Brothers, assured shareholders at a meeting that "*the worst ... [is] ... behind us*".[871] Some sources asserted that Fuld's personal experience was mainly as a bond trader, and that he had little technical understanding of such new financial instruments as CDOs and CDSs. Moreover, the majority of Lehman's board of directors had no specialized financial expertise: nine of them were retired, four of them over 75 years old, one was a theater producer, another a former Navy admiral… in fact only two had direct experience in the financial services industry.[872] Even after the bankruptcy of Lehman Brothers—for which he voted—Fuld insisted that "*There was no capital hole at Lehman Brothers. At the end of Lehman's third quarter [of 2008], we had US$28.4 billion of equity capital*".[873] There was a similar situation in AIG: executives at the insurance firm told the FCIC commission that "*they did not even know about these terms of the [credit default] swaps until the collateral calls started rolling*" in July 2007.[874] Not even the office of Thrift Supervision, the regulators who supervised AIG on a consolidated basis, knew the true level of risk the company was underwriting.[875] By the fall of 2007, AIG management certainly knew where things

[869]http://www.bloomberg.com/news/2011-11-04/berkshire-earnings-decline-24-on-derivatives.html.

[870]The Financial Crisis Inquiry Report, The Financial Crisis Inquiry Commission, Washington, D.C., Jan. 2011, p. xx.

[871]Lorraine Woellert, Yalman Onaran, Fuld Targeted by Lawmakers as Surrogate for Wall Street Excess, Bloomberg, Oct. 6, 2008.

[872]Larry McDonald, Patrick Robinson, A Colossal Failure of Common Sense: The Incredible Inside Story of the Collapse of Lehman, Random House, Nov. 24, 2009, p. 91, 226, 234–236.

[873]The Financial Crisis Inquiry Report, The Financial Crisis Inquiry Commission, Washington, D.C., Jan. 2011, p. 235.

[874]Ibid, p. 243.

[875]The Financial Crisis Inquiry Report, The Financial Crisis Inquiry Commission, Washington, D.C., Jan. 2011, p. 243.

were heading—and despite this, they too continued to convince investors that "*the risk we have taken in the U.S. residential housing sector is supported by sound analysis and a risk management structure ... we believe the probability that it will sustain an economic loss is close to zero ... We are confident in our marks and the reasonableness of our valuation methods ... [AIG has] active and strong risk management*".[876] In September 2008, the US government took over AIG in a US$85 billion bailout, because of AIG's liquidity shortage on credit default swap positions.

In addition, there was no unified regulator gathering information to build up a holistic picture of the risks involved in the housing bubble and the securitization pipeline. John Snow, US Secretary of the Treasury from 2003 to 2006, testified that regulators tended not to see a problem at their own institutions: "*Nobody had a full 360-degree view. The basic reaction from financial regulators was, 'Well, there may be a problem. But it's not in my field of view*'".[877] One member of the FCIC commission observed that "*it appears that market participants were unprepared for the destructiveness of this bubble's collapse because of a chronic lack of information about the composition of the mortgage market. Information about the composition of the mortgage market was simply not known when the bubble began to deflate*".[878] After the crash, Federal Reserve Chairman Ben Bernanke admitted that he had missed the systemic risks: "*Prospective subprime losses were clearly not large enough on their own to account for the magnitude of the crisis*".[879] In 2006, property prices peaked and Bear Stearns investment bank was found problematic during the following year, but regulators stated that it was a "*relatively unique*" case. They continued to convince the financial community that there was "*comfort about the capital cushions*" at the big investment banks until the collapse of Bear Stearns in March 2008.[880] Henry Paulson, US Secretary of the Treasury during the crisis, had been CEO of Goldman Sachs—one of the key players of the securitization pipeline—from 1999 until 2006. He warned in October 2007 that the burst of the housing bubble was "*the most significant risk to our economy*".[881] Despite his warning, and the occurrence of US$100 billion mortgage-related losses in 2007, the government did not act decisively to assess the real situation of the financial institutions, or to mitigate the consequences of a possible crisis, until the autumn of 2008. Because nobody could really see the whole picture, few could guess the real magnitude of the approaching calamity—even in the last few months before the government takeover of "Fannie Mae" (the Federal National

[876]Ibid, p. 272.

[877]Ibid, p. 172.

[878]Ibid, p. 465.

[879]Ibid, p. 27.

[880]Ibid, p. 288.

[881]Edmund L. Andrews, Housing Slump 'Unfolding,' Treasury Chief Says, The New York Times, Oct. 16, 2007.

Mortgage Association) and "Freddie Mac" (the Federal Home Loan Mortgage Corporation) in August 2008.

"Why Are There no Jail Sentences for Wall Street CEOs?"

When the crisis occurred, a great number of Americans asked each other: "*Why are there no jail sentences for Wall Street CEOs?*" There was plenty of evidence of fraud, conspiracy and lies, but no criminal prosecutions of the executives of investment banks and auditors as there had been when the cases of Enron, WorldCom or Madoff's 'Ponzi' scheme came to light. The answer is likely to be quite simple: this crisis was in significant part created as an unintended consequence of a close cooperation between the US government and private financial institutions—which were giving billions of dollars in contributions to the campaign coffers of both the Republican and Democratic parties. Nobody wanted to dig into the dirty laundry. The situation could be summarized by the following aphorism, which summarizes the generic problem underlying sound risk management: "*No one sees any pressing need to ask hard questions about the sources of profits when things are doing well*".

Subprime Mortgage Crisis: Why Risks Were Concealed

- **Deregulation** was implemented in the (mistaken) pursuit of long-term improvement in the efficient allocation of resources. In this respect, the transient triumph of the Efficient Market and Rational Expectations Hypotheses created an intellectual environment that rationalized and legitimized policy initiatives that created the opportunity for massive, unregulated pursuit of short-terms profits by all the intermediaries in the financial supply chain. So, **the captains of finance got carte blanche from the government to take further risks with derivatives**—and to conceal the risks they were taking—with near impunity.
- **Government representatives, and the executives and board members of financial institutions, did not fully understand the complexity of innovative financial instruments** and the potential consequences of deregulating the financial sector. **Government control over these complex systems was too weak** in the absence of a "mega-regulator", and there was only **fragmentary perception of the whole picture of risks** among representatives of the government and the top managers of companies in the mortgage pipeline.
- **Wishful thinking among borrowers, investors and the media**—they preferred to believe only what they wanted to believe and in particular in the illusion of a "perpetual money machine" promising endless wealth

and prosperity for everyone based on the sure thing, the never ending growth of real-estate prices, and ignored known facts and early warnings about the real estate bubble and the low quality of CDOs.

- **Government executives were reluctant to admit mistakes in previous deregulation efforts,** which, together with the policy of low interest rates in 2002–2003, had help create the real estate bubble. Any admission of oversight would massively reduce the value of assets and lower US economic figures. So, government decision-makers preferred not to respond to clear evidence of risk before the collapse of Bear Stearns and Lehman Brothers.

2.3 Military, Social and Natural Disasters

2.3.1 Unreadiness of the Soviet Red Army for the Nazi Invasion (1941)

On June 22, 1941 at 3:30 a.m., the Nazi German armed forces (the Wehrmacht) together with Italian, Romanian, Finnish, Hungarian, and Slovakian forces invaded the Soviet Union. It was the most powerful invasion in world history in terms of the number of soldiers: more than 5.5 million fighters were amassed in 192 divisions for the Eastern campaign. The forces had more than 4300 tanks, 5000 military airplanes and 47,200 artillery guns and mortars.[882] The Soviet Red Army actually had numerical superiority over the Wehrmacht, but could not make use of

[882] "The Great Patriotic War of the Soviet Union, 1941–45" article, Great Soviet Encyclopedia, Moscow, 3rd edn., Vol. 4, 1971.

it because of its unreadiness for the sudden attack. During the first day, the Wehrmacht penetrated between 25 and 50 km into Soviet territory. By the end of the first week, Minsk, the capital of the Soviet Republic of Belarus, was taken. By the third week, the depth of the invasion exceeded 600 km and the Wehrmacht was close to Leningrad (the former St. Petersburg) and Kiev (the capital of the Soviet Republic of the Ukraine). After $3^1/_2$ months of fierce battles, the Nazis had advanced up to 1000 km and reached the suburbs of Moscow, the capital of the Soviet Union.

The first months of the war on the Eastern Front—a major part of the Second World War—turned out to be a military catastrophe for the Red Army: more than 850,000 soldiers died, more than 1 million soldiers were captured, and nearly 3500 military airplanes and 6000 tanks were lost. The Wehrmacht seized territory that normally produced up to 40 % of Soviet GDP.[883] In doing so, they lost 100,000 soldiers, 950 airplanes and 1700 tanks.

From December 1941 until May 1945, the Red Army successfully counterattacked the Wehrmacht near Moscow, destroyed Nazi divisions near Stalingrad and Kursk, liberated Eastern Europe from the Nazis and reached Berlin.

Nevertheless, the military failure experienced by the Red Army in those first few months was unprecedented. It remains one of the most tragic examples of mismanagement in world history. As we will document below, the widespread concealment of risks—at all levels of the Soviet military and political hierarchy—led to the failure of Soviet political and military executives to assess the reality adequately right before the invasion.

2.3.1.1 Risk Concealment Before the (Red Army) Disaster

Confrontation Between Soviet Politicians and Red Army Executives

There was a covert struggle extending over decades between the political and military executives of the Soviet Union. The fact is that Joseph Stalin rose to become the almighty leader of the Soviet Union and successor to Vladimir Lenin, the leader of the Communist Revolution of 1917, by systematic political struggle within the Communist Party and by his constant presence in Moscow, where all political decisions were made. He had huge political influence but did not have a military background and, during the Russian Civil War of 1918–1920, his military achievements were modest.[884] By contrast, there was a small group of outstanding generals, who had won the Russian Civil war for the Bolsheviks. These officers

---

[883]Shamil Munchayev, Viktor Ustinov, History of Russia. Textbook for high schools, Moscow, 3rd edn., 2003, p. 194.

[884]Reference of the Commission of the Presidium of the Central Committee of Communist Party "Inspection of charges brought against Tukhachevsky, Yakir, Uborevich and other military leaders for treason, terrorism and military conspiracy by the judicial and the Party organs in 1937". Russian Military Archives. 1993. #1. pp. 4–113, Military Historical Archive. 1998. #12. pp. 3–81.

had started to develop the Red Army from an initial small force of a few thousand soldiers in St. Petersburg and Moscow to end up creating a force that won control of the largest country in the world, and that defeated the well-equipped anti-communist forces of 14 countries, all trying to seize territory from a Russia weakened by the revolution. As the heroes of the Russian Civil War, they became the most prominent and popular people in the Soviet Union. To reduce their political influence and mitigate the possibility of possible military coups, Stalin appointed Kliment Voroshilov, his most loyal supporter, as People's Commissar for Defense (defense minister) of the Soviet Union in 1925, a position he retained until 1940. Like Stalin, Voroshilov did not have a military background and had little knowledge of modern military strategy, but he was able to control the ambitious, well-educated and self-determined generals.

Georgy Zhukov, one of the notable generals of World War II and Head of General Staff of the Red Army in 1941, remembered an episode which demonstrated the relationship between Voroshilov and the generals: "*I must say that Voroshilov, later People's Commissar for Defense, was incompetent in his position. To the end of his days, he remained a dilettante in military matters and never knew them deeply and seriously ... Responsibility for military questions fell to [Mikhail] Tukhachevsky, who was really a military expert [and was Voroshilov's First Deputy]. He had permanent struggles with Voroshilov and there were hostile relations between them. Voroshilov did not like Tukhachevsky... During the development of the [Red Army's] articles of war, there was an episode... Tukhachevsky, as Chairman of the articles of war commission, reported to Voroshilov. I was present at that moment. Voroshilov focused on some of the articles and began to express dissatisfaction and suggested changing something. Tukhachevsky, after listening to him, said: 'Comrade Commissar, the Commission can not accept your amendments'. Voroshilov asked 'Why?'. Tukhachevsky: 'Because your amendments are incompetent, Comrade Commissar'*".[885] Tukhachevsky was perceived as the informal leader and innovator among the officers of the Red Army. He realized that further war with Germany was unavoidable, and proposed fundamental changes in the Red Army, turning it from an equestrian armed force to a mechanized one through massive production of tanks, aircrafts, missiles, and so on, and through the complete re-training of soldiers to use this new equipment.[886] Obviously, he wanted to remove Voroshilov from the top military position because of the latter's inability to implement the required changes. However, Stalin saw in Tukhachevsky's opposition a threat, not only to his loyal ally Voroshilov, but also to his own political position.

Ultimately, a severe purge of senior and middle-ranking officers of the Red Army began in 1937, with the intention of eliminating any opposition from the military to the politicians and giving the Politburo total control over the Red Army.

---

885 Konstantin Simonov, The point of view of the person of my generation. Reflections on Joseph Stalin, Novosti Press Agency Publishing House, Moscow 1989, p. 338.

886 M.N. Tukhachevsky Selected Works in 2 volumes, Moscow, Military Publishing House, 1964, Volume 1, pp. 24, 162, 180.

Between 1936 and 1938, the NKVD—the Soviet secret police and ancestors of the KGB—eliminated more than 44,000 officers, 7 % of the total number of commanders in the Red Army.[887] Voroshilov personally signed orders for the executions of opposing officers.[888] The majority of these officers, including Tukhachevsky, were executed on pro-German espionage charges. The question of whether any senior officers of the Red Army were actually involved in pro-German espionage remains open to this day. Nevertheless, by this severe purge, Stalin dealt with two problems: he eliminated any opposition to the central government among the Red Army command and put an end to any espionage activity within the Red Army ahead of the Second World War. One NKVD officer, arrested in 1939, testified in a statement that was eventually published in the late 1980s: "*The mass repression of decision makers could be explained by Stalin's dictatorial methods of running the country, he makes all decisions in the country by himself, he does not tolerate objections and ignores the opinion of others and organizes massive operations [repressions] against those individuals who contradict (criticize) him*".[889] Stalin personally controlled the investigation process, received interrogation reports of the arrested officers and participated in deciding what to charge them with.[890] Over several years, the entire regional command of the army—all commanders of regional army divisions, and of military training and administrative establishments—along with 90 % of the deputies and chiefs of arms, 80 % of the leadership at division level, and 91 % of regimental commanders and their deputies, were replaced.[891] Many witnesses confirmed cases after the repressions where junior commanders (captains) became colonels, and were appointed to command whole regiments, because there were no superior officers left. From 1937 to 1942, a military commissariat in the Red Army was created to ensure political oversight of the military command and to re-educate or indoctrinate personnel with pro-Communist ideology. Ultimately, Stalin obtained a new executive officer corps, who were devoted to him and the Communist Party.

---

[887]N.M. Ramanichev, People paid a high price for the victory, Military History Magazine, Moscow, 1991, № 12, pp. 2–9.

[888]Reference of the Commission of the Presidium of the Central Committee of Communist Party "Inspection of charges brought against Tukhachevsky, Yakir, Uborevich and other military leaders for treason, terrorism and military conspiracy by the judicial and the Party organs in 1937". Russian Military Archives. 1993. #1. pp. 4–113, Military Historical Archive. 1998. #12. pp. 3–81.

[889]Reference of the Commission of the Presidium of the Central Committee of Communist Party "Inspection of charges brought against Tukhachevsky, Yakir, Uborevich and other military leaders for treason, terrorism and military conspiracy by the judicial and the Party organs in 1937". Russian Military Archives. 1993. #1. pp. 4–113, Military Historical Archive. 1998. #12. pp. 3–81.

[890]Reference of the Commission of the Presidium of the Central Committee of Communist Party "Inspection of charges brought against Tukhachevsky, Yakir, Uborevich and other military leaders for treason, terrorism and military conspiracy by the judicial and the Party organs in 1937". Russian Military Archives. 1993. #1. pp. 4–113, Military Historical Archive. 1998. #12. pp. 3–81.

[891]N.M. Ramanichev, People paid a high price for the victory, Military History Magazine, Moscow, 1991, № 12, pp. 2–9.

Distorted Red Army Causalities of the Finish Campaign

In spite of the repressions, the modernization plans of the former military command were being implemented intensively. At the end of the 1920s, the Red Army had only 89 tanks and 1394 military aircraft imported from Europe but, by 1941, the army already had 20,000 tanks and 22,000 military aircraft, designed and manufactured in the Soviet Union.[892] In order to test the modernized Red Army in a real war, between November 1939 and March 1940, the Soviet Union tried to bring Finland back under Russian control. Finland had been part of the Russian Empire from 1809 to 1917, but was lost during the Russian Civil War. Stalin and the renewed military executive were counting on the power of mechanized armed forces, confident that the huge amount of modern military equipment would ensure victory. Voroshilov reported to Stalin before the Finnish campaign that "*everything is good, everything is fine, everything is ready [for a successful military operation]*".[893]

However, the Red Army made modest military progress, gaining back for Russia only 11 % of Finnish territory; originally Stalin had hoped for total reunification with Finland. Against the Red Army's military operation, Finnish generals—the majority of whom were former officers of the Russian Empire, just like the Soviet senior officers executed in 1937–1938—mounted a sophisticated defense strategy, adapting well to the terrain and carefully coordinating units of well-trained soldiers. As a result, the Red Army formally defeated Finland, but in reality failed to achieve its ambitious goal, and the two countries signed a peace treaty. During and after that war, Red Army commanders at all levels began to embellish the real situation in their reports to superiors, because of the fear of further repression from Stalin after such a poor performance from what was—despite its tremendous wealth of military equipment—an under-trained army with young and inexperienced commanders.

This embellishment manifested itself especially in the falsification of figures for war casualties. Red Army officers tried to underplay their own losses, and exaggerate those of the enemy, in their reports to Stalin and the General Headquarters of the Red Army. Thus—according to the report received by the Politburo and the Supreme Council of the USSR—48,475 Soviet soldiers were killed and 158,863 were injured during the Finnish campaign, but the Finnish Defense Forces lost more than 70,000 soldiers and 250,000 were injured. Decades later, historians found out that the Red Army had actually lost 95,200 soldiers, and the Finnish Defense Forces had lost 23,500.[894] In other words, Red Army officers halved their own losses and exaggerated the losses of the enemy by a factor of three.

---

[892]Ibid.

[893]Konstantin Simonov, The point of view of the person of my generation. Reflections on Joseph Stalin, Novosti Press Agency Publishing House, Moscow 1989, p. 307.

[894]Alexander Pravdin, Casualties in ground battles of the Winter War, http://www.proza.ru/2011/08/05/1121.

After the war, the Finns declared that the main defect of the Red Army was the weakness of its command. At debriefs, this statement was eventually admitted by the Red Army generals. They accepted that the troops did not suffer from a lack of equipment, but from an abundance of equipment and the inability of commanders of the infantry, tank divisions, the air force and the navy to interact effectively with each other: "*The war... showed that the weakest link in the chain was the level of training of commanders, who could not make the full use of... the personnel subordinated to them*".[895] Nevertheless, nobody had the courage to inform Stalin and the politicians, and to state openly that the main cause of the low level of training of Red Army commanders was the previous purges of senior and middle military management. Everybody was afraid of further repression. As a result, Stalin began to receive the information he wanted to hear and not an honest appraisal of the condition of the Red Army. In 1937, Hitler had also eliminated opposition among generals of the Wehrmacht, who objected to his plans for further conquests. However, only 60 generals were replaced and these were retired. Hitler also surrounded himself with generals who did not criticize his maniacal desire to expand the Third Reich.

The poor level of training of the Red Army commanders was one of the key factors in Hitler's decision to open an Eastern Front. In January 1941 at an executive military meeting with the commanders of his armies, he stated that "*The Russian Armed Forces are like a headless colossus with feet of clay but we cannot with certainty foresee what they might become in the future. The Russians must not be underestimated. All available resources must therefore be used in the German attack*".[896] In May 1941, Colonel-General Halder noted in his diary: "*The Russian officer corps is exceptionally bad. It produces a worse impression than [the officer corps] in 1933. It will take about 20 years until it reaches the same level [as 1933]*".[897] After the Second World War and Stalin's death, Alexander Vasilevsky, Marshall of the Soviet Union and Head of the General Staff of the Red Army (1942–1945), declared that "*without [the repressions of] the thirty-seventh year, there might not have been any war in the forty-first year. [When] Hitler decided to start the war in the forty-first year, the assessment about the degree of destruction of the military command [that had] occurred in the USSR played a significant role*".[898]

After the Finnish campaign, the General Staff of the Red Army ordered their subordinates to improve the quality of military training of personnel. The subordinates began to send assuaging reports to Moscow about the supposed serious progress in training. In turn, the military command tried to reassure Stalin that the

[895]Boris Sokolov, Secrets of the Finnish War, Veche Publishing House, Moscow, 2000, p. 384.

[896]R. H. S. Stolfi, Hitler's Panzers East: World War II Reinterpreted, University of Oklahoma Press, 1993, p. 20.

[897]Shamil Munchayev, Viktor Ustinov, History of Russia. Textbook for high schools, Moscow, 3rd edn., 2003, p. 194.

[898]Konstantin Simonov, The point of view of the person of my generation. Reflections on Joseph Stalin, Novosti Press Agency Publishing House, Moscow 1989, p. 338.

Red Army had already overcome the shortcomings identified during the Finnish war and that it was ready for any war. In May 1941, inspired by the apparent progress, Stalin declared that fundamental restructuring of the army was over, that they now had 300 divisions with 10,000–13,000 soldiers in each, and that one third of the divisions were mechanized.

Stalin's Self-deception

Stalin did not believe that Hitler would attack the Soviet Union. There were several reasons for this view.

Firstly, Germany had a history of military failures caused by trying to wage a war simultaneously on two battlefronts, and Stalin expected that Hitler would not be keen to repeat this experience, but would wait for the total fall of Great Britain—especially since he had completely defeated France in only $1^1/_2$ months in the summer of 1940.

Secondly, Stalin was relying on the Ribbentrop–Molotov Pact or Nazi–Soviet Pact, a mutual non-aggression pact signed in 1939, which determined agreed spheres of interests for both parties in Eastern Europe and declared the absence of aggressive motives.

Thirdly, if Hitler intended to attack the Soviet Union, the Wehrmacht would have to prepare for a period of winter warfare. Napoleon's "unbeatable" army was catastrophically defeated in Russia in the late autumn months of 1812 because they had not prepared for the long Russian winter. During 1940–1941, there was no intelligence evidence that the Wehrmacht was prepared for a winter campaign. Stalin always believed in taking a logical approach, and did not expect reckless actions from Hitler like launching an Eastern Front campaign with only summer equipment.

Fourthly, Russian military production output had risen sharply and this created the illusion that, as the Red Army had numerical superiority over any army in Europe, it would be in a favorable position to protect the Soviet Union from any external attack.

Fifthly, when the Red Army military intelligence service began to register serious movement of the Wehrmacht in Poland close to the western borders of the Soviet Union, Stalin suggested that it was a provocation from Hitler, and ordered the Red Army to give no reaction—to abstain from returning fire, or from shooting down German spy planes, and so on—in order to minimize the probability of any accusation of starting a war between the Soviet Union and Germany by mobilization of the Red Army. A quarter of a century earlier, on July 31, 1914, the Russian Empire had ordered the full mobilization of its troops in response to the attack by Austria-Hungary against Serbia, a close Russian ally; the next day, August 1, 1914, Germany had declared war on the Russian Empire and the First World War had started. With this dreadful precedent in mind, Stalin preferred not to mobilize the Red Army in order to avoid any blame for starting another war.

Sixthly, Stalin emphasized that all the foreign intelligence reports that had been warning him of Hitler's imminent invasion since late 1940 were merely provocations fabricated by Great Britain in its great distress, in order to draw the Soviet Union into the war with Germany. There was a meeting in the Kremlin on May 14, 1941, when the Soviet military command informed the Politburo about the concentration of German troops near the border of the Soviet Union, but Stalin categorically rejected the conclusions of the Soviet military: "*Germany is stuck up to its ears in the war on the West, and I believe that Hitler would not dare to create for himself a second front by attacking the Soviet Union. Hitler is not such fool as not to realize that the Soviet Union – is not Poland, it is not France, and it is not even England and all of them put together... Do you [generals] propose to mobilize the whole country, raise troops and move them to the western borders? This is war! Do you [military] understand it? ... Comrade Zhukov, tell us, why is your information about deployment of German army correct? – Comrade Stalin, all conclusions were refined by aerial reconnaissance and confirmed through a network of agents. – A network of agents? Whose? Ours or English? Our agents send me every week a new date for the commencement of hostilities, but nothing happens... Have you come to frighten us by war or do you want war? Do you have shortage of awards or titles? Stop generating nonsense!*".[899] Stalin was finally convinced than he was right a week later, when Nazi Germany launched an invasion on Crete on May 20, 1941, which was a base for the Allies (the United Kingdom, New Zealand, Australia and Greece). Moreover, Hitler assured Stalin in personal correspondence that he had only ordered the concentration of the Wehrmacht's forces in Poland to reduce their losses from air raids by British bombers, which were occurring in France and Germany.[900]

Subordinates Provide Calming Reports to Stalin

In their turn, the Red Army command did not want to be seen by Stalin as British or German spies, so they did not try to transmit fully objective information from the western border. They preferred to change the "facts" to please Stalin. For example, Dmitry Pavlov, commander of the key Soviet Western Front in 1941, received extensive reports from subordinates about the Wehrmacht's prewar frontier activity, but in a call with Stalin, he said: "*No, Comrade Stalin, that's not true! I just came back from defensive positions. There is no concentration of German troops on the border, and my intelligence works well. I checked again, but I think it's just a provocation*". After the call, he commented to a colleague: "*Some bastards are trying to convince him [Stalin] that the Germans are concentrating*

[899]Georgy Zhukov, Memories and Reflections, OLMA Media Group, Moscow, Vol. 1, pp. 258–259.

[900]Sergey Smirnov, Marshal Zhukov, Moscow, Politizdat, 1988, p. 100.

troops on our border".[901] Moscow also tried to interpret German preparations as a provocation and "*assured [subordinates] that everything is in order, and [they have to] be quiet and not panic*".[902] As a result, the order to bring troops to full combat readiness was sent from Moscow only a few hours before the invasion started.

Pavlov and several other generals were accused of "*failure to perform their duties*" and executed one month after the invasion. Nobody among the Politburo and the Red Army command blamed Stalin for mismanaging the situation, refusing to listen to warnings and, above all, for having set up a system of absolute fear, which prevented reasonable criticism and the communication of objective information about risks. Stalin received the usual servile replies from his subordinates to anything he said: "'*Yes, Comrade Stalin', 'Of course, Comrade Stalin', 'Quite right, Comrade Stalin', 'You have made a wise decision, Comrade Stalin...*'".[903] After the war and Stalin's death, Georgy Zhukov stated that "... *it was the responsibility of the Red Army military executives that we did persistently demand to bring the army to full combat readiness [before the war] and urge early implementation of the necessary measures in case of war... Of course, we should be realistic about the possible consequences of any objections to Stalin about his assessment of the general political situation. Everyone remembers recent years during which, if anybody would speak aloud that Stalin was not right, it meant that this person immediately had to meet with the NKVD repression system*".[904] Only once Stalin confessed his own mistakes—several days after the invasion, when he said angrily to his close subordinates: "*Lenin founded our state, and we've pissed it away*".[905]

2.3.1.2 Risk Concealment After the Disaster

During the first days of the invasion, Stalin and the General Staff of the Red Army had little information about the real situation on the battlefronts. In the midst of such a disaster, few officers at any level of the Red Army would have dared to

[901]Contents of the call published after the war in memoirs of Alexander Golovanov, chief of long-range aviation of the Red Army (1942–1944).
Alexander Golovanov, Long-distance bomber aviation, Moscow, Centrpoligraf publishing house, 2007, p. 56.

[902]Protocol of the closed hearing of the Military Collegium of the USSR Supreme Court concerning cases of Pavlov D.G., Klimovskikh V.E, Grigoriev A.T. and Korobkov A.A., Moscow, 22 July 1941. V.Pakhomov, The Great Patriotic War in the documents and testimonies of contemporaries, Samara, 1995, p. 145.

[903]Konstantin Simonov, The point of view of the person of my generation. Reflections on Joseph Stalin, Novosti Press Agency Publishing House, Moscow 1989, p. 360.

[904]Konstantin Simonov, The point of view of the person of my generation. Reflections on Joseph Stalin, Novosti Press Agency Publishing House, Moscow 1989, p. 357.

[905]Nikita Khrushchev, Time. People. Power, Moscow, Moscow News, Book I, 1999, p. 695.

admit their own inability to resist the Wehrmacht because of fears of repression. Therefore, they distorted all figures about actual casualties, and tried to convince superiors that a counterattack was possible in the near future. Moreover, because communication channels had been destroyed, the General Staff lost contact with many army units. According to the memoirs of senior army officers and members of the Politburo, in the first weeks after the invasion, Stalin thought the enemy could be defeated in a very short time, because of reports from the front claiming minimal losses for the Red Army and serious damage to the Wehrmacht: "*[The battlefront reports] instilled confidence in him that [the enemy could not continue for long to sustain such losses] and soon the enemy would be defeated*".[906] The General Staff of the Red Army were misinforming Stalin from the very first weeks: they consistently reduced casualty figures and concealed deplorable facts. For example, they informed Stalin that 700 aircraft had been lost in the first day because the order for full combat readiness came too late, but the real number of damaged aircraft exceeded 1200.

Later, Stalin began to understand that information had been withheld or distorted and shouted furiously: "*You [the military] are just afraid to tell us [the Politburo] the truth*".[907] On the 12th day after the invasion, Stalin called Alexander Golovanov, long range aviation chief for the Red Army, and ordered the use of high-altitude bombers to collect information about the German forces and their own army: "*We are not well informed about the situation at the front. We do not even know exactly where our military units and their staff are located and do not know where the enemy is. You have the most experienced flight crews. We need credible data... It will be your main concern. All gathered information must be immediately transmitted to us*".[908]

Concentration of Political and Military Power in Stalin's Hands for Effective Decision-Making

Stalin soon recognized that only direct and simultaneous leadership of the Soviet economy and the Red Army, and deep immersion in the military decision-making process, could guarantee an effective response to the most challenging crisis in Russian history. Therefore, six weeks after the invasion, Stalin became Supreme Commander of the Armed Forces of the USSR, and combined political and military power in his hands. Georgy Zhukov remembered: "*Stalin had his own method of conducting a military operation... Before preparing the operation, he was calling the officers of the General Staff – majors, lieutenant colonels, who oversaw the relevant operational areas. He called them for one-to-one meetings, worked with*

[906]Yuri Emelianov, Stalin: Generalissimus of the Great Victory, Yauza Publishing House, Moscow, 2008, p. 233.

[907]Ivan Stadnyuk, The war, Moscow, Military Publishing, 1987.

[908]Alexander Golovanov, Long-distance bomber aviation, Moscow, Centrpoligraf publishing house, 2007, p. 67.

them for $1^1/_2$–2 hours in each specified situation. He was so prepared that he sometimes surprised the commanders of the battlefronts by his detailed knowledge... It was impossible to visit Stalin with a report with maps that had some "white spots" and reveal incomplete or exaggerated data. Stalin could not tolerate random answers, demanded exhaustive completeness and clarity. He had a special flair for weaknesses in the reports and documents. He immediately showed them and demanded exact clarification of the fuzzy information... Therefore, we tried to prepare reports very carefully".[909] This allowed Stalin to understand the reality of the situation at the battlefronts, competently discuss the planning of military operations with senior officers and increase the speed of the decision-making process. According to Georgy Zhukov: "*At the beginning of the war, Stalin understood poorly the matters of operational art. ... He could conduct operational issues well... in the last period of the Battle of Stalingrad [winter of 1943] and the Battle of Kursk [summer 1943]... He began to rely on objective reality. [His previous] viewpoint summarized by 'I decided about something and it must be done in any case' evolved into a sober attitude based on a more objective assessment of reality... [More importantly] his mind and talent enabled him [to conduct military operations] not worse, and sometimes even better than his subordinates [professional battlefront commanders]*".[910]

Ultimately, the war of the Soviet Union with the Nazis continued for 1418 days and resulted in the deaths of up to 27 million Soviet people. The Red Army made a major contribution to the defeat of Nazi Germany and to the victory of Allied forces in the Second World War.

Comments from a Modern Russian Historian About an Alternative Account of the Nazi German Invasion of the Soviet Union

After the preparation of this case, we sent it for comment to experts in the field. We received a very interesting assessment from Russian historian Sergey Nefedov, Senior Fellow in History at the Ural Federal University, about alternative views on the history of the Nazi German invasion of the Soviet Union—in particular addressing the question "*What if Stalin had ordered the Red Army to prepare for defense months before the actual time of the invasion?*" The following is Nefedov's conclusion: even if Stalin had not conducted repressions and the highest officer corps had been saved; even if Stalin had received reliable information about the state of the Red Army during 1940 and 1941; even if Stalin had trusted the intelligence data clearly warning him about a huge concentration of German forces in Poland in the summer of 1941, and about the high likelihood of an attack on the Soviet Union in June 1941; even if Stalin had declared a total mobilization

[909]Konstantin Simonov, The point of view of the person of my generation. Reflections on Joseph Stalin, Novosti Press Agency Publishing House, Moscow 1989, p. 328.

[910]Ibid.

of the USSR and ordered preparations for defense in the spring of 1941; even then the USSR would have suffered huge losses, and German troops would have been deep into Soviet territory in the late autumn of 1941.

The main problem was not that the USSR did not have fully mobilized forces and was not ready for defense, but the fact that the Soviet General Staff was preparing for the previous war (WWI)—much like the French military in 1940. They could not imagine that the Germans had developed far superior military strategy and tactics with their active use of "Blitzkrieg": fast mixed attacks by hundreds of tanks, attack planes and motorized infantry, deep into enemy territory through the weakest points in its defense lines. Such blitzkrieg attacks could not be stopped by defensive lines like the famous Maginot Line: in 1940 the Germans bypassed it through Dutch and Belgian territory, advancing suddenly with 1250 armored vehicles and outflanking the French troops to attack them from the rear. This operation was the main cause of their remarkable swift defeat of France in WWII. On the Eastern front too, the Wehrmacht found weaknesses in the thousand-kilometer defensive line of the Red Army; General Guderian's tank divisions slipped into the rear of the Red Army through these weak points, surrounding Red Army divisions and destroying them. Thus they advanced, gradually but inexorably, towards Moscow…

What stopped the Wehrmacht from continuing its unique military strategy into the late autumn of 1941, and caused its eventual defeat in the battle for Moscow, was the notorious Russian winter—nicknamed "General Frost"—which had also helped the Red Army's ancestors to destroy the "unbeatable" army of Napoleon in 1812. Even during November 1941, the temperature near Moscow was around −10 °C (−14 °F), well below freezing—but on December 4, the temperature dropped to −35 °C (−31 °F). The German tanks and motorized troops were completely frozen: with no adequate winter clothing or equipment, the Wehrmacht soldiers lost any motivation for further military breakthroughs in their desperate search for sources of heat. On 6 December, fresh Red Army units arrived, transferred from Siberia, and began a successful counterattack on Wehrmacht troops near Moscow. This turned the tide, and laid the foundation for their ultimate victory over Nazism three and half years later. In these later years, the German forces used blitzkrieg attacks less frequently, because the Red Army had learned how to anticipate such attacks and prevent them from developing. Dr. Nefedov published an article about his findings in a Russian history periodical.[911]

Thus, according to Dr. Nefedov's assessment, Stalin and the Soviet Union had only one chance to avoid the tragic defeat that befell them in the first months of 1941, and save the lives of millions of its citizens: by a preventive attack on the Nazis in Poland. But in this case, the Soviet Union would have been recognized as the aggressor, violating the Ribbentrop-Molotov pact. Consequently, Great Britain and the United States would have been obliged to help Germany in a war against the Communist Russia. Harry S. Truman—later to be president of the

[911]Nefedov S.A., 1941: The road to Moscow, Questions of History, 2012, № 7, http://book.uraic.ru/elib/Authors/NEFEDOV/Science/Russia/Guderian.htm.

US—made the cynical position of the American establishment very clear on June 24, 1941, just three days after the Nazi invasion of Communist Russia: "*If we see that Germany is winning, we ought to help Russia and if Russia is winning we ought to help Germany, and that way let them kill as many as possible, although I don't want to see Hitler victorious under any circumstances*".[912]

Unreadiness of the Soviet Army for the Nazi Invasion: Why Risks Were Concealed

- **The wishful thinking/overconfidence/self-suggestion/self-deception** of Stalin, who convinced himself in 1941 that an attack on the Soviet Union by Nazi Germany was impossible.
- **A prevailing culture of "success at any price" and "no bad news"**: the fear among Soviet army officers of being punished (dismissed, criminally prosecuted or executed) for communicating any information about the situation on the battlefronts that did not match Stalin's perception and expectations.

2.3.2 Worldwide Spanish Flu and SARS Outbreaks (1918–1919, 2003)

Sunlight is the best disinfectant

William O. Douglas

Severe acute respiratory syndrome (SARS) originated in Guangdong Province in China in November 2002. The Chinese authorities suppressed news of the outbreak of an unknown disease, concealing it both from residents of the province and specialists of the World Health Organization (WHO). As a result, large-scale preventive measures were delayed for four months. The WHO issued a global warning only in mid-March 2003. A unique collaboration of governmental organizations and research centers throughout the world made it possible to halt the last human chain of the transmission of SARS on 5 July 2003. But, by that time, the international spread of SARS had resulted in 8098 cases in 26 countries, with 774 deaths.[913] Early preventive action could have reduced these numbers many times.

[912]Harry S. Truman's statement in The New York Times, June 24, 1941.

[913]WHO guidelines for the global surveillance of severe acute respiratory syndrome (SARS), WHO (Department of Communicable Disease Surveillance and Response) Updated recommendations, October 2004, p. 6.

2.3.2.1 Risk Concealment Before and After the Disaster

"Spanish Flu" Global Pandemic

To understand what motivated the Chinese authorities to attempt to hide information about the disease, we will start by looking at the actions of the American and European governments during the global pandemic of 1918–1919 (also known as the "Spanish flu"), when around 500–600 million people—a third of the world's population at that time—were infected, and nearly 50 million lost their lives (some estimates put the figure at nearer 100 million casualties).[914] The pandemic took five times more lives than the First World War. The H1N1 influenza virus originally came from birds, but then appeared in pigs before crossing the species barrier again to ignite the pandemic among humans. The first cases of the unknown disease were registered in Kansas, America, in January 1918. By March 1918, more than 100 soldiers fell ill at the US army camp in Funston, Haskell County, where more than 5000 recruits were training for further military operations on the European battlefronts of the First World War. Most of the recruits were farmers, had regular contact with domestic animals and were less resistant to viruses than recruits from cities. The high concentration of personnel in the camp simplified human-to-human transmission. At that time, viruses were not known to medicine, and some doctors had not even accepted the idea that microorganisms could cause disease.[915] Later, the personnel of Funston camp were transferred to Europe by ship, and during the long transatlantic crossing the virus spread among soldiers coming from other parts of the USA. Upon arriving in Europe, American soldiers

[914]Jeffery Taubenberger, David Morens, 1918 Influenza: the Mother of All Pandemics, Emerging Infectious Diseases, 2006, Vol. 12, No. 1, p. 15.

[915]John M. Barry, The Great Influenza: The Epic Story of the Deadliest Plague in History, New York, Penguin, 2004, p. 51.

infected British and French forces, which in their turn infected German forces in hand-to-hand combat. When Woodrow Wilson, President of the United States from 1913 to 1921, began to receive reports about a severe epidemic among American forces, he made no public acknowledgement of the disease.[916] Moreover, other governments involved in the war made similar decisions—censorship, lies, and even active propaganda—to keep up morale, allowing the disease to continue to spread without any preventive measures. The pandemic was named "Spanish flu" because Spain was a neutral country during the First World War and did not suppress the media, so it was only Spanish newspapers that published honest articles about the severity of the disease—despite the fact that it had originated in the USA and spread initially among American soldiers in the absence of a proper response by the US government. This lack of response was probably due to the US strategic goal of developing a strong political influence in the post-WWI peace process that was to shape international politics in the following decades.

In the USA alone, the disease claimed more than 650,000 lives during 1918–1919. One remarkable incident occurred in September 1918 in Philadelphia, during preparations for the Liberty Loan Parade. City officials received warnings about the flu threat from soldiers participating in the parade, but did not cancel it because the parade was expected to be the largest in the city's history and help sell millions of dollars in bonds to finance the war.[917]

SARS Outbreak

According to some virologists, the "Spanish flu" was "*the mother of all pandemics in the 20th century*".[918] The H2N2 virus, which caused the Asian influenza pandemic in 1957, and the H3N2, which caused Hong Kong influenza in 1968, were both closely related to H1N1.[919] In 1997 in Hong Kong—which was by then a Special Administrative Region of China—a highly pathogenic avian influenza A (H5N1) virus appeared, which crossed the species barrier between birds and humans in a local live poultry market. During the outbreak, 18 people fell ill and 6 perished; more than 1.5 million chickens were slaughtered to stop the spread of the virus. The consequences of the outbreak were severe for Hong Kong's economy and for its reputation as a center of international tourism in South East Asia, because of negative reports about Hong Kong in the world media.[920] The

[916]Ibid, p. 302.

[917]John M. Barry, The Great Influenza: The Epic Story of the Deadliest Plague in History, New York, Penguin, 2004, p. 203.

[918]Jeffery Taubenberger, David Morens, 1918 Influenza: the Mother of All Pandemics, Emerging Infectious Diseases, 2006, Vol. 12, No. 1, p. 15.

[919]Edwin D. Kilbourne, Influenza Pandemics of the 20th Century, Emerging Infectious Diseases, 2006, Vol. 12, No. 1, p. 11.

[920]Gilbert Wong, Nina Hansen, Vanessa N. Clark, Crisis Communication: The Asian Bird Flu, Centre for Asian Business Cases, School of Business, University of Hong Kong, 1998.

frequency of aerial flu outbreaks originating in South East Asia led to a widespread opinion among the world virology community that the next high-mortality flu pandemic would start in South China, because of the high concentration of people who permanently live close to domestic animals.[921] Obviously, based on these historical facts, the Chinese government understands the possible consequences of a third of the population of China becoming infected: based on the mortality statistics for "Spanish flu", such an epidemic would mean the possible death of at least 40 million people, with inevitable catastrophic political and economic consequences.

In the middle of November 2002, the first case of SARS was registered in Foshan hospital, in Guangdong province, South East China, 130 km from Hong Kong. Staff of the hospital informed a local anti-epidemic station about a "*strange disease*" by mid-December. Representatives of Chinese Ministry of Health landed in Guangdong province for a detailed investigation in mid-January 2003. On January 27, they issued a "*top secret*" report, which was distributed to executives of the provincial health bureau and to the Ministry of Health in Beijing. By the beginning of February, hospitals across Guangdong province were alerted, but the majority of staff did not receive any warnings because of the Chinese New Year holiday, which began on February 1 and continued for 15 days. No public warnings were issued.[922]

The unauthorized transmission of health–related information is prohibited in China until there has been an announcement by "*the Ministry of Health or organs authorized by the Ministry*", in order to "*avoid confusion and panic*". Before such official acknowledgement, all data about infectious diseases is classified as a state secret: any doctor or journalist who reveals information about the development of a disease will be prosecuted for leaking state secrets.[923] So, the provincial authorities were not allowed to discuss the SARS problem openly until the national authorities had authorized it.[924] In the absence of any government statement, the growing number of apparently infected people led to rumors that were widely spreading among the Chinese concerning a "*deadly flu*". The rumors were carried by word-of-mouth, texting and conversing on social networks.[925] By mid-February, nearly 50 % of people interviewed in Guangzhou City confirmed that they had heard about the disease from friends, relatives or the overseas media. The Great

[921]Documentary "SARS: the True Story", BBC, 2003 and documentary "The Silent killer: SARS", Eurovision Science with support of European Commission, Films Media Group, 2009.

[922]Yanzhong Huang, The SARS epidemic and its aftermath in China: a political perspective, Institute of Medicine (US) Forum on Microbial Threats, Washington, D.C., National Academies Press, 2004, pp. 116–136.

[923]Ibid.

[924]China's Chernobyl? A health minister fired. A cover-up admitted to. Is China headed in a new direction? The Ecomonist, April 24, 2003.

[925]Yanzhong Huang, The SARS epidemic and its aftermath in China: a political perspective, Institute of Medicine (US) Forum on Microbial Threats, Washington, D.C., National Academies Press, 2004, pp. 116–136.

Chinese Firewall—a countrywide surveillance system—began to block the sending of SMS and Skype messages about SARS.[926] Later, 93 people were arrested for spreading rumors.[927]

On February 10, 2003, Guangdong health officials reported atypical pneumonia cases in the province—emphasizing that the disease was controllable and treatable. The government played down the risk of the illness.[928] The WHO was first officially informed of the outbreak on February 11, 2003, when the Chinese Ministry of Health reported 305 cases of an unknown disease with five deaths in Guangdong Province. Transmission of the disease was largely confined to health care workers—105 doctors, nurses, and other medical workers were infected—and the friends or families of patients.[929] The Chinese health authorities did not impose strict anti-epidemic measures (respiratory precautions to protect medical staff, taking people's temperature on entry to the hospital, and so on). In spite of clear evidence that hospital staff were the main source of transmission of the virus, Liu Jianlun, a Chinese doctor who had treated cases in Guangdong and had symptoms of SARS already from February 15, was allowed to travel to Hong Kong to attend a wedding ceremony. On February 21, he stayed in the Metropole Hotel and infected several people, who transmitted the virus to Canada, Vietnam, Singapore over the following few days, initiating a global SARS outbreak. By the end of February, 806 people were ill in China and 34 had died.[930]

On March 12, 2003, the WHO issued a global health alert about the mysterious pneumonia and recommended that people avoid traveling to Hong Kong and Vietnam. After the WHO warning, Hong Kong airport lost 18 percent of flights.[931] From February 11 to April 25, the Hang Seng Index—the market index in Hong Kong—lost 8.55 %, while the Dow Jones and NASDAQ Composite rose by 7–10 %.[932]

Meanwhile, on March 1, SARS had been recorded in Beijing. Nevertheless, the Beijing municipal authorities hid this fact from the national authorities because they wanted to convince superiors that Beijing was implementing all necessary anti-epidemic measures, and there was no reason for canceling the upcoming National People's Congress meeting, which was planned for March 5–18 and to

[926]Tim Richardson, China snoops on text messages Stamping out 'false political rumours, The Register, July 2, 2004.

[927]Pete Sweeney, Michael Martina, China detains 10 for bird flu rumors, death toll at 9, Reuters, April 10, 2013.

[928]Yanzhong Huang, The SARS epidemic and its aftermath in China: a political perspective, Institute of Medicine (US) Forum on Microbial Threats, Washington, D.C., National Academies Press, 2004, pp. 116–136.

[929]Severe Acute Respiratory Syndrome (SARS) multi-country outbreak—Update 6, WHO, 21 March 2003.

[930]David Cyranoski, China joins investigation of mystery pneumonia, Nature, April 3, 2003.

[931]WHO targets SARS "super spreaders", CNN, April 6, 2003.

[932]Comparing of trade results between Hang Seng Index, Dow Jones and NASDAQ Composite for period February 11 to April 25, 2003.

which up to 3000 representatives from every province of China were invited. Beijing's municipal government also convinced superiors that it had all the resources to deal with the situation, and refused any assistance from the central government. Many other local and municipal authorities demonstrated the same behavior: "*officials at all levels tended to distort the information they passed up to their political masters in order to place themselves in a good light*".[933] Moreover, military hospitals, which were not obliged to reveal information about their activity to the civilian authorities, also tried to deny the existence of infected soldiers. As a result, the national authorities received encouraging statistics on SARS development and were convinced there was no need for urgent nationwide preventive measures. During March, they reassured themselves that the SARS outbreak was limited to Guangdong Province and Hong Kong. It was only on March 25 that the central government confirmed the spread of SARS outside of Guangdong Province. Moreover, when the Chinese authorities invited WHO experts to China, they would only allow them access to Guangdong Province until April 2. The decision to allow them wider access was only made under tremendous international pressure on China, when it became obvious that the Chinese authorities were covering up what was actually a national SARS outbreak. On April 5, Chinese Vice Premier Wu Yi announced "*the immediate establishment of a national medical emergency mechanism, with emphasis placed on public health information and an early warning reporting mechanism*".[934] On April 6, China apologized for its slow reporting on the outbreak[935] and the Chinese Health Ministry reported 19 cases and four deaths in Beijing.[936] However, Dr. Jiang Yanyong, retired chief of surgery for a Beijing military hospital, had previously stated on TV channels that he was aware of a hundred cases and six deaths in his hospital—more than five times the number of cases announced by the authorities.[937] So on April 9, WHO experts gained permission to inspect military hospitals in Beijing as well. After clear evidence that the real size of the outbreak was being hushed up, Chinese Communist Party executives launched their own investigation into the matter. On April 20, Health Minister Zhang Wenkang and Beijing mayor Meng Xuenong were fired for their mismanagement of the crisis.[938] During May, more than 120 health execu-

[933]Yanzhong Huang, The SARS epidemic and its aftermath in China: a political perspective, Institute of Medicine (US) Forum on Microbial Threats, Washington, D.C., National Academies Press, 2004, pp. 116–136.

[934]WHO targets SARS "super spreaders", CNN, April 6, 2003.

[935]Ibid.

[936]Howard Kunreuther, Michael Useem, Learning from Catastrophes: Strategies for Reaction and Response, Pearson Prentice Hall, 2009, p. 196.

[937]Andrew Scobell, Larry M. Wortzel, Chinese National Security Decision-making Under Stress, Diane Publishing, 2005, pp. 108–115.

[938]China's Chernobyl? A health minister fired. A cover-up admitted to. Is China headed in a new direction? The Ecomonist, April 24, 2003.

tives were sacked and nearly 1000 government officials were reprimanded for their "*slack*" response to the outbreak.[939]

On April 16, an international network of 11 leading laboratories under the leadership of the WHO discovered the etiological agent of SARS and suggested potential treatments. The cause of the Severe Acute Respiratory Syndrome was determined as a coronavirus (subsequently named the SARS coronavirus, or SARSCoV) originating from bats. Virologists breathed a sigh of relief, as coronaviruses are less dangerous than flu viruses. Had SARS been a kind of influenza, the consequences for the world population could have likely been devastating.[940] In such a scenario, the mismanagement and risk concealment demonstrated by the Chinese health system at the early stages of the epidemic could have cost millions of lives throughout the world—as it did in the case of the "Spanish flu" pandemic because of the inadequate response of the US government.

Worldwide Spanish Flu: Why Risks Were Concealed

- The military requirement to keep up the morale of the US nation caused deliberate suppression of any information about the disease. Such **secrecy on the grounds of "national security" was common during the war period**.
- **The absence of scientific knowledge about viruses, the principles of their transmission and the associated risks meant** that decision-makers underestimated the need for urgent and decisive action.
- American (and other allied countries) politicians apparently gave **priority to their political interests over the lives of hundreds of thousands of their own citizens, and millions of people around the World**.

SARS Outbreak: Why Risks Were Concealed

- National security concerns: the Chinese authorities were **afraid of massive panic, and were worried about the threat to social stability and continued economic growth** if SARS caused a similar death rate as the Spanish flu pandemic.
- **The Chinese provincial authorities wanted to be seen in a good light by the central government**, which in turn tacitly approved of the **"no bad news" culture** that existed within the Chinese communist party.

---

[939] Yanzhong Huang, The SARS epidemic and its aftermath in China: a political perspective, Institute of Medicine (US) Forum on Microbial Threats, Washington, D.C., National Academies Press, 2004, pp. 116–136.

[940] Documentary "SARS: the True Story", BBC, 2003 and documentary "The Silent killer: SARS", Eurovision Science with support of European Commission, Films Media Group, 2009.

2.3.3 *Great Wildfires in the European Part of Russia (Russia, 2010)*

In July 2010, gigantic wildfires and a drought occurred in the western part of Russia caused by a record-breaking heat wave. Fifty-four people perished and 458 were injured in the wildfires themselves and, according to Munich Re estimates, around 56,000 people died from the effects of the smog and heat wave cause by the fires.[941] More than 2000 buildings were destroyed and more than 9 million hectares of crops were lost. Total damages from the wildfires and drought were estimated at between US$15 billion and $50 billion.[942] In June 2010, temperatures exceeded the previous Russian maximum 36 times, and in July 124 times.[943] In July, maximum temperatures were recorded in all regions of the European part of Russia for the first time since records began.

After the wildfires were extinguished, the Russian Minister of Emergency Situations described the causes of the severe wildfires. Firstly, there had been no warnings of a wildfire threat from Russian regional authorities. Regional bureaucrats had simply repeated the same message to the federal authorities: "*We have enough forces and ability to deal with any situation*". For example, the government of the Nizhny Novgorod region refused federal help.[944] Less than three days afterwards, fire coming out of the forest destroyed a whole village of 341 dwellings. Secondly, intervention by the federal forces should have been organized

[941]Natural catastrophes 2010. Analyses, assessments, positions, Munich Re, Feb. 2011, p. 27.

[942]A. Shapovalov, D. Butrin, $15 billion lost in Russian fires, Kommersant, Aug. 10, 2010.

[943]Anomalous phenomena don't exempt from responsibility, Parliamentary newspaper, Sep. 14, 2010.

[944]Russian Prime Minister Vladimir Putin chaired a conference on measures to reduce the number of wildfires, July 27, 2010, Website of the Government of the Russian Federation, http://archive.government.ru/docs/11511.

much earlier, despite the absence of requests from the regions and the misinformation about the real situation from local and regional authorities. Thirdly, federal authorities should have ignored the existing regional framework for granting legal permission to fight the fires, and unilaterally declared a state of emergency in the regions. Fourthly, aerial support should have been provided earlier in the development of the wildfires. Fifthly, local authorities should have taken the necessary firefighting measures in advance, and extinguished peat fields preventively. Sixthly, regional leaders were not paying attention to pessimistic weather forecasts or taking the necessary action over a whole month of heat waves, which later caused the wildfires.

This disaster confirms again that the distortion of information occurring on one level of the managerial hierarchy of an organization or a country leads to a situation where senior management receive information about risks too late to allow them to deploy serious force in time to respond adequately to foreseeable risks, resulting in a much larger-scale disaster.

2.3.3.1 Risk Concealment Before and During the Disaster

Russia accounts for around 20 % of all the forested area on the planet. In the Soviet Union, a centralized system of forest protection allowed firefighting forces from one region to be redeployed to another. The last giant wildfire disaster occurred in the European part of Russia in 1972, when tens of thousands of Red Army soldiers with military engineering equipment helped federal forest firefighters to extinguish the fires. However, by the 2000s, this event had been forgotten, and the Federal Forest Service had been eliminated as part of a wider deregulation movement and to reduce public expenditure. In 2007, a new Forest Code was passed that provided no federal forest protection whatsoever. State control was transferred to regional authorities who were expected to work with private logging operators to manage forest fires and fight wildfires. However, the regional authorities did not have the budget to do what was required: for example, Mordovia, a region the size of the state of Massachusetts and with 800,000 residents, had just 50 forest firefighters and 2 specialized fire trucks. In their turn, private logging operators were only concerned with the short-term profitability of business and were reluctant to invest in fire protection activity. The overall financing of Russian forest management dropped to US$0.55 per hectare per year, although nearby Kazakhstan spent US$1.05, the USA $4 and Belarus—with forests contaminated by radionuclides from Chernobyl—$7.45.[945] In addition, the new provisions of the Forest Code did not regulate clearly the separate jurisdiction of federal, regional and local authorities during interregional wildfires, and the responsibility of the different authorities during fires moving from forests to settlements and conversely. Ultimately, after these legal changes, the area of forest destroyed by fire

[945]Wildfires 2010: The miser pays twice, Greenpeace (Russia), press release, August 6, 2010.

doubled. The public did not see the consequences of the absence of federal control over forests, until an unprecedented natural phenomenon occurred, perhaps associated with the phenomenon of global change due to anthropogenic causes: a two-month long anticyclone hovered over the European part of Russia and caused heat waves with temperatures up to +40 °C, hot wind storms with speeds of up to 30 m/s and a total absence of rainfall.

The heat and drought of summer 2010 were predicted by NASA analysts in December 2009 and confirmed by Russian scientists. On March 23, 2010, the Russian meteorological service Roshydromet issued a weather forecast for the coming summer season. It predicted an increased wildfire hazard in the European part of the Russian and Ural regions. Despite this warning, large-scale preventive actions were carried out neither by federal authorities (because there was no legal framework for intervention) nor by regional and local authorities, (because of the lack of money and resources). Moreover, none of the officials could imagine the possible consequences of two months of extreme heat and the resulting wildfires, because since 1972 there had been no similar disaster. In June 2010, when the wildfires started, regional agencies began to pass the responsibility for fire prevention and fighting to each other because it was not defined clearly enough in the existing legislative base. As a result, they just spent precious time playing bureaucratic games and lost the initiative in preventive action. In addition, the failure to manage forest firefighting was aggravated by the Russian political system. In December 2004, Vladimir Putin, President of Russia, had changed the rules for appointing regional leaders: instead of being elected by popular vote, all heads of Russian regions were appointed by the President. This led to a situation where their performance was evaluated in Moscow, rather than in their regions by the citizens they were supposed to be serving. Eager to make a favorable impression on Putin and ensure their continuation in power, regional leaders preferred to send only reassuring reports to the central government. They always tried to convince the federal authorities that they could handle any situation but, in reality, they did not have the power or the resources to deal with a crisis of this scale. Some of governors were continuing to send soothing reports to Moscow and reassure the residents of their regions that "*everything is under control*". Some, having delegated authority to their deputies, flew out from the regions for 2–3 weeks on planned vacations: such distasteful cases occurred in the Moscow, Vladimir and Voronezh regions. Without proper management, the situation was thus getting worse.

When the fires started to reach settlements in many of the regions, which were supposedly "*under control*", and Russian newspapers and social networks were filled with horrific details of deaths from fire and burned houses, the federal authorities realized the true scale of the disaster. Moreover, when smog from the giant wildfires covered Moscow—the information and control center of Russia—and all nearby regions, questions about the adequacy of the crisis response became the main agenda of both the Russian media and foreign correspondents. From this point, the swift extinction of the fires was seen as a test of the ability of Vladimir Putin and the federal government to manage emergency situations. This was perhaps ironic in that the crisis was created in the first place by the federal government's shortsightedness in deregulating forest management. The Russian political culture, which motivated subordinates to conceal risks, only aggravated the magnitude of the disaster.

In the end, the inaction of the regional authorities was balanced by the vigorous intervention of the federal authorities. Vladimir Putin and the Minister of Emergency Situations were presiding over the response forces directly from the Government Emergency Center, and were present in person in the most damaged places. Putin set a record in terms of the number of visits to disaster-stricken areas: at no other time in his political career has he attended affected regions six times to make decisions about comprehensive state assistance to the victims. Some remarkable facts include the following: Dmitry Medvedev, the President of Russia from 2008 to 2012, visited the ambulance control center in Moscow in place of the mayor, who had left the city on vacation during the worst of the smog, and would duly be dismissed after the disaster; and Vladimir Putin (who was President of Russia from 2000 to 2008, Prime Minster at the time of the disaster, and President again from 2012 till the time of writing [summer 2014]), personally flew a multipurpose amphibious firefighting aircraft to demonstrate his direct involvement and control over the situation to the public. Such actions gave these politicians additional credits to their ratings and drew the attention of Russian and world media, but also revealed the weakness of the state management hierarchy, which could not adequately and promptly react to risks without the intervention of federal government officials.

After the disaster, the Minister of Emergency Situations described the behavior of the regional authorities as follows: "*Half of the governors sat and waited until the fires died out themselves. Another half continued to relax, waiting for the Minister of Emergency Situations and the Prime Minister to come, who like the magicians on the blue helicopter could extinguish the fires*".[946] The governors refused to admit that they could not cope with the fires, and that they needed external help, until the bitter end.

Massive Wildfires in the European Part of Russia: Why Risks Were Concealed

- **This propensity for hiding bad news** resulted in part from the change of the rules for appointing regional leaders: instead of being elected by popular vote, all heads of Russian regions were appointed by the President. This led to a situation where their performance was evaluated in Moscow, rather than in their regions by the citizens they were supposed to be serving. **Eager to make a favorable impression on Putin and ensure their**

[946] Anomalous phenomena don't exempt from responsibility, Parliamentary newspaper, September 14, 2010.

continuation in power, regional leaders preferred to send only reassuring reports to the central government. They always tried to convince the federal authorities that they could handle any situation. This led to massive distortion of information about the real situation concerning wildfires in several Russian regions and, as a result, delayed the reaction of Russian federal government to the threat. There is a prevalent Russian political culture, which motivated subordinates to conceal risks.

- This was further reinforce by the federal government's **shortsightedness in deregulating forest management**, leading to confusing and badly designed attribution of responsibilities among involved parties from local government to the private sector.

2.3.4 Krymsk Flooding (Russia, 2012)

On July 7, 2012 from 2 until 4 a.m., a powerful flash flood with a 6.8-m water surge occurred in the Krymsk district of Southwestern Russia. Krymsk is in the Krasnodar region, just 30 km from the coast of the Black Sea and 200 km from Sochi, where the Winter Olympic Games took place in 2014. For two days before the disaster, the volume of rainfall exceeded the monthly average by three to five times. The torrential rain caused a sharp rise in the water level of rivers flowing from the nearby Caucasus Mountains, which led to the flooding of several districts and cities. However, it was only in the Krymsk district that the consequences of the flooding were dreadful. The disaster affected 34,650 people, 171 people died—153 in the Krymsk district—and 2225 people (including 496 children) were injured. More than 7200 residential and public buildings in the district were destroyed or damaged by the flood.[947]

The magnitude of the disaster—a large number of deaths and injuries—was caused by the absence of emergency information about the coming flood from local authorities, despite the fact that local officials received information about a dangerous rise in river levels more than 36 h before the flooding. Moreover, 12 h after the disaster, Vasily Krut'ko, the head of Krymsk district, reported directly to Vladimir Putin—who had immediately come to the affected district from the nearby presidential residence with a group of journalists—that local officials had declared a state of emergency in good time and taken action to inform the population 5 h before the peak of the flash floods. He claimed that the crisis information campaign had included personal visits to private houses in potential flooding areas, continuous coverage on local television and emergency SMS messages. However, investigations that following revealed that Krut'ko had lied about the decision by local officials to declare a state of emergency—it was actually issued 2 h after the gigantic flash flood occurred—and that the majority of the above actions were never implemented in reality. As a result, only 52 out of the 60,000 residents of Krymsk confirmed that they received warnings about the coming flood.[948] The majority of the dead were elderly people who drowned while sleeping, or when it was too late to escape from their houses. Krut'ko was arrested after the investigation, on charges of negligence resulting in the death of two or more people. On March 2013, during the trial, he partially admitted his guilt. In August 2013, he and other former local officials were sentenced to between $3^1/_2$ and 6 years in prison.

As in many other cases, this bottom line manager tried to convince his superiors that he had responded adequately, in spite of the presence of thousands of witnesses and a dozen journalists with TV cameras, and the likelihood of criminal prosecution for document forgery and bluffing. The Russian investigation committee declared: "*In this situation, the [local] officials worried more how they could justify their inaction by any means rather than how to minimize the consequences [magnitude] of disaster. [Local executives] instructed their subordinates to retroactively make official documents containing false information about the meeting of the Emergency Commission, declaration of state of emergency and warnings of the residents*".[949]

Facts and documents published after the disaster clearly demonstrated that the actual scale of the flooding was unpredictable in advance by any meteorological or engineering service—let alone by local officials who had no specialist training or experience of hydrology—until the floods came. If only they had been honest and straightforward about their activities before and during the disaster, the local

[947] 2012: the year the disaster, Interfax, Dec. 24, 2012.

[948] Falsification of emergency, Interfax, July 25, 2012.

[949] The criminal case hearings concerning flooding in July 2012 in Krymsk city, Krasnodar region, The Investigative Committee of the Russian Federation, May 15, 2013.

managers would probably never have become the main defendants following the disaster. Instead, their undeniable distortion of the facts about what they had done in the attempt to prove their competence by any means led them to prison…

2.3.4.1 Risk Concealment Before and After the Disaster

From the 1970s onwards, private houses began to be constructed without the permission of the authorities in flood zones on the banks of the River Adagum in Krymsk district. Local officials did not consistently oppose this activity: they ruled out the possibility of a disaster for the simple reason that, in the Soviet Union, the beds of rivers flowing from the Caucasus Mountains had been cleaned up on a regular basis. Flood-protection measures had been effective enough. But in post-Soviet times, the funding for cleaning riverbeds and limiting the deforestation of mountain areas was terminated,[950] due to lack of financial resources and the dismantling of the integrated system of state control over the political and economic matters in 1991. Almost every year after 1995, Krymsk district faced minor floods of the River Adagum, with water surges of up to 1.5 m. In August 2002, after heavy rains in the Krasnodar region and the collapse of a dam reservoir, 7000 homes and commercial buildings were flooded, 4968 homes damaged and 447 destroyed. Two villages in Krymsk district were completely washed away, killing 62 people. An investigation initiated by the Parliament after these floods found that they were caused by the widespread unauthorized construction of private houses and the silting of riverbeds. In both cases, local authorities were found to have been negligent. The investigation commission stated: "*[Local and regional] officials ... were not recognizing their mistakes and are still trying to reduce the degree of [their] own guilt and amount of damage by misinforming federal authorities about the actual size [of the disaster]... The residents and tourists were not informed about the coming disaster... Responsible services did not take the necessary steps to prepare for a disaster: [there was no] drainage clearing, riverbeds were not maintained in a good state*".[951] After the investigation, two local officials were sentenced to $3^1/_2$ years on probation. According to one witness who survived the flood, "*those who were in the flood zone got from [local and regional] authorities significant compensations and everyone has forgotten about the incident*".[952] After the flood, the Krasnodar regional authorities ordered hydrological studies of the affected districts, but did not subsequently implement the recommendations of scientists. None of the regional and local officials worried about complex anti-flooding measures because their had been no serious floods since 2002. Therefore, only 12 % of the budget intended for flood prevention measures—up to

[950]Y. Vorobev, V. Akimov, Y. Sokolov, Catastrophic floods of beginning of XXI century. Lessons and conclusions, Moscow, DAX-Press, 2003, p. 72.

[951]C. Baimukhametov, The country of unlearned lessons. Why is there no hydraulic security in Russia? Moscow Pravda, July 12, 2012.

[952]Ibid.

US$40 million had been funded by federal government—was actually spent for this purpose.[953]

Another problem was the high frequency of warnings about dangerous weather events issued by the Russian metrological service (Roshydromet). In 2010, Roshydromet released 2700 storm warnings throughout Russia; the average accuracy was 92 %.[954] In 2011, the service released 1800 storm warnings throughout the country (average accuracy was 91 %) and 700 warnings about dangerous weather events (average accuracy was 88 %). More than 100 storm warnings were registered in Krasnodar region alone in 2011 (roughly one warning every three days). In the five days before the disaster, Roshydromet issued 18 weather forecasts[955] and two storm warnings. Local authorities in Krymsk district received the first warning 36 h before the flash flood: "*in the next 3–6 hours [and during the whole of the next day], heavy thunderstorms, severe hailstorms and strong winds up to 20 m/s are expected*".[956] Six hours before the flash flood, they got a storm warning about possible natural hazards referring to the serious threat of flooding, severe rise of water levels and so on. During the investigation, it was revealed that, after receiving the second warning, local executives under the leadership of Vasily Krut'ko began to collect information about the current water level in the River Adagum by personal observation of the banks of the river. At 11:30 p.m. on 6 July 2012, about 2$^1/_2$ h before the flash flood, the level of the Adagum in the city of Krymsk had only risen by 24 cm (later investigators declared that the flash flood raised the water level in the river by up to 6.8 m!).[957] Local officials were communicating regularly with each other by cell phones. Vasily Krut'ko, who had been the head of Krymsk district for the previous seven years, was reassuring subordinates: he emphasized that Roshydromet's storm warnings were always similar, such warnings came to him weekly and they had never been followed by serious flooding.[958] After the disaster, Russian meteorologists calculated that the volume of rainfalls on the day of the flood was three times the average daily maximum, and the river discharge exceeded the previous maximum recorded flow rate by a factor of 2.5; the last record was during the 2002 flooding. The likelihood of such anomalous levels of precipitation has been estimated to correspond to one such event every

---

[953]Ibid.

[954]S. Shilova, Little rainfall and heat are expected in summer in most parts of the country, Russian newspaper, March 23, 2011.

[955]Boris Pasternak, Caucasian Prisoners. Chief Meteorologist spoke about the causes of floods in Kuban, Moscow News newspaper, July 13, 2012.

[956]Transcript of the crisis response meeting under leadership of Vladimir Putin in Gelendzhik, The Kremlin, July 25, 2012.

[957]Open letter of Vladimir Ulanovsky, former mayor of Krymsk city, Site of Krymsk radio station Electron FM, 2013, http://www.electron-fm.com/%D0%BE%D1%82%D0%BA%D1%80%D1%8B%D1%82%D0%BE%D0%B5-%D0%BF%D0%B8%D1%81%D1%8C%D0%BC%D0%BE.

[958]Ibid.

200–300 years.[959] Despite this, Roshydromet's storm warnings did not point out that this was an exceptional event, unprecedented by any previous dangerous weather events in the Krasnodar region. Ultimately, local officials did not declare a state of emergency and did not wake up this district of 60,000 residents before the flash flood. All 52 witnesses who confirmed that they had heard an emergency signal had received it from three loudspeakers during the second hour after the flash flood reached Krymsk city[960]: by this time, the abnormal level of water was obvious to local officials, and they turned on the emergency loudspeakers. Later, scientists found the cause of the sudden 4-m wave of water described by many witnesses: a huge amount of debris from previous flooding (wood, silt, stones, and so on) had accumulated on the uncleaned bed and banks of the River Adagum over the previous ten years, which blocked the river flow under railway and road bridges located near the city, forming in this way temporary "dams". When a critical amount of water had built up, the bridges were destroyed and a wave-like flash flood rushed down to Krymsk city.[961] This explains why local officials did not record an unusual water surge within the city a couple of hours before the disaster.

Detailed investigations later proved the innocence of local executives, but their criminal activity after the disaster—forging documents and misleading residents and federal executives—guaranteed them prison terms and strong rebuke from the majority of Russians.

Even after the disaster, regional and some federal officials continued the practice of sending falsely reassuring reports to Vladimir Putin and the public. For example, in Krymsk district, the flood killed about 40,000 animals. When the temperature exceeded 30 °C, there was a high risk of the spread of disease if the carcasses were not eliminated immediately. Regional officials continued to assure the federal authorities that they had all the resources they needed to clean up Krymsk city. However, clearing the city of dead animals began only after the deployment of army units on the 8th day after the disaster. During the week after the flooding, the regional authorities were still sending messages to Moscow that "*everything is fine, we have everything [we need] to control the situation*".[962] The Federal Minister of Health and Social Development assured Vladimir Putin immediately after the flood that all necessary vaccinations (tetanus, hepatitis and dysentery) would be made promptly. Nevertheless, a week later, the President said after his visit to affected areas: "*For me, it is strange now to hear that people were not vaccinated*". Victims complained to the President about difficulties in the state

[959]Ibid.

[960]Transcript of the crisis response meeting under leadership of Vladimir Putin in Gelendzhik, The Kremlin, July 25, 2012.

[961]A catastrophic flood in the river basin of Adagum 6–7 July 2012 and its causes, State Hydrological Institute, Krasnodar Regional Center for Hydrometeorology and Environmental Monitoring Department and RosHydromet's departments in South Federal District and North Caucasus Federal District, 2012, pp. 13–16.

[962]Transcript of the crisis response meeting under leadership of Vladimir Putin in Krymsk, The Kremlin, July 15, 2012.

compensation process and the slow work of regional officials. Putin was furious about that: "*People complain that it is hard to reach the sources of state financial assistance. The queues are large. These queues crush. They say that 'We have already suffered. We need to deal with housing and children, but [clerks] force us to stand in queues'. It is necessary to change the situation radically. Next time [on the president's next visit to Krymsk city], I will come and stand myself in the queue. Is it so difficult to deploy mobile assistance offices throughout the city for comfortable handling?*".[963] The phrase of Vladimir Putin—"*I will come and stand myself in a queue*"—is reminiscent of his piloting of a firefighting aircraft during the wildfires of 2010, when the then Prime Minister tried to sway his subordinates in order to ensure that crisis response actions were provided well.

The conclusion is simple: whenever a senior executive hears reports from subordinates in a crisis situation that everything is under control, generally the situation is much worse than reports state and the executive should personally go to the place of the disaster to check the actual conditions there. During the Krymsk tragedy, the Russian leader made three emergency visits in 18 days for crisis response meetings that he held personally, and two additional trips in the year after the flood to check on the reconstruction of the city. Obviously, such frequent visits and ongoing control of subordinates originated from previous leadership mistakes. For instance, during the Kursk submarine accident, Russian naval officials gave distorted information to Putin and the public, reassuring them that the Northern Fleet of Russia could mount a successful rescue operation, when Vladimir Putin was on vacation. And when hurricane Katrina struck the coast off Louisiana, George Bush was on vacation at his Texas ranch. In the midst of an uncoordinated response from the Federal Emergency Management Agency, Louisiana state and local government, the National Guards and the US Army Corps of Engineers, Bush visited the Gulf Coast only on the fifth day after the disaster.

Krymsk Flooding: Why Risks Were Concealed

- **Habituation/false reassurance/overconfidence/self-deception** among representatives of the local authorities about the low probability of a catastrophic flash flood in Krymsk district.
- **Regional authorities were unwilling to investigate the causes of previous flash floods in detail**, since this would inevitably lead to the lengthy and embarrassing process of passing on the lessons learnt and making recommendations to subordinates.
- **The high frequency of flood and severe weather warnings** previously received by local authorities, which were often not realized, **leading to a "crying wolf" psychological response and growing complacency.**

---

[963]Transcript of the crisis response meeting under leadership of Vladimir Putin in Krymsk, The Kremlin, July 15, 2012.

- **Russian regional bureaucrats and federal ministers wanted to appear in a good light in the eyes of the Russian president**. This led to massive distortion of information about the timeliness of the state of emergency during the disaster and the adequacy of crisis response measures after the disaster.

2.4 Retail Production Industry

Maybe it's shameful ... [but w]e live in a capitalist world

Jean-Claude Mas

2.4.1 Nature of the Industry

This sector produces consumer goods: foods and drinks, drugs, cosmetics, electronic gadgets, cars and so on. The majority of risk concealment cases in the sector are similar because of its specific business practice. Mutually competing manufacturers tend to produce similar goods in every product segment because each manufacturer is continually watching competitors for any innovations, which will be implemented as fast as possible in the products of all manufacturers. Prof. Leveson assessed the problem very clearly: "*At the same time that the development of new technology has sprinted forward, the time to market for new products has greatly decreased, and strong pressures exist to decrease this time even further. The average time to translate a basic technical discovery into a commercial product in the early part of this century was thirty years. Today our technologies get to market in two to three years and may be obsolete in five. We no longer have the luxury of carefully testing systems and designs to understand all the potential behaviors and risks before commercial or scientific use. [This leads to] reduced ability to learn from experience*".[964] Therefore, manufacturers try to launch new products as swiftly as possible to gain a competitive advantage during the first few months, and sometimes ignore defects in the design of innovative production.

Such problems have occurred in many cases with complex innovative products. The retail sector has seen the Ford-Firestone tire controversy (1990), the Intel Pentium FDIV Bug Crisis (1994), and problems with the antenna of the Apple iPhone 4 (2010) and with the brakes of the Toyota Prius (2010–2013). In the

[964]Nancy G. Leveson, Engineering a Safer World: Systems Thinking Applied to Safety, MIT Press, 2011, p. 3.

industrial sector, notorious cases include the lithium ion batteries on the Boeing 787 Dreamliner (2012–2013)[965] and the chassis of the Sukhoi SuperJet (2013).[966]

To take just one of these examples: the management of Apple was aware of problems with the quality of signal reception of the iPhone 4 long before it was released, but Apple's co-founder Steve Jobs liked the design of the new phone so much that he personally gave an order to launch it into mass production without redesigning the antenna. He also cancelled real-world testing before the launch – the testing process usually takes a minimum of 14 weeks.[967] Within three weeks of the launch, Steve Jobs and Apple were denying that the new phone had flaws. This position angered many people and attracted media attention to the problem. Ultimately, the company had no choice but to admit the problem, issue a temporary solution for the 25 million customers who had already bought the phone (a free case for the phone) and update the software.

2.4.2 *Complexity, Cost Reductions, Arrogance and The Toyota Problems (USA–Japan, 2000s)*

With recalls of millions of Toyotas in 2010, the automaker has been facing heavy criticism over the design of its cars, about concealment of information concerning known defects and on the matter of the arrogance of middle-managers in

[965]Boeing 787 Dreamliner battery problems, http://en.wikipedia.org/wiki/Boeing_787_Dreamliner_battery_problems.

[966]State of emergency with the aircraft Sukhoi SuperJet 100 during 2011–2014, RIA Novosti, Feb. 25, 2014.

[967]Yukari Iwatani Kane, Niraj Sheth, Apple Knew of iPhone Issue, The Wall Street Journal, July 16, 2010.

dealing with customers' complaints. The general perception in the West is that a main cause of this apparent drift away from the Toyota Way is the cost reduction war between carmakers. Interestingly, the typical view of Japanese business persons and researchers represented also by the Japanese media is different, emphasizing that the accidents happened because Toyota tried to produce too complicated products and did not manage the raising complexity imposed by social pressure and market demand.

2.4.2.1 Toyota's Cost Reduction Challenge

The typical narrative of the West draws attention to the need for companies to monitor the activity of competitors, while the market forces of supply and demand lead to a situation where prices on similar products become equalized. Usually, the race to keep prices competitive with similar products results in the reduction of production costs by any means necessary. Thus, Carlos Ghosn, Chairman and CEO of the Renault-Nissan Alliance, initiated an aggressive downsizing campaign in the late 1990s to rescue Nissan, the second largest car producer in Japan, from imminent bankruptcy. He was so successful with huge cost reductions—10 % down on supplier parts and US$2.25 billion in savings—that he was nicknamed "*le cost killer*" and "*Mr. Fix It*".[968] Toyota was forced to respond and adopt a similar action: the company launched a cost-cutting strategy called "*Construction of Cost Competitiveness for the 21st Century*" (CCC21) and may have perhaps broken its own cardinal rule that "*customers are always prior to profits*".[969] CCC21 had an impact on many spheres of Toyota production. After its implementation, Toyota faced extensive recalls of cars because of flaws in design and supplied parts; between 2004 and 2006, the company recalled more vehicles than ever before in its history. In 2006, Japanese investigators found out that, since 1996, Toyota had been aware of a defective relay rod on the Toyota Hilux Surf, but recalled the model only after a lawsuit was initiated by relatives of people killed in a crash caused by the defective rod. Before the crash, Toyota managers thought that problems occurred only in "*unusual and extreme conditions*".[970] Information about flaws in car designs, shortcomings in manufacturing processes and production defects in parts from other suppliers was not communicated well within the world's largest car-maker: according to insiders from the American branch of Toyota, "*working for the company is like working for the Central Intelligence Agency, where information is shared only on a 'need to know' basis*".[971]

---

[968]David Austen-Smith, Daniel Diermeier, Eitan Zemel, Unintended Acceleration: Toyota's Recall Crisis, The Kellogg School of Management, Northwestern University, 2011.

[969]Ken Bensinger, Toyota tried to cut costs on recalls. Los Angeles Times, February 22, 2010.

[970]David Austen-Smith, Daniel Diermeier, Eitan Zemel, Unintended Acceleration: Toyota's Recall Crisis, The Kellogg School of Management, Northwestern University, 2011.

[971]William J. Holstein, Toyota Recall Highlights Deep Organizational Failures, CBS (Money Watch), Feb. 9, 2010.

2.4.2.2 The Acceleration Pedal Problem

During the crisis of the uncontrolled acceleration of certain Toyota vehicles (2009–2010), investigators published internal correspondence between Toyota managers, confirming that 37,900 customer contact reports of unintended acceleration had been logged in Toyota's complaint coding system[972] since the early 2000s. This problem had led to 34 deaths and thousands of customer complaints to the National Highway Traffic Safety Administration (NHTSA).[973] Toyota knew about problems with gas pedals made by an Indiana-based supplier, and that the unintended acceleration problem only existed on US-made cars.[974] Nevertheless, the company did not launch a detailed investigation of the problem and made only a limited recall, because of the significant incurred expenses: in 2007, they saved up to $100 million on a limited recall of floor mats in Camry and Lexus ES sedans.[975] But, in autumn 2009, the American media released a recording of a conversation between a 911 call center employee and the driver of a Lexus ES 350 in San Diego, California, who was asking for help with a stuck accelerator pedal. During the call, the car crashed, and all four occupants were killed.[976] This case led to severe accusations of Toyota by the American media and attracted the attention of regulators. According to Jim Press, Toyota's former U.S. chief, "*The root cause of their problems is that the company was hijacked, some years ago, by anti-(Toyoda) family, financially oriented pirates... [These executives] didn't have the character necessary to maintain a customer first focus*".[977]

The Japanese view emphasizing the role of complexity, and of mistakes in managing it, is well represented by economic professor Takahiro Fujimoto from the University of Tokyo, a leading authority on the Toyota production system and automotive product development. Toyota has been a frontrunner in making very complex products, like hybrids or luxury cars, at very high volume and then growing the volume. In addition, constraints on cars have grown tremendously in the last decade: "*What was okay 10 years ago is not okay now... some of the things that are part of the Toyota problem now were not a big problem 20 years ago. So customers and society are fussier and fussier about what they expect from car... For example, with the Prius recall, the problem resulted because Toyota tried to*

[972] Letter to James E. Lentz, President and CEO, Toyota Motor Sales, USA, Inc., Congress of the United States, Committee on energy and commerce, from Chairmen Waxman and Stupak, June 29, 2010.

[973] Brady Dennis, Toyota hit by new surge of reported fatalities in vehicles, The Washington Post, Feb. 16, 2010.

[974] William J. Holstein, Toyota Recall Highlights Deep Organizational Failures, CBS (Money Watch), Feb. 9, 2010.

[975] Ken Bensinger, Toyota tried to cut costs on recalls. Las Angeles Times, Feb. 22, 2010.

[976] 'There's no brakes... hold on and pray': Last words of man before he and his family died in Toyota Lexus crash, The Daily Mail, Feb. 3, 2010.

[977] Alan Ohnsman, Jeff Green and Kae Inoue, Toyota Recall Crisis Said to Lie in Cost Cuts, Growth Ambitions, Bloomberg, February 26, 2010.

improve fuel efficiency and safety and quietness at the same time through a nice combination of very powerful regenerating brakes, plus the latest antilock brake system, plus the hydraulic braking system. But the relationship between the three kinds of brakes changed with the new design, and then drivers could have an uneasy experience when there was switching between the different brakes a little bit… Toyota failed to see this problem in the right way, at least in the beginning.[978]

Akio Toyoda, the president of Toyota, grandson of the company's founder, in his testimony before the US Congress in February 24, 2010, confirmed this view: "*At times, we do find defects. But in such situations, we always stop, strive to understand the problem, and make changes to improve further. In the name of the company, its longstanding tradition and pride, we never run away from our problems or pretend we don't notice them. Toyota has, for the past few years, been expanding its business rapidly. Quite frankly, I fear the pace at which we have grown may have been too quick [to train adequately our personal]. I would like to point out here that Toyota's priorities have traditionally been the following: first, safety; second, quality; third, volume. These priorities became confused. And we were not able to stop, think, and make improvements as much as we were able to before, and our basic stance to listen to our customers' voices to make better products has weakened somewhat. We pursued growth over the speed at which we were able to develop our people and our organization, and we should sincerely be mindful of that*".[979]

Professor Fujimoto emphasized that all the problems associated with the Toyota recalls were design quality problems rather than manufacturing quality problems. He is adamant in pointing out that "*to my knowledge they are not trying to hide the problems. But when a very complex problem happened, they were not sure to what extent this was a responsibility for the company, and to what extent other parties were responsible. So their attitude was, 'Wait a minute. This is complicated'. They were sure that they were not the only party responsible for this problem… But it is also obvious that Toyota was at least partly responsible for many problems that were popping up one after another. Probably what they should have done was to deal with it as quickly as possible – [such as send] a senior person to America as quickly as possible and then have [the company] apologize for whatever [it felt was their] responsibility. So a partial but thorough apology, and definitely a quick apology, was what they had to do. But they probably hesitated to come to the U.S. because they were not sure to what extent they were responsible for those*

[978]Interview of Professor Takahiro Fujimoto by Professor MacDuffie, Under the Hood of Toyota's Recall: 'A Tremendous Expansion of Complexity', Knowledge@Wharton, March 31, 2010, http://knowledge.wharton.upenn.edu/article/under-the-hood-of-toyotas-recall-a-tremendous-expansion-of-complexity.

[979]Toyota gas pedals: is the public at risk? Hearing before the Committee on oversight and government reform House of Representatives one hundred eleventh Congress, second session, serial No. 111–75, February 24, 2010, http://www.gpo.gov/fdsys/pkg/CHRG-111hhrg58346/html/CHRG-111hhrg58346.htm.

problems. Then people saw that as, 'Gee, Toyota is escaping from responsibility for this problem'. This is not what Toyota meant – but the way they handled the initial problem was very bad, I think.[980]

It is thus the combination of growing complexity, pushed by competition as well as society demands, together with overconfidence in being able to deal with the problems that led to Toyota's crisis. Overconfidence led Toyota managers to underestimate the difficulties in handling the novel kinds of arising complexities, and to believe that Toyota would do better than other companies: "*Ironically, as a result they probably took in way too much complexity. It was [beyond] their capacity*".[981] According to professor Fujimoto, arrogance is the number one enemy of the Toyota philosophy: "*But they didn't take this seriously until big problems happened. I would probably say middle managers, particularly at headquarters, started to deviate from the Toyota Way by being arrogant, being overconfident, and also they started not to listen to the problems that customers raised*".[982]

2.4.2.3 Public Perception Versus Reality

The Big Three US motor manufacturers Ford, General Motors, and Chrysler, who were in a complicated financial situation at the time, benefited from the accusations of the more successful and profitable Toyota.[983] Only then, under public pressure and the threat of losing the American market,[984] did Toyota finally decide to recall 8.5 million vehicles worldwide to solve the problem of accelerator pedal entrapment by loose floor mats, and other pedal problems. Fortunately for Toyota, suggestions that flaws with their electronic throttle control systems could be the main cause of unintended acceleration were not confirmed by a year-long NHTSA investigation.[985] The National Highway Traffic Safety Administration (NHTSA) aided by NASA Engineering and Safety Center concluded the following[986,987]:

---

[980]Ibid.

[981]Ibid.

[982]Ibid.

[983]David Bailey and Kevin Krolicki, Toyota U.S. sales reel from crisis; GM, Ford surge, Reuters, Feb. 2, 2010.

[984]William J. Holstein, Toyota Recall Highlights Deep Organizational Failures, CBS (Money Watch), Feb. 9, 2010.

[985]Peter Valdes-Dapena, Pedals, drivers blamed for out of control Toyotas, CNN, February 8, 2011.

[986]National Highway Traffic Safety Administration (NHTSA), U.S. Department of Transportation, Technical assessment of Toyota electronic throttle control (ETC) systems, February 2011, http://www.nhtsa.gov/UA.

[987]NASA Engineering and Safety Center, National highway traffic safety administration Toyota unintended acceleration investigation, Technical support to the National Highway Traffic Safety Administration (NHTSA) on the reported Toyota motor corporation (TMC) unintended acceleration (UA) investigation available at http://www.nhtsa.gov/UA.

"After conducting the most exacting study of a motor vehicle electronic control system ever performed by a government agency, NASA did not find that the ETC *electronics are a likely cause of a large throttle openings in Toyota vehicles as described in consumers' complaints to NHTSA... NASA found no evidence that any failures of the* ETC *system had an effect on the performance of the braking system*". Remarkably, they declared that it was mainly "*the publicity surrounding NHTSA's investigations, related recalls, and Congressional hearings [that] was the major contributor to the timing and volume of complaints*". Both also noted that "*the vast majority of complaints involved incidents that originated when the vehicle was stationary or at very low speeds and contained allegations of very wide throttle openings, often with allegations that brakes were not effective*". NHTSA's analysis indicated that "*the most likely cause of the acceleration was actually pedal misapplication (i.e., the driver's unintended application of the accelerator rather than, or in addition to, the brake)*".[988]

The hysteria in the US against Toyota is an example of the "social proof" mechanism that Robert Cialdini has abundantly documented and dissected to describe group actions under social influence.[989,990] Social influence describes the fact that people will do things that they see other people are doing, such as the famous example of someone looking up into the sky, leading to bystanders then looking up into the sky to see what he was seeing. This may partly explain the initial reluctance of Toyota management to acknowledge the responsibility of their product whose reliability they trusted so much. Thus, we can conclude that Toyota made essentially no mistakes in producing automobiles but made serious blunders in its management of the public perception and in dealing with the politics in Washington, as reported by US[991] as well as Japanese media. As Mr. Yoshimi Inaba, the president of Toyota North America, said: "*there is sometimes a lack of communication because of the language differences, because of the cultural differences... [about] this sticky pedal situation... yes, we knew that probably a year ago in Europe. And I say that had not been shared enough well on this side. So we did not hide it, but it was not properly shared. We need to do a much better job in sharing. Whatever is happening in Europe should be known in the United States so we are all alert*".[992] Toyota had thus a problem with managing the cultural difference in driving as well as the different expectations of their customers. Since then, imitating the major US firms that sell extensively to the government or depend on

[988]Ibid.

[989]Cialdini, R. B., Influence: The Psychology of Persuasion, Collins; Revised edition, 1998.

[990]Cialdini, R. B., The science of persuasion. Scientific American, 284, 76–81, 2001.

[991]NHTSA report clears up mystery—and hysteria—on Toyota cars, The Washington Post, February 9, 2011, http://www.washingtonpost.com/opinions/slowing-the-toyota-furor/2011/02/08/AB35mZF_story.html.

[992]Toyota gas pedals: is the public at risk? Hearing before the Committee on oversight and government reform House of Representatives one hundred eleventh Congress, second session, serial No. 111-75, February 24, 2010, http://www.gpo.gov/fdsys/pkg/CHRG-111hhrg58346/html/CHRG-111hhrg58346.htm.

it for regulations, and must of necessity engage in lobbying (which is protected by the US constitution as an expression of free speech), Toyota has taken steps to strengthen its connections to US politicians and their cronies in Washington so as not to repeat the same mistake.

The recall cost more than $2 billion.[993] Moreover, Toyota was ordered to pay a record US$1.2 billion to settle a criminal investigation into safety issues—the largest penalty ever levied by the US authorities on an auto company.[994] These expenses are very significant compared with the previous cost-cutting measures that were an important contributor to the crisis, as suggested by some observers such as Rep. Dennis Kucinich during the testimony of Akio Toyoda before the US Congress in February 24, 2010.[995]

2.4.2.4 Genuine Cases of Concealment of Defects in Automotive Industry

In 2014, information concealment practices were revealed within General Motors, which for ten years had been hiding a defect with a tiny part in the ignition switch. For the sake of a part worth just 57 cents, the concealment caused 13 traffic deaths, and 2.6 million cars had to be recalled.[996] According to the more recent report of May 22, 2015 by The New York Times,[997] "*in February 2014, the automaker began recalling 2.6 million Chevrolet Cobalts and other small cars with faulty ignitions that could unexpectedly turn off the engine, disabling power steering, power brakes and the airbags. The switch crisis prompted a wave of additional recalls by G.M. for various safety issues. All told, G.M. recalled more than 30 million vehicles worldwide last year, a record for the automaker.*" And "*Justice Department investigators have identified criminal wrongdoing in General Motors' failure to disclose a defect tied to at least 104 deaths, and are negotiating what is expected to be a record penalty, according to people briefed on the inquiry.*"

The recent Takata case concerning defective inflator and propellent devices that may deploy improperly in the event of a car crash is also a clear-cut case of risk

---

[993] Margaret Cronin Fisk, Toyota Recall Cost to Exceed $2 Billion, Lawyers Say (Update2), Bloomberg, Feb. 9, 2010.

[994] Aruna Viswanatha, David Ingram, Ben Klayman, Toyota's $1.2 billion settlement may be model for U.S. probe into GM, March 19, 2014.

[995] Toyota gas pedals: is the public at risk? Hearing before the Committee on oversight and government reform House of Representatives one hundred eleventh Congress, second session, serial No. 111-75, February 24, 2010, http://www.gpo.gov/fdsys/pkg/CHRG-111hhrg58346/html/CHRG-111hhrg58346.htm.

[996] GM recall linked to 57-cent ignition switch component, The Associated Press, April 01, 2014.

[997] Ivory, D., B. Protess and B. Vlasic, G.M. Inquiry Said to Find Criminal Wrongdoing, The New York Times, 22 May 2015 (www.nytimes.com/2015/05/23/business/gm-inquiry-said-to-find-criminal-wrongdoing.html).

information concealment. The available evidence suggests too much cost-cutting, early realization of problems by Takata (one of the world's leading suppliers of advanced automotive safety systems and products) that were intentionally hidden for years,[998] eventually obliging many car makers to issue massive car recalls.[999] "*Under pressure from safety regulators,* [Takata] *agreed* [on Tuesday May 19, 2015] *to declare nearly 34 million vehicles defective, doubling the size of its recall in the United States and making it the largest automotive recall in American history*".[1000]

2.4.3 The 17-Year Poly Implant Prothese Fraud (France, 1993–2010)

The French company Poly Implant Prothese (PIP) was founded by Jean-Claude Mas in 1991. PIP specialized in breast implants. In 1993, Jean-Claude Mas secretly decided to change the formula of the silicon gel used in the implants to reduce production costs. Instead of the expensive medically approved silicon at US$45/liter, he ordered the use of industrial silicon, which was seven times cheaper at US$6.5/liter. PIP implants would now consist of 25 % medical gel and 75 % industrial gel, so the cost of producing the new implant was US$50, instead of US$68 for an implant filled with 100 % medical gel; and the implants sold for about US$400 to French cosmetic surgeons and US$130 to Latin American surgeons. Thus, using the new gel formula allowed PIP to save up to US$1.6 million per year.

Due to shortcomings in the deregulated legislation of the cosmetic industry in France and the European Union, it had become possible to hide changes in the gel formula for decades. Every year, the company was selling up to 100,000 implants in 60 countries worldwide and, nowadays, more than 300,000 women carry PIP implants[1001]; a significant amount of them live in Latin America and Asia, where regulation and government control over the cosmetic industry are poor. Seventy five percent of all the implants produced by PIP were made with the unapproved gel formula.[1002]

---

[998]Hiroko Tabuchi, Takata Saw and Hid Risk in Airbags in 2004, Former Workers Say, The New York Times, Nov. 6, 2014.

[999]Clifford Atiyeh, Massive Takata Airbag Recall: Everything You Need to Know, Including Full List of Affected Vehicles, Dec. 2, 2014, http://blog.caranddriver.com/massive-takata-airbag-recall-everything-you-need-to-know-including-full-list-of-affected-vehicles.

[1000]Ivory, D., B. Protess and B. Vlasic, G.M. Inquiry Said to Find Criminal Wrongdoing, The New York Times, 22 May 2015 (www.nytimes.com/2015/05/23/business/gm-inquiry-said-to-find-criminal-wrongdoing.html).

[1001]Jean-François Rosnoblet, Makers of fraudulent breast implants on trial in France, Reuters, Apr. 17, 2013.

[1002]Alexandria Sage, Natalie Huet, Jean-Francois Rosnoblet, Special Report: The French breast implant scandal, Reuters, Feb. 2, 2012.

The gel formula scam only came to light accidentally, when French regulator AFSSAPS (Agence Française de Sécurité Sanitaire des Produits de Santé) initiated an inspection of PIP's plants after numerous complaints from French plastic surgeons and customers about the abnormal rupture rate of PIP implants compared with those of other producers. Jean-Claude Mas was arrested and confessed the fraud: "*I knew that this gel wasn't officially registered, but I did it knowingly because the PIP gel was less expensive ... and of much higher quality... The material was better than that used to make the officially authorized gel*". However, British surgeons subsequently stated, after 453 patient examinations, that PIP breast implants ruptured in at least 16 % of patients fitted with them—while the usual failure rate does not exceed 1 %.[1003] Twenty cases of different types of cancer were connected with flawed PIP's implants.[1004] A former PIP employee testified that there were "*unscientific tests of product quality, such as judging silicone gel by sticking a finger*". Yves Haddad, a lawyer who represented Jean-Claude Mas, declared to journalists: "*Maybe it's shameful ... [but w]e live in a capitalist world*".[1005]

Three factors helped Jean-Claude Mas to conceal for decades the risk of flawed implants.

Firstly, staff at the company was under orders to "*hide the truth*". Some of them were convinced by Mas that the industrial gel they were using was "*better than that used to make the officially authorized gel... We organized everything to escape being*

[1003]Sarah-Kate Templeton and John Follain, French breast implant ruptures '16 times worse', The Australian, Jan. 09, 2012.

[1004]Jean Yves Henry, Les prothèses mammaires PIP, http://www.medecine-integree.com/les-protheses-mammaires-pip/#.VN0uXCixXrY.

[1005]Alexandria Sage, Natalie Huet, Jean-Francois Rosnoblet, Special Report: The French breast implant scandal, Reuters, Feb. 2, 2012.

monitored". Some of the staff "*kept quiet because they were worried about their jobs*". Some testified: "*Mas would tell them we used the silicone oil for creams, certainly not breast implants… We were very uncomfortable and let Mas do all the talking*". Others were simply not aware of changes in the gel formula. As a result, Mas was able to declare that they had been working with it for decades without any problems.[1006]

Secondly, NuSil, the American gel that was medically approved worldwide, was purchased from the California-based NuSil Technology. The founder of NuSil Technology was Donald McGhan—who also distributed PIP's saline implants in the USA. McGhan had started his career at a Dow Corning laboratory where the first breast implants were made in the early 1960s, but, in 2009, he was given a 10-year jail sentence for wire fraud: he had been illegally using money from clients of a real estate company in an attempt to build yet another implant business.[1007] Jean-Claude Mas and Donald McGhan had a mutual business interest in working together: McGhan distributed PIP's saline implants on the American market and, in return, his company was able to export its silicon gel for decades without fear of possible competition from other medical gel suppliers.

Thirdly, flaws in legislation helped Mas mislead the regulator and the certifying agency. AFSSAPS was the regulator of PIP, but did not certify cosmetic products—which, by the way, are much less strictly certified than pharmaceuticals. Jean-Claude Mas invited TUV Rheinland, a private German company, to certify PIP's factory and implants. In all documents, PIP was demonstrating the use of NuSil, the American medical gel. Because of the absence of legal requirements about unannounced inspections from AFSSAPS and TUV Rheinland, PIP was informed of upcoming checks ten days in advance according to European guidelines. Moreover, during yearly audits, TUV Rheinland did not make any random on-site lab tests of the implants. Mas testified: "*Since 1997, we automatically hid the products that allowed us to make the PIP gel… because I knew they weren't abiding to regulation*".[1008] After the truth was revealed, the French regulator declared that the fraud was so sophisticated that "*it's not evident that an inspection, even an unannounced one, could have been effective*".[1009] Also, PIP was focusing on expansion to the emerging markets of Latin America and Asia, where regulation of the cosmetic industry was weaker than in France.[1010] Regulators in these countries relied on previous validation of the quality of the implants issued by European institutions, and did not carefully test the implants.

In the first few years, the low quality of the PIP implants was not obvious to surgeons throughout the world: statistics about the rupture of PIP's implants

[1006]Alexandria Sage, Natalie Huet, Jean-Francois Rosnoblet, Special Report: The French breast implant scandal, Reuters, Feb. 2, 2012.

[1007]Tom Hals, Exclusive: The troubled history of PIP's implants man in America, Reuters, Jan. 10, 2012.

[1008]Alexandria Sage, Natalie Huet, Jean-Francois Rosnoblet, Special Report: The French breast implant scandal, Reuters, Feb. 2, 2012.

[1009]Ibid.

[1010]Ibid.

became alarming only during years after implantation.[1011] In many emerging countries, such statistics were totally absent. In 2000, the FDA (US Food and Drug Administration) made an inspection of a PIP factory in La Seyne-sur-Mer in Southern France after the regulator had received 1810 reports about problems with PIP's saline implants. Immediately after the inspection, FDA prohibited the sale of PIP's saline implants in the USA, because of evidence of the use of "*adulterated*" material in the production of saline implants.[1012] Nevertheless, there is no confirmation that the FDA shared information about its findings with French colleagues from AFSSAPS.[1013] It was only 10 years later in March 2010, after numerous complaints from French surgeons and customers on the low quality of PIP's silicon implants, that the AFSSAPS made a comprehensive inspection of the PIP factory: the inspectors accidentally found six discarded plastic containers of Silopren industrial gel. It took 17 (!) years from the first attempts of Jean-Claude Mas to conceal what was going on at PIP to reveal a world-wide fraud, which affected more than 300,000 patients!

Dow Corning and other corporations knew that silicon breast implants were porous when they marketed them. In 1994, these manufacturers agreed to pay $4.75 billion to 60,000 affected women.[1014,1015] Throughout the 1980s and 1990s, class-action lawsuits claimed that Dow Corning's silicone breast implants caused systemic health problems. The claims first focused on breast cancer and then drifted to a range of autoimmune diseases, including lupus, rheumatoid arthritis and various neurological problems. This led to numerous lawsuits beginning in 1984 and culminating in a 1998 multibillion-dollar class action settlement. As a result, Dow Corning was in bankruptcy protection for nine years, ending in June 2004 during which time it largely withdrew from clinical market.[1016]

2.4.4 Other Cases with Risk Information Concealment: Tobacco and Food Industries

Misguided and ultimately counterproductive attempts to save money and increase profitability can be seen in other retail industries too. Of course the most

[1011]Dr. Joseph Mele, PIP Silicone Breast Implants—Not in the US, http://www.sanfranciscoplasticsurgeryblog.com/pip-silicone-breast-implants-not-in-the-us.

[1012]Tom Hals, Exclusive: The troubled history of PIP's implants man in America, Reuters, Jan. 10, 2012.

[1013]Dr. Joseph Mele, PIP Silicone Breast Implants—Not in the US. http://www.sanfranciscoplasticsurgeryblog.com/pip-silicone-breast-implants-not-in-the-us.

[1014]http://www.nytimes.com/1997/08/19/us/dow-chemical-deceived-women-on-breast-implants-jury-decides.html.

[1015]http://www.singerpubs.com/ethikos/html/dowcorning.html.

[1016]http://en.wikipedia.org/wiki/Dow_Corning.

notorious, flagrant and—for decades at least—successful attempt to conceal the truth from the public took place within the world tobacco industry, causing up to 6 million deaths per year from smoking worldwide.[1017] In 1992, Judge H. Lee Sarokin, during the tobacco-related case of Haines v. Liggett Group, had this to say: "*All too often in the choice between the physical health of consumers and the financial well-being of business, concealment is chosen over disclosure, sales over safety, and money over morality. Who are these persons who knowingly and secretly decide to put the buying public at risk solely for the purpose of making profits and who believe that illness and death of consumers is an apparent cost of their own prosperity? As the following facts disclose, despite some rising pretenders, the tobacco industry may be the king of concealment and disinformation*".[1018]

Like the tobacco industry, the world food industry has decades of experience in concealing the truth from the public, most notably about the high sugar content of many processed foods. Some examples of concealment within the food industry on a national scale follow. In the middle of the 1980s, because the domestic meat industry was highly profitable, the British Government concealed facts at all stages and even corrupted evidence about the mad cow disease,[1019] which led to more than 200 deaths during the following decades.[1020] In 2013, the meat adulteration scandal in 14 European countries (the undeclared or improperly declared

[1017]Tobacco, The World Health Organization, July 2013.

[1018]R. Kluger, Ashes to Ashes—America's Hundred-Year Cigarette War, the Public Health, and the Unabashed Triumph of Philip Morris, New York, Alfred A. Knopf Publishing House, 1996, p. 676.

[1019]Richard Lacey, Mad Cow Disease: The History of BSE in Britain, Cypsela, 1994, p. xx.

[1020]Variant Creutzfeldt-Jakob disease, The World Health Organization, Feb. 2012.

horse meat) revealed a notable failing in the traceability of the food supply chain, raising fear that other harmful additives could be incorporated as well without the disclosure to the consumers.[1021] In 2006, an outbreak of salmonella occurred in Cadbury's chocolate, affecting up to 37 people, after the existence of the salmonella bacteria at Cadbury Schweppes's factory in Marlbrook (UK) had been concealed for six months.[1022,1023] Milk scandals occurred in China in 2004 and 2008, when it was revealed that companies producing infant formula had reduced its nutritional value—causing 13 baby deaths from malnutrition—and added inexpensive melamine and other compounds such as cyanuric acid, ammeline and ammelide to infant formula to increase their apparent protein content—affecting more than 300,000 babies, of whom 54,000 were hospitalized and six died.[1024,1025] After investigating the case, a World Health Organization (WHO) expert stated: "*It was a large-scale intentional activity to deceive consumers for simple, basic, short-term profits*".[1026]

Retail Production Industry: Why Risks Were Concealed

- **Companies prioritised short-term profitability** and used all means necessary to gain a competitive advantage by launching products as quickly and cheaply as possible, at the expense of the quality of their products and the long-term health and loyalty of customers.
- This happened in some cases as the path of least (short-term) effort to **respond to the pressure from emerging competition or other appearing stressors**.
- In a capitalistic free market system, a narrow view is that firms aim at **maximizing shareholder value and nothing else counts**. In such rational optimization framework, additional considerations involving the physical health of consumers, if not directly impacting the financial well-being of business, will be relegated, ignored or simply negated. Of course, this is a short-term view, but humans tend to be biased towards short-term preferences.

1021 http://en.wikipedia.org/wiki/2013_meat_adulteration_scandal.

1022 Fiona Walsh, Salmonella outbreak costs Cadbury £20 m, The Guardian, Aug. 3, 2006.

1023 Nick Britten, Cadbury's let salmonella get into bars, Telegraph, July 14, 2007.

1024 2008 Chinese milk scandal. http://en.wikipedia.org/wiki/2008_Chinese_milk_scandal.

1025 Chinese protein adulteration. http://en.wikipedia.org/wiki/Chinese_protein_adulteration.

1026 China's Melamine Milk Crisis Creates Crisis Of Confidence, Voice of America, Nov. 1, 2009.

Chapter 3
Causes of Risk Information Concealment

There is not a crime, there is not a dodge,
there is not a trick, there is not a swindle,
there is not a vice which does not live by secrecy

Joseph Pulitzer

That men do not learn very much from the lessons of history
is the most important of all the lessons that history has to
teach

Aldous Huxley

Only a fool learns from his own mistakes. The wise man learns
from the mistakes of others

Otto von Bismarck

In what follows, we identify and synthesize the 30 main factors that compelled organizations and personnel, in previously mentioned cases, to hide risks. This is based on the observation that people have clearly acted in similar ways to conceal risks in different disaster situations, across a very broad range of industry and contexts. Our analysis can be viewed as a significant generalization of the findings of Scott Sagan, in his study of the safety of the US nuclear weapons command organizations[1] in which he provided numerous examples of some detrimental factors to the safety of the operations: he documented serious barriers to learning and improvement due to political infighting, organized deception, normalization of errors, reclassification of failure as success, and conflicts over short-term interests. In a recent book, Eric Schlosser goes further by reporting in details on known accidents with nuclear weapons that have been regularly taking place since 1945.[2] Centering on the 1980 Damascus accident of the explosion in a Titan II Inter-Continental Ballistic Missile housed in Damascus, Schlosser documents a litany of

[1]Scott D. Sagan, The Limits of Safety. Princeton University Press (1995).
[2]Eric Schlosser, Command and Control: Nuclear Weapons, the Damascus Accident, and the Illusion of Safety, Penguin Books; Reprint edition (August 26, 2014).

D. Chernov and D. Sornette, *Man-made Catastrophes and Risk Information Concealment*, DOI 10.1007/978-3-319-24301-6_3

nuclear accidents revealing the past, present and future vulnerability of the exceedingly complicated technical systems that are nuclear weapons, embedded within layers of bureaucracy and subjected to the continuously changing nuclear policies of the politicians. These two studies exemplify the existence of and interactions between the external environment and the internal processes of the organization. The present chapter builds on this distinction to organize the presentation of the main factors promoting crises via information concealment.

In the many examples that we discussed, supplemented by a detailed investigation that follows, we demonstrate that, when human beings distort information about risks before or during a disaster, their behavior is determined not only by personal factors but also by the existence of a flawed organizational environment that promotes secretive patterns of behavior by members of an organization. Underlying the observed fallibility are both the person and the system approaches[3]: "*the person approach focuses on the errors of individuals, blaming them for forgetfulness, inattention, or moral weakness. The system approach concentrates on the conditions under which individuals work and tries to build defenses to avert errors or mitigate their effects... High reliability organisations recognise that human variability is a force to harness in averting errors, but they work hard to focus that variability and are constantly preoccupied with the possibility of failure.*" With this insight, the following areas should be analyzed in order to identify the reasons why organizations and particular employees conceal risks:

- the nature of the external environment surrounding an organization and the incentives that it provides;
- the corporate objectives and strategy of the organization and internal managerial practices;
- the conditions of the internal system for communicating and gathering information about risks within the organization (formal and informal channels);
- the internal practices for managing risk assessment;
- the psychological characteristics of employees within an organization that hides risks.

The figure below presents a synthesis of the main factors catalyzing concealment of information, which led to the previously reviewed disasters. In the following sections, we review in details each of these five factors and delineate the specific conditions and circumstances under which they occur.

[3]James Reason, Human error: models and management. BMJ 320, 768–770 (2000).

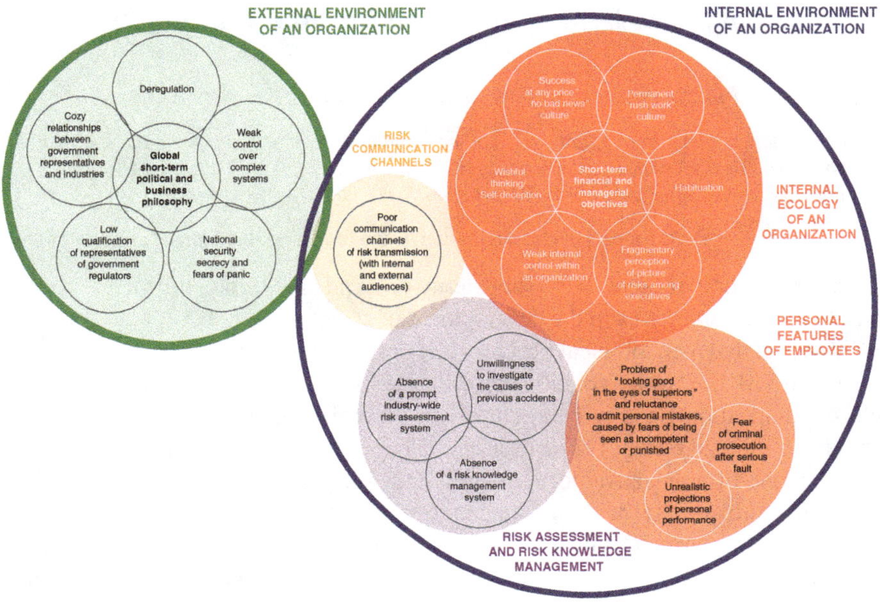

An essential help we can provide managers with is in understanding the reasons that motivated participants of the 25 elaborated cases of risk concealment. Having this information, managers can be prepared and implement specific and unique measures to reduce such practice within their particular organization and environment. More specifically, we recommend that managers use the catalog we provide below from Sects. 3.1 to 3.5 as a kind of checklist to sense whether their organizations may be prone to one or several of these weaknesses.

Strengthened by the knowledge of the 30 main causes of information concealment, an organization can continuously tests for their possible presence and degrees. And the managers together with the staff can design their own solutions and procedures in order to improve the transmission of risk information within their organizations, making them more resilient and sustainable.[4]

3.1 External Environment of an Organization

The fiduciary principle: No Man Can Serve Two Masters

John C. Bogle, Founder and former chief executive of The Vanguard Group

These metaphors [Black swans and perfect storms] have been used as excuses to wait for an accident to happen before taking risk management measures

Elisabeth Paté-Cornell (Risk Analysis, 2012)

[4]Tatyana Kovalenko, Didier Sornette, Dynamical Diagnosis and Solutions for Resilient Natural and Social Systems, Planet@ Risk 1 (1), 7–33 (2013) Davos, Global Risk Forum (GRF) Davos, http://arxiv.org/abs/1211.1949.

3.1.1 Global Short-Term Political and Business Philosophy

(*This occurred in the majority of cases elaborated above*)

> *History shows that where ethics and economics come in conflict, victory is always with economics. Vested interests have never been known to have willingly divested themselves unless there was sufficient force to compel them*
>
> B. R. Ambedkar (Indian Politician and founder of the Indian Constitution)

There are several reasons why short-term political and economic goals prevail over long-term objectives in the development of society.

Firstly, global technological progress had dramatically reduced the length of the organizational life-cycle: some complex systems (power plants, heavy industry, telecommunications, and infrastructure) are becoming outdated and less effective already in the early stages of a project's lifetime, because of momentous global development of innovative technologies and rapid information exchange. The unpredictability of the medium- and long-term future and the limited planning horizon exerted pressure on decision makers to avoid long-term projects whose income could be put into jeopardy by another breakthrough in knowledge and technology.

Secondly, globalization has expanded opportunities for investment and increased competition for investment resources. In this fight, the winners are those companies that can show high profitability with very short turnaround and can keep production costs low. Cost reductions often come at the expense of material quality, of putting safety systems in place to protect employees or of installing purification systems against environmentally harmful emissions. Cost reductions may also increase profitability by pressuring staff to over-work and develop risky behaviors. As a result, managers of many complex facilities are ordered by headquarters to run those facilities at the limits of their performance, while minimizing costs on maintenance and staff compensation. These forces business to focus on continuous cost reduction at the expense of the long-term firm (and national) interests, and to avoid investments in long-term projects due to the unpredictability of global developments. In addition, the increasingly complicated financial situation of many developed countries creates uncertainty about the stable income of citizens, the earnings of corporations and the tax base of governments in the long run, which leaves business reluctant to implement risky long-term investments. Indeed, maximization of short-term revenues involves the existence of an optimization process whose solution is often found at the extreme boundaries of the physical and legal constraints. One can refer to the following general result: as the number of constraints increases, the optimal solution of most optimization problem is found on the boundaries delineated by the constraints. This implies stretching the capacities of the system to the maximum as well as operating at the margin of legality. This is the unavoidable conclusion adopted by any rational utility maximizer.

Thirdly, the existence of competition between different systems at different levels, from individual and organizational emulation to the struggle between nations and political systems, puts pressure on decision makers to direct their subordinates towards achieving noticeable results in a very short period of time in order to demonstrate the superiority of own organization in comparison with others. Such approach is usually performed at the expense of the long-term individual, organizational, or societal objectives.

Fourthly, the dominance of democratic electoral procedures in many countries forces politicians to promise voters that they will achieve visible results in a very short period of time. This has a great impact on their choice of foreground national projects. Thus, short-term high-impact projects—which can immediately revive economic growth and provide jobs—often win out over longer-term strategic initiatives, whose results are only seen over decades. Since politicians often begin to prepare for the next election campaign immediately after the start of a new term, they generally favor populist measures to fulfill the immediate desires of their voters; understandably, they are reluctant to handle the painful and unpopular (but necessary) reforms, or develop the strategic programs, which are important for the survival and resilience of a society. Since few voters want to hear about hard times and decades of hard work, politicians and the economic elite proclaim what voters want to hear—they focus on specific and tangible results in the near future, but keep quiet about the fact that these short-term results may be harmful to the survival of a state in the long term.

Fifthly, fundamental changes have taken place in urbanized society. Indeed, during a long active life, a person may graduate from several educational institutions, practice a number of professions, live and work in a variety of countries and have a succession of partners. These changes do not promote the development of long-term relationship skills, but rather encourage individuals to maximize their own benefit in a limited period of time—sometimes even consciously harming and exploiting other people who they may not need to meet again.

3.1.2 Deregulation

(Barings bank collapse|Enron bankruptcy|Subprime mortgage crisis|Great wildfires in the European part of Russia|Deepwater Horizon oil spill)

Before the collapse of the Soviet Union and of the communist ideology in 1991, the socialist and capitalist systems competed with one another, and each was aware that any reckless efforts or shortsighted decisions could weaken their own system and give a competitive advantage to the other. Thus, both players tried to conduct a balanced long-term policy; long-term objectives for any society include its continuing survival with current borders without violation of its sovereignty, and peaceful, safe and prosperous development over centuries.

When capitalism and liberal values triumphed in the competition with socialism, the victor of this battle lost a crucial stabilizing factor: the opposition of socialism had previously limited the unbridled development of capitalism. In addition, those who wished to limit the power of government to protect the long-term interests of society, by keeping a controlling hand on private business, had a perfect argument: that the collapse of socialism had been caused by the excessive power of government in modern society. Politicians—financed by private business—tried to convince voters that reducing the power of government was the call of the times: reduction of budget spending on bureaucrats could release resources for social programs; without government interference, private business could bring faster growth, create new jobs and contribute to the public good through increased tax revenues.

However, the catalogue of industrial and business disasters we have described above shows that, in the absence of regulators, the management of private industries—with the approval of shareholders—prefer to focus on short-term financial performance, local growth of capitalization, and high annual bonuses at the expense of long-term cost reduction and sustainability. This ultimately destroys the tax base to the detriment of national interests in the long run. The expansion of globalization—the free flow of capital and goods between countries—was accompanied by a worldwide trend towards the thoughtless deregulation of markets, often causing serious corruption of political systems and weakening national governments, whose job is to represent and act on the opinion (and for the benefit) of the vast majority of the people, and not on that of a small fraction of the population in the form of big business.

In addition, because governments want to gain a competitive advantage over other countries through innovative development, they often encourage, permit the development or adopt new technologies without a proper assessment of the potential impacts of these technologies on society in the long term, and without a rigid system of government control over their testing and implementation. For example, deregulation of deepwater oil field drilling has raised oil production in the United States—but the uncontrolled growth of the sector led to the largest maritime oil spill in the world, caused by an avoidable technological accident. Deregulation of the derivatives market gave the United States significant economic growth for decades—but ultimately led to huge losses of economic output and financial wealth for many states in the world, including the US itself.[5]

Government interest in the implementation of advanced but risky technologies, and the desire to accelerate economic growth by developing them, leads to a weakening state control over industries and a climate of impunity for executives who use or authorize dangerous working practices. This policy is never officially commented upon but, informally, executives get in fact a carte blanche from the authorities to ignore risks and conduct high-risk operations, the true details of which are disclosed neither to regulators nor to employees or external

[5]Tyler Atkinson, David Luttrell and Harvey Rosenblum, How Bad Was It? The Costs and Consequences of the 2007–2009 Financial Crisis, Staff Papers is published by the Federal Reserve Bank of Dallas, No. 20, 1–22, July 2013.

audiences—investors, contractors, and so on. In fact, government officials seem to be saying to industry executives: "*We are interested in immediate results, so that the implementation of dangerous technologies will be approved by voters in spite of possible long-term losses, which other politicians and generations will have to deal with*".

In Chap. 4, where on-going cases of risk concealment are presented, there are two innovative industries, which now grow in such an environment of weak government control: the development of shale gas and oil production and the use of genetically modified organisms (GMOs). The technology of gas and oil extraction from shale formations requires the use of either water or propane gel but, since water is far cheaper than propane gel, the shale hydrocarbon companies—with the tacit approval of the US government—began to pump billions of cubic meters of water mixed with harmful chemicals into the ground. The USA intensified domestic gas and oil production—reviving economic growth, reducing the cost of electricity and allowing government to declare the imminent achievement of US energy independence from imported hydrocarbons. However, from the perspective of long-term national interests and possible damage to the nation's health, there is no comprehensive research to evaluate the damage caused by polluting huge underground water resources—damage which, although it did not deter all parties from going ahead with shale energy extraction, should at least be added to the declared production costs. A similar approach can be recognized during the promotion of GMOs: the aspiration of some governments to increase their own agricultural output as quickly as possible is pushing aside the barriers against genetically modified crops without much considerations for the safety of this type of crop for the health of the present and future generations, which has not yet been clearly confirmed by the world scientific community.

3.1.3 Cozy Relationships Between Government Representatives and Representatives of Industries

(*Vajont Dam disaster|Exxon Valdez oil spill|Challenger Shuttle disaster|Chernobyl disaster|Raspadskaya coal mine burnout|Deepwater Horizon oil spill|Fukushima-Daiichi nuclear disaster|Barings bank collapse|Enron bankruptcy|Subprime mortgage crisis*)

Active cooperation between governments and private industries in the development of state economic policy, the widespread practice of getting a job in industries after a career in government, legitimate corporate financing of election campaigns and sometimes outright corruption—all these contribute to a convergence of interests between the political and business elites. After business has poured vast sums of money into cultivating and supporting politicians, representatives of government all too readily agree to promote the deregulation of markets and industries. The experience of the disasters elaborated above shows that corporations, which had actively lobbied for deregulation measures and weakened public control over their activity, lost powerful and objective external controllers. Before deregulation, such controllers could prevent the implementation of risky and reckless management decisions by strict legislation of industries, and through continuous external control of common business practice by highly skilled government representatives. However, business in general chooses to get rid of independent regulators—and in some cases such activity has led to catastrophic consequences. The Exxon Valdez and Deepwater Horizon oil spills, the Subprime mortgage crisis and the Fukushima-Daiichi nuclear disaster did not just cause public outcry: even owners, executives and investors obviously regretted that they had persistently supported previous deregulation initiatives within the industries. The chastening experience of disasters like these shows that organizations working with dangerous technologies could, after all, have an interest in maintaining the existence of strong, independent regulators, who may be annoying in the everyday operations but could help those organizations to mitigate risks at the early stages.

3.1.4 Low Qualification and Unattractive Wages of Representatives of Government Regulators

(*Bhopal disaster|Exxon Valdez oil spill|Deepwater Horizon oil spill|Enron bankruptcy|Subprime mortgage crisis*)

Under the pretext of cutting costs on bureaucracy—and with determined lobbying from private industries—politicians, often with the support of voters, cut the salaries of government representatives and the budgets of public-sector organizations. As a result, regulators cannot hire highly educated and experienced staff on their limited budget. Moreover, pay levels in private industry are several times higher than for government jobs, which lead to a decline in the prestige of public service and an absence of skilled government officials. In a confrontation with executives who have a strong interest in pulling the wool over their eyes, representatives of the regulators do not have sufficient education and experience to allow them to understand new technologies being implemented in the industries they are overseeing, and to identify major risks that these industries want to hide from regulators and the public. Moreover, the loyalty of regulatory representatives can either be bought by private business—through employment guarantees after the end of the government job, or just simple bribery—or extorted by threats of dismissal following a word in the ear of the right government official.

3.1.5 Weak Control Over Complex Systems

(*Barings bank collapse|Enron bankruptcy|Subprime mortgage crisis*)

The deregulation of industries and subsequent fierce competition push many organizations towards the unification of previously competing businesses, in order to withstand competitive pressure by means of vertical and horizontal integration of businesses, and cross-subsidization within the new larger organization. As a result, continuous mergers and acquisitions establish complex and large-scale systems; and the risks from the activities of these giants are not always obvious even to their top executives because the multi-level transmission of information, and the variety of businesses, have become too complicated. Moreover, the speed of mergers stays ahead of changes in the government regulatory framework: the new corporate giants are still regulated by obsolete and uncoordinated regulatory measures, with no interaction between different regulators. In other words, the development of gigantic multi-industry mergers is not matched by the parallel development of a "mega-regulator". In addition, there is weak coordination between regulators from different countries: there is in general little or no exchange of information about the activity of transnational corporations, or sharing of best practice [exceptions include the World Health Organization (WHO) and the International Atomic Energy Agency (IAEA)]. Consequently, both regulators and executives have a fragmented picture of risks, which does not allow them to understand all the risks associated with such large and complex systems.

3.1.6 Political Instability and Struggle Between Political Camps

(*Vajont Dam disaster|Chernobyl disaster|Fukushima-Daiichi nuclear disaster|Enron business and Californian electricity crisis|Hurricane Katrina disaster*)

The practice of crisis management shows that the magnitude of a crisis, and the quality of the crisis response, can be influenced by political instability in a country and the permanent confrontation of political forces. The fact is that industries sponsor certain political parties, which in return lobby for the interests of these industries. In a critical situation, the damaged industry then becomes a hostage of the political confrontation, obtaining support only from the political parties they have supported and receiving blame and (often unfounded) criticism from opposing political camps. The struggle between political factions leads to deliberate distortion of information among decision makers concerning the real scale of a disaster and its possible consequences, in order to use the disaster to achieve a political gain.

3.1.7 National Arrogance

(*Chernobyl disaster|Deepwater Horizon oil spill|Fukushima-Daiichi nuclear disaster|Toyota pedal crisis*)

This problem was revealed in four disastrous cases. The brightest one is the Fukushima-Daiichi nuclear disaster, when Japanese demonstrated an example of national self-delusion. The Japanese nuclear industry arrogantly refused to learn from the experience of other countries, which had previously faced several accidents at nuclear power plants (NPPs). Japanese nuclear scientists believed so fervently in their country's technological superiority over the rest of the world that they even began to falsify data about minor shortcomings at plants, in the belief that minor defects would always be compensated by the Japanese attitude towards work. Ignoring global trends in the continuous updating of NPP security systems left Japanese plants far behind global best practices. Previously, in a similar way, the lessons of the Three Mile Island accident were not learned by Soviet nuclear specialists. In the case of the Deepwater Horizon oil spill, both US regulators and American oil companies—which saw themselves as the pioneers of shelf oil extraction—also demonstrated an unwillingness to follow international experience in the sphere of deepwater drilling safety. After decades of trouble-free deepwater drilling, they believed a major accident on an American platform was impossible—even though Mexican, Canadian, Norwegian and British companies had all faced such accidents.

Ultimately, national arrogance—a worldwide phenomenon, of course—leads to a situation where nobody uses a "learn-from-history" approach. When assessing the probability of accidents in hazardous industries, managers generally ignore previous accidents in their own country, and also pay no attention to the statistics of international incidents. Regulators and corporations rarely develop a database of dangerous industrial events worldwide,[6] and have weak exchange procedures with international colleagues. There is indeed little correlation between national safety standards and international cutting-edge legislation created after severe disasters. There is no common practice of visiting related accident sites around the world. There are few international industrial conferences or translations of foreign internal reports on the causes of events.

[6]Exceptions include the Energy-related Severe Accident Database (ENSAD) (see http://www.psi.ch/ta/risk-assessment) at the Paul Scherrer Institute, ETH, Switzerland. ENSAD currently comprises 32705 accident records. Of these 83.2 % are classified as man-made, 16.3 % as natural disasters, and 0.5 % as conflicts. Among man-made accidents, 20245 are attributable to the energy sector, and of these 93.8 % occurred in the years 1970–2008 (see Burgherr, P. and Hirschberg, S., Comparative risk assessment of severe accidents in the energy sector, Energy Policy http://dx.doi.org/10.1016/j.enpol.2014.01.035i, 2014). Another example is the Energy Infrastructure Attack Database (EIAD), a large dataset that uniquely categorizes reported incidents where non-state actors target energy infrastructure. It was jointly developed by the Crisis & Risk Network (CRN) research group at the Center for Security Studies (CSS) at ETH Zurich and the Technology Assessment (TA) Group of the Laboratory for Energy Systems (LEA) at the Paul Scherrer Institute (PSI). See Center for Security Studies (CSS), ETH Zurich (2012). Energy Infrastructure Attack Database (EIAD). http://www.css.ethz.ch/research/research_projects/index/EIAD.

3.1.8 Fear of Massive Panic

(Three Mile Island nuclear accident|Chernobyl disaster|Fukushima-Daiichi nuclear disaster|Worldwide Spanish flu and SARS outbreaks)

Radiation is the most dangerous threat in the public perception. Therefore, in three civil nuclear accidents, the practice of dosing the information about the events was adopted in order to avoid panic among the civilian population, on the rational that the casualties resulting from hysteria and irrational self-evacuation might exceed mortality stemming directly from the accident. The opposite unintended effect occurred as a result of the uncertainty created by such crisis management.

In the case of the Three Mile Island accident, 17 people were killed in the nearby city, while, at the time of the accident, no one died or was irradiated. Locals rushed to their cars to run away from the neighborhood and, in the process, 17 people died in car crashes.[7] In the case of Chernobyl, "the dosing of information" led to a massive falsification of the real radiation data in the first months after the disaster and engendered a strong distrust of the public with respect to the Politburo's statements. Consequently, the death rate from psychosomatic disorders exceeded that resulting from radiation.[8] The absence of reliable information emanating from the Japanese government about the conditions of the reactors on the Fukushima-Daiichi plant during the first weeks after the disaster ended up provoking terror in Japan and on the Far East of Russia, leading to hundreds of thousands of people leaving their homes and thousands to fly abroad. As a result of the trauma of the evacuation, approximately 1000 Japanese died,[9] while no loss resulted from radiation during the first weeks after the accident on the Fukushima-Daiichi plant.

3.1.9 National Security Secrecy

(Worldwide Spanish Flu and SARS outbreaks|Unreadiness of the Soviet Army for the Nazi invasion|Challenger Shuttle disaster|Chernobyl disaster)

National security requirements impose stringent regulations on the disclosure of information within an organization, and to civil authorities and the public. Typically, such information is provided only in an abbreviated or completely

---

[7]Valery Legasov, Interview to Ales Adamovich, Record from cassette #5, 1986–1988.

[8]25 years of the Chernobyl accident (1986–2011). Results and Prospects overcoming its consequences in Russia, Ministry for Civil Defense, Emergencies and Disaster Management of the Russian Federation, Moscow, 2011, p. 20.

[9]Fukushima Accident, World Nuclear Association, Updated February 2015, http://www.world-nuclear.org/info/Safety-and-Security/Safety-of-Plants/Fukushima-Accident/.

distorted form in order to mislead a potential enemy. Ironically, this secrecy can lead to information on risks not reaching key decision makers in time, because it is too difficult to quickly transmit emergency information through special communication channels and across different levels of a secrecy-prone organization.

3.2 Internal Ecology of an Organization

The very fact that knowledge is itself the basis for civilization points directly to openness as the way to overcome the present crisis

Niels Bohr, 1950

3.2.1 Short-Term Financial and Managerial Objectives and Unrealistic Projections of Future Development

(*This occurred in the majority of cases elaborated*)

The focus of investors on quick turnover projects, the volatility of a globalized world market, the orientation of politicians towards achieving a quick fix to please voters: all these factors lead to a tendency for companies to develop short-term strategies. The objectives of shareholders and politicians to achieve short-term business development—and thus economic progress through revenue growth and tax payments—are transformed into ambitious and often unrealistic business strategies. Building on these strategies, companies develop operational plans that create enormous psychological stress among personnel, forcing them to ignore risks to achieve unrealistic results, and to distort information about risks in order to prove their competence in the eyes of superiors: otherwise, they could be blamed by management for being unable to handle a challenging environment and subsequently dismissed. Those who disagree with such practice, as a rule, do not stay in the organization for long. Because of short-term goals and unrealistic plans, managers promote a risk-taking approach, the organization operates facilities close to their ultimate limits while at the same time trying to save on costs and labor… and the whole situation is often heading for disaster.

In addition, a major component of managerial compensation all around the world is the awarding of annual bonuses, which plays an important role in the implementation of short-term business strategies precisely because it motivates executives to show short-term results by any means. Obviously, the experience of major business and industrial disasters shows that managers should rather be rewarded according to an organization's progress over at least five years, and—for industries with a longer turn-around—after ten years of successful and stable operation.

3.2.2 Permanent "Rush Work" Culture

(*Challenger Shuttle disaster|Chernobyl disaster (in aspect of RBMK development)|Ufa train disaster (during construction of NGL pipeline)|Sayano-Shushenskaya hydropower station (on construction phase)|Deepwater Horizon oil spill|Apple iPhone 4 antenna|Intel Pentium FDIV bug crisis|Boeing 787 Dreamliner lithium-ion batteries|Sukhoi Superjet*)

Unrealistic targets, set under pressure to achieve short-term results or to compete in the market, force managers to demand constant haste from their subordinates in implementing new projects. Because of the perpetual rush, subordinates do not have time to sit down with other departments and external contractors to fully evaluate potential risks, so they have to rely on their own experience to assess risks, without alerting other units about possible threats. If an employee decides to conduct a comprehensive risk assessment and this assessment does not ultimately confirm their fears, they would be blamed by superiors and colleagues for delaying the schedule and wasting company money and time. Moreover, the rush to implement projects requires compromises with the quality of project work: experience shows that the production of half-baked and barely tested solutions motivates an organization to conceal information about shortcomings, which if widely known would cause the outrage of customers and regulators, and force the company to recall the flawed product.

3.2.3 "Success at Any Price" and "No Bad News" Culture

(*This occurred in the majority of cases elaborated*)

Unrealistic plans and the desire to increase productivity within an organization compel managers to shift the responsibility for implementing plans onto their subordinates, promoting the principle of "*no bad news*" by encouraging only those who produce tangible short-term results. This means that employees have to find "*their own solution*", and take the initiative without bothering managers. This approach creates a corporate culture where employees are afraid of layoffs, afraid of being publicly accused of incompetence if they fail, and under pressure to set unrealistic goals. Such a corporate culture obliges employees to distort information about their own success, even to the extent of falsification and fraud, to delay informing managers about risks and shortcomings until there is no choice, to reject personal guilt and to tell managers only what they want to hear.

Prof. Leveson goes further in analyzing the negative impact of blaming and punishing as follows: "*Blame is the enemy of safety. Focus should be on understanding how the system behavior as a whole contributed to the loss and not on who or what to blame for it... A culture of blame creates a climate of fear that makes people reluctant to share information. It also hampers the potential to learn from incidents; people may even tamper with safety recording devices, turning them off, for example. A culture of blame interferes with regulatory work and the investigation of accidents because people and organizations are less willing to cooperate ... In such cultures, risk assessment is unrealistic and credible warnings are dismissed without appropriate action. Management only wants to hear good news and may ensure that is what they hear by punishing bad news, sometimes in a subtle way and other times not so subtly ... The mistake and any harm from it should be acknowledged, but the response should be to lay out the opportunities for reducing such mistakes by everyone (not just this particular person), and the responsibilities for making changes so that the probability of it happening again is reduced. This approach allows people and organizations to move forward to prevent mistakes in the future and not just focus on punishing past behavior... Punishment is usually not a long-term deterrent for mistakes if the system in which the person operates has not changed the reason for the mistake ... Overcoming our cultural bias to punish people for their mistakes and the common belief that punishment is the only way to change behavior can be very difficult. But the payoff is enormous if we want to significantly reduce accident rates. Trust is a critical requirement for encouraging people to share their mistakes and safety problems with others so something can be done before major losses occur".*[10]

3.2.4 "Ivory Tower Syndrome" or Fragmentary Perception of the Whole Picture of Risks Among Top Managers

(*This occurred in the majority of cases elaborated*)

The complexity of modern technologies, the constant process of mergers and acquisitions, and the selection of top management not predominantly on the basis of industry knowledge and experience, but on the basis of financial management skills—their ability to reduce costs, increase profits and enhance the company's position in the market—all lead to companies being headed by executives who have little understanding of the complexity of the organization and even less grasp of the whole picture of risks involved with that business or industry. There have been instances where management teams were replaced at hazardous industrial sites: instead of skilled managers with technical background and experience in industry, shareholders put in their place financiers and economists who considered the plant they were supposed to be managing not as a complex socio-technical system, but only as a means of generating profits, where financial security was more important than safety and the interests of the broader society. The imperative of maximizing income alone does not require top management to learn the details of a business—so the people in charge of potentially dangerous projects or facilities may have only a fragmented perception of the real situation in an organization, or an unbalanced picture of the condition of its different units.

[10]Nancy G. Leveson, Engineering a Safer World: Systems Thinking Applied to Safety, MIT Press, 2011, pp. 56–57, 428, 432.

3.2.5 *Absence of Specific Knowledge and Experience Among Members of Boards of Directors*

(*Enron bankruptcy|Lehman Brothers bankruptcy|Subprime mortgage crisis*)

In order to convince investors that a company is in good hands, major shareholders and executives invite revered and well-known people from various industries onto the board of directors—people who often lack specific knowledge and experience of company business. Most of them immediately get an impressive compensation package, which sometimes is linked to the share value of the company. This makes board members financially dependent on the current state of the company and motivates them not to ask tricky questions, but to trust information from managers about what they have done to increase the profitability of the company and the performance of its shares. In fact, such a board of directors becomes unable to perform its primary function—to control top managers of a company. Rather than being a restraining hand, an incompetent and financially interested board may even provoke managers to take risks, eventually leading to disaster.

3.2.6 *Weak Internal Control Within an Organization*

(*Barings bank collapse|Enron bankruptcy|Lehman Brothers bankruptcy|Subprime mortgage crisis|Chernobyl disaster|Bhopal disaster|Fukushima-Daiichi nuclear disaster*)

To executives under pressure to achieve impressive results in a short time, the weakening of internal control seems to be in their best interest. A comprehensive, highly professional and independent control department, which collects information about all activities of both staff and managers and produces impartial assessments, constitutes a dangerous witness that can be exploited by regulators and government investigators in the event of disaster. Therefore, it is not surprising that, in many of the cases we have investigated, internal regulatory departments were either eliminated or made up of incompetent and corrupted employees who could not—or would not—carry out their duties properly.

3.2.7 *Frequent Labor Turnover*

(*Bhopal disaster|Enron bankruptcy|Subprime mortgage crisis*)

A short-term strategy to maximize revenue and reduce costs dictates significant wage reductions, encouraging the most competent staff to look elsewhere. This practice leads to a loss of knowledge about complicated or risky aspects of the organization's work. On complex technological sites, such a loss of expertise is not acceptable and always leads to accidents. However, in the financial sector—for example at Enron and in the companies most implicated in the subprime mortgage bubble—a high staff turnover was deliberately encouraged in order not to allow employees to understand the true scale of the risks the organizations were taking. These companies mostly selected young people with little experience—people who were loyal, ambitious and willing to take cynical actions to achieve results at any cost on the short-term.

3.2.8 Habituation (Loss of Fresh Vision on Problems and Risks Because Nothing Has Gone Wrong in the Past)

(*This occurred in the majority of cases elaborated*)

When a managerial team has been running an organization or an industrial facility without accidents for many years, people often become complacent, creating an environment where risks are not taken seriously anymore. Among the managers of such projects, a permeating sentiment of confidence is progressively generated that the plant (or the market, or the world economy…) is fundamentally reliable, and that even unusual deviations in their work will never actually lead to a catastrophe. Therefore, identified operational risks are not transmitted within the organization anymore or subjected to a comprehensive assessment. If a risk is judged to be insignificant by only a single manager, the matter will go no further—although a more appropriate response strategy would be for risks to be assessed by a pool of external experts with diverse experience and knowledge.

Some statistics illustrate this point. Over 12 years of oil transportation through Prince William Sound, there was no major oil spill until the Exxon Valdez case. BP has been involved in deepwater drilling for more than 20 years and never encountered a huge oil spill until the Deepwater Horizon case. The Sayano-Shuschenskaya hydropower plant generated electricity without major incidents for more than 30 years. The Fukushima-1 nuclear power plants worked for 40 years before the largest technological accident in Japan's history occurred. Barings bank faced major damage from securities trading and lived with the risk of default for about 100 years before its complete collapse in 1995—and by this time the bank had been operating for over 230 (!) years. And the list can go on and on. These examples illustrate the fallacious conception that, because an entity was (apparently) sound yesterday, it will be so today and therefore tomorrow. As a consequence, operational disruptions and crises come as a surprise to the naïve and unprepared. But any structure or system on Earth is born one day, then develops, flourishes and eventually exits in one way or another. The Center Research on Security Prices (CRSP)[11] of the Booth School of Business at the University of

[11]http://www.crsp.com.

Chicago. CRSP maintains one of the largest and most comprehensive proprietary historical database in stock market research. In particular, the database on US firms from 1927 to present includes close to 30,000 firms, most of which are now either defunct or have been acquired or have merged. One discovers in this extensive database that the half-life of a typical firm is less than 14 years. We tend to think of large firms as part of our industrial and commercial landscape but the reality is that they live typically much shorter than humans. Remember the big banks in 2007–2008, General Motors and many others, which were once thought to be there for the long-term… until some development proved otherwise.

Moreover, the phenomena of risk compensation and risk reflexivity describe the fact that people tend to take more risks when they feel more protected, in safer environment, or when they are cognitively unaware of the real risks. By their behavior, they actually invite more risks, thus leading to lower benefits than expected and sometimes to the exact opposite of the intended goal of risk reduction programs.[12]

Another important prejudice is that the absence of catastrophes for many years leads to reductions in expenses on risk mitigation. For instance, Professor N. Leveson from MIT documented this practice within NASA: "*Shuttle management also had a belief that less safety, reliability, and quality assurance activity would be required during routine Shuttle operations [when number of launches exceeded several dozens flights]. Therefore, after the successful completion of the orbital test phase and the declaration of the Shuttle as 'operational', several safety, reliability, and quality assurance groups were reorganized and reduced in size. Some safety panels, which were providing safety review, went out of existence entirely or were merged… William Readdy, head of the NASA Manned Space Program, for example, in 2001 wrote that 'The safety of the Space Shuttle has been dramatically improved by reducing risk by more than a factor of five'. It is difficult to imagine where this number came from as safety upgrades and improvements had been deferred while, at the same time, the infrastructure continued to erode. The unrealistic risk assessment was also reflected in the 1995 Kraft report, which concluded that 'the Shuttle is a mature and reliable system, about as safe as today's technology will provide'. A recommendation of the Kraft report was that NASA should "restructure and reduce overall safety, reliability, and quality assurance elements*".[13] In 2003, Columbia disaster occurred. Professor N. Leveson also concluded: "*Often, ironically, our successful efforts to eliminate or reduce accidents contribute to the march toward higher risk. Perception of the risk associated with an activity often decreases over a period of time when no losses occur even though the real risk has not changed at all. This misperception leads to reducing the very factors that are preventing accidents*

[12]Adams, John, Risk. Routledge, London/New York (1995).

[13]Nancy G. Leveson, MIT, Technical and Managerial Factors in the NASA Challenger and Columbia Losses: Looking Forward to the Future, published within Kleinman, Cloud-Hansen, Matta, and Handelsman, Controveries in Science and Technology Vol. 2, Mary Ann Liebert Press, 2008.

because they are seen as no longer needed and available to trade off with other needs. The result is that risk increases until a major loss occurs … In the absence of accurate information about the state of the process, risk perception may be reevaluated downward as time passes without an accident. In fact, risk probably has not changed, only our perception of it. In this trap, risk is assumed to be reflected by a lack of accidents or incidents and not by the state of the safety control structure".[14]

3.2.9 Wishful Thinking/Self-Suggestion/Self-Deception Among Decision Makers

(*This occurred in the majority of cases elaborated*)

Self-deception among those receiving information about risks is one of the main obstacles to quickly identifying risks, assessing them and communicating them to others. Instead of looking at a situation realistically, studying the facts, searching for primary sources and independently assessing information, people tend to convince themselves of what they want to believe. Instead of maintaining a skeptical and critical attitude towards information they receive, the majority of people prefer to trust it unconditionally. Moreover, a group mechanism starts to work when people are trying to convince each other of what they all want to believe. Such behavior contributes to the formation of bubbles on financial markets—when a market does not grow in response to objective factors, but to a widespread euphoria about a fictional future—and to the unfounded confidence of executives in the loyalty of their subordinates, or the naive reliance of subordinates on their executives. Wishful thinking is a sure path to the inadequate perception of reality, and consequently to inadequate action, or inaction, towards essential changes.

3.2.10 The Remoteness of Units/Facilities

(*Bhopal disaster|Exxon Valdez oil spill|Barings bank collapse|SARS outbreak|Sayano-Shushenskaya hydropower station*)

Due to their remoteness from headquarters, some units are not always governed under the same organizational standards as the others. Especially if these remote units are in addition unprofitable, or bring insignificant revenue compared with other parts of the organization located geographically closer to headquarters, then executives will tend to lose interest in what is going on there. Usually, this creates

[14]Nancy G. Leveson, Engineering a Safer World: Systems Thinking Applied to Safety, MIT Press, 2011, p. 419, 423.

the illusion that "no news is good news": if the managers of a distant outpost have not said anything to headquarters about their actions, then everything must be going fine there and it is not worth wasting time and effort on regular and accurate updates. In spite of the rapid development of modern communication, geographical remoteness still has a negative impact on how quickly and thoroughly information about risks is passed on. Experience has shown that lack of control over remote locations, to the extent of almost ignoring their existence, can cause a crisis that takes top management by surprise.

3.3 Risk Communication Channels

3.3.1 *Long Chains of Communication for Risk Information. Absence of a Direct, Urgent 24-7-365 Channel Between Field Staff and Executives. Field Staff who do not have Authority to Immediately Stop a Process if they Suspect Evidence of Risk*

(*Ufa train disaster|Fukushima-Daiichi nuclear disaster|Chernobyl disaster|Challenger Shuttle disaster*)

In many pre-crisis situations that developed into disasters, emergency communication between facility operators and executives was difficult. Even when there are clear signs that something may be going wrong, getting emergency powers to shut down operations often involves a long chain of confirmations. In many organizations that produce goods or provide services 24-7-365, suddenly stopping the process would automatically violate supply agreements and may incur damage compensation for the shutdown of other dependent facilities downstream in the supply chain. Therefore, operators often do not have the authority to stop facilities preventively. Moreover, the procedure for making such a decision is very complicated. Generally, the fact that local staff is not authorized to make independent decisions in a difficult situation is one of the main reasons why severe accidents are not prevented as soon as there are alarming signs.

Synthetizing the experience obtained from many disasters, Prof. Leveson also concluded: "*Decentralized decision making is, of course, required in some time-critical situations. But like all safety-critical decision making, the decentralized decisions must be made in the context of system-level information and from a total systems perspective in order to be effective in reducing accidents... Formal methods of operation and strict hierarchies can limit communication. When information is passed up hierarchies, it may be distorted, depending on the interests of managers and the way they interpret the information. Concerns about safety may even be completely silenced as it passes up the chain of command. Employees may not feel comfortable going around a superior who does not respond to their concerns. The result may be a misperception of risk, leading to inadequate control actions to enforce the safety constraints*".[15]

3.3.2 No Internal or External Incentives for Whistleblowers

(*Challenger Shuttle disaster|Chernobyl disaster|Exxon Valdez oil spill|Fukushima-Daiichi nuclear disaster|Enron bankruptcy*)

Only a few countries have comprehensive witness protection programs for whistleblowers—active citizens who are ready to inform regulators and the public about violations of laws or safety regulations. Usually, such programs are part of independent government investigating agencies, with the power to suspend the work of any organization that systematically conceals risks and breaks the law. However, the analysis of multiple disasters reveals the existence of cozy ties between regulators and industry executives, who are equally interested in the concealment of risky working practice in industry. Since regulators have often explicit or implicit instructions from politicians to turn a blind eye to risks in order to maintain economic growth, whistleblowers are a threat not just to industries, but to the regulators who tacitly approve of dangerous working practices. In addition, the majority of state legal systems do not encourage disclosure of information by whistleblowers; and employees who are known to be active within their trade union or their local community are often regarded as dangerous within their company, and within the industry as a whole. After they have put their necks on the line by disclosing risk information, it becomes impossible for whistleblowers to pursue their professional careers within their previous industry and/or to obtain wages that are no lower than in their previous jobs. Therefore, most employees are forced to keep silent about risky practices at work, to avoid being subjected to condemnation or even ostracism by colleagues and losing their income. Experience shows that most whistleblowers who revealed risks within their organizations did so solely out of personal moral principles—but unfortunately, in spite of public approval, their careers were ruined and they paid a high price in their personal and professional lives.

Only scientists and operational staff with deep knowledge coming from many years experience of the exploitation of their objects can transmit appropriately and adequately information on the risks of their objects to their superiors. We have seen in the cases of Chernobyl, for instance, that the presence of KGB agents did not help at all, only in adding to the confusion of the Politburo. This strongly bolsters the need for executives to provide sufficient stimulus and incentives to their qualified subordinates to share information freely.

[15]Nancy G. Leveson, Engineering a Safer World: Systems Thinking Applied to Safety, MIT Press, 2011, pp. 44, 425.

3.3.3 Poor Inter-organization Risk Transmission

(*Three Mile Island Nuclear Accident|Bhopal disaster|Chernobyl disaster|Ufa train disaster|Deepwater Horizon oil spill|Fukushima-Daiichi nuclear disaster|Great wildfires in the European part of Russia|Hurricane Katrina disaster*)

The analysis of major accidents shows that the involved organizations had insufficient mutual exchanges of risk-related information with other institutions such as contractors, representatives of other critical infrastructure objects, local authorities, police, fire and medical services, local military units, and so on. As a result, when it came to a crisis, these organizations did not understand the risks born by each of the involved structures, they had no idea about the real severity of the accident in absence of suitable assessments of existed risks prior to the disaster, and no one had adequate infrastructure or trained personnel for the unexpected scale of the disaster. For example, emergency services officers in Bhopal understood neither the level of the public threat from an accident at the chemical plant nor its nature, so when tragedy struck, no one knew which chemical compounds were being released or about possible antidotes or simple recommendations for the local population. And, in the case of Chernobyl, firefighters arriving at the station after the initial incident did not know that feeding water into the burning reactor was prohibited, due to the danger associated with the resulting mixture of water, uranium and graphite. Later, it would take great effort and the loss of further lives to pump out the large injected amounts of water from beneath the reactor. Before the Ufa train disaster, the railway dispatchers did not understand the fatal threat from the poorly maintained nearby NGL pipeline. In the case of the Macondo deepwater oil well, the service contactor did not communicate about the failure of the cement tests to BP, its client. Representatives of TEPCO did not asked help from local military units, notwithstanding the fact that these units had available experienced staff that could have organized a prompt organization of the reactors cooling right after the tsunami. And so on.

3.3.4 Absence of Direct Horizontal Communication Between Departments of an Organization (Communication Between Units Only Occurs Through Superiors)

(*Three Mile Island Nuclear Accident|Bhopal disaster|Chernobyl disaster|Piper Alpha platform disaster|Sayano-Shushenskaya hydropower station accident*)

There is a general problem of communication between units that are running similar projects or facilities. In most cases, they only communicate via managers at headquarters, so there is no exchange of experience about risks—and even when something goes wrong at one unit, managers at others do not find out in detail what caused the incident. In the case of Bhopal, information about the causes of several accidents at Union Carbide Corporation plants in the USA were withheld from managers of the company's Indian division (and probably from executives at other factories scattered around the world). The same practice was common in the Soviet power industry: after accidents at nuclear and hydro stations, staff at other stations would not generally be informed—which left them prone to repeating the same mistakes.

Generally, managers are making a mistake with possible dramatic consequences if they omit to develop a unified database of near-miss cases and of any accidents occurring within an organization/industry (which should include a detailed elaboration of their causes) and to give free access to it for executive staff of all units and departments. The silo approach, where risk may be monitored in each individual division but not consolidated globally, allows the uncontrolled maturation of the overall risk.

3.4 Risk Assessment and Risk Knowledge Management

It is not what we know, but what we do not know which we must always address,

to avoid major failures, catastrophes and panics

Richard Feynman

Most of the greatest evils that man has inflicted upon man have come through people feeling quite certain about something which, in fact, was false

Bertrand Russell

3.4.1 Absence of a Prompt Industry-Wide Risk Assessment System

(Three Mile Island Nuclear Accident|Chernobyl disaster|Bhopal disaster|Fukushima-Daiichi nuclear disaster|Deepwater Horizon oil spill|Krymsk flooding)

It is rare that a disaster occurs unexpectedly. As a rule with very few exceptions, a disaster is preceded by unusual deviations from the normal functioning of equipment. Such deviations are often noticed as being suspicious by operators. Typically, they then diligently seek advice to assess the possible risks, but often this process can take weeks. After an accident, the response team also needs prompt assessment of the effectiveness of different types of response actions. However, in the majority of cases we have studied, there was no mechanism for the immediate meeting of industry experts, who could remotely simulate risks and make recommendations to operators and the response team.

For example, after the Three Mile Island nuclear accident, the Institute for Nuclear Power Operations (INPO) was established in response to an "*immediate need to provide the onsite crew with prompt, reliable access to offsite expertise to assist them in diagnosing and responding to a potentially serious accident situation... [M]any of the important diagnoses in an emergency can best be made by individuals not embroiled in the hectic atmosphere of the control room... At TMI, difficulties in communications, management, and logistics contributed to the failure to bring available expertise to bear*".[16] A similar necessity for external expertise occurred during the first days after the Chernobyl accident, no one knew exactly how to deal effectively with the consequences of the disaster: what kind of materials to use to extinguish the reactor, how to assess which areas should be permanently evacuated and which temporarily, and so on. In response to these problems in dealing with Chernobyl, a network of 24-7-365 crisis centers has been established within the Soviet/Russian nuclear industry. Operators of the centers are ready to make immediate forecasts about the likely spread of radioactive emissions after any nuclear accident, and to create a dynamic model of any nuclear accident scenario. The centers are equipped with automated systems, which in case of disaster integrate data from the distressed nuclear power station, from communications by atomic scientists, meteorologists and local authorities. Usually, three shifts of two operators work day and night. After receiving information about a possible emergency situation, operators immediately call in a pool of a dozen experts from various scientific fields to evaluate the risks; these scientists are informed about the emergency meeting through an automated system by SMS, phone call, and e-mail. Experts must reach the center within an hour after the emergency call—or if they cannot attend the emergency meeting in person, they must immediately begin to evaluate the risks remotely via video conferencing, mobile or fixed telephone lines.[17] Similar systems are now operating in most of the developed nations with nuclear power plants.

3.4.2 Unwillingness to Investigate in Details the Causes of an Accident and Absence of Permanent Risk Assessment Systems Within Organizations (Recording, Evaluating and Ranking Risks Over Decades)

(*This occurred in the majority of cases elaborated*)

The unwillingness of executives to carry out a detailed investigation after an accident, and to publish detailed reports about its causes, is a very common corporate problem. Usually, details of an accident remain in the archives of investigation bodies, and these bodies do not produce summary reports for the further use of industry specialists. Moreover, organizations do not welcome enquiry or

[16]Report of the President's Commission on the Accident at Three Mile Island: The Need for Change: The Legacy of TMI, October 1979, pp. 103.

[17]Technical Crisis Center of Russian Safety Institute of Atomic Energy Sciences. http://www.youtube.com/watch?v=7Vlj9vmGPBE.

discussion about the experience of near-miss incidents, because even news about an accident that did not happen can be perceived by the public and regulators as a very worrisome sign of trouble within an organization. So, the practice of concealing the details of accidents prevents anyone in the organization from drawing valuable conclusions from the tragic experience. Consequently, 10–20 years after an accident or a near-miss incident, new managers repeat the mistakes of their predecessors without realizing they are facing similar risks.

Many companies faced with severe accidents did not have a system for regularly collecting information about risks and for ranking the revealed risks according to severity. Managers themselves are often uninterested in collecting such information, because the existence of such a system, detailing all the shortcomings of the equipment they are operating, will show investigators that they knew about the risks before an accident, but took no action.

3.4.3 High Frequency of Unconfirmed Alerts

(*Krymsk flooding and also in cases, which were not mentioned in the book, like Mumbai terrorist attacks 2008, Oslo bombing 2011 and 9-11*)

The near-impossibility for managers of assessing the likelihood of a catastrophic scenario was highlighted in cases where leaders faced a vast amount of information about possible threats—such as weather conditions or the actions of terrorists—which were impossible to verify quickly enough to understand how to direct response efforts. It is important to remember that an abundance of unranked risk information slows down an organization's response and reduces the quality of that response.

Herbert Alexander Simon (1916–2001), Nobel Memorial Prize winner in Economics for his pioneering research into the decision-making process within economic organizations, declared once that "*in an information-rich world, the wealth of information means a dearth of something else: a scarcity of whatever it is that information consumes. What information consumes is rather obvious: it consumes the attention of its recipients. Hence a wealth of information creates a poverty of attention and a need to allocate that attention efficiently among the overabundance of information sources that might consume it*".[18]

[18]Herbert Simon, "Designing Organizations for an Information-Rich World", in: Martin Greenberger: Computers, Communication, and the Public Interest. Baltimore, MD: The Johns Hopkins Press (1971) pp. 40–41.

3.4.4 Ignorance Among Critical Personnel and Managers of Other Accidents or Near-Miss Cases Within the Organization, the Industry and Abroad. Absence of a Risk Knowledge Management System (Accumulation, Systematization, and Transmission)

(*Three Mile Island Nuclear Accident|Chernobyl disaster|Fukushima-Daiichi nuclear disaster|Deepwater Horizon oil spill|Sayano-Shushenskaya hydropower station accident|Bhopal disaster|Krymsk flooding*)

Many managers believe that their problems and risks are unique, leading them to try to find their own solutions, but experience demonstrates that they are simply ignorant of the experience of other departments, companies, industries or countries, because there is no accurate, systematized, detailed knowledge of previous accidents. Unfortunately, many organizations—and even government ministries or think-tanks for a given industry—do not systematize existing sector risks, and do not collect information about near-miss cases on an ongoing basis; so, basically no one describes and studies in detail the causes of accidents elsewhere in an industry, or abroad. In addition, there are few articles in internal corporate journals related to the experiences of other departments of the organization, competitors and foreign enterprises.

3.5 Personal Features of Managers and Employees

James Dean died at 24 year old. His last known words, uttered right before impact when Wütherich told Dean to slow down when they saw the Ford coupe in front of them about to drive into their lane, were said to have been: "That guy's gotta stop… He'll see us."

The trouble with the world is that the stupid
are cocksure and the intelligent are full of doubt

Bertrand Russell

People only accept change when they are faced with necessity, and only
recognize necessity when a crisis is upon them

Jean Monnet

The characteristics of human nature, which drive individual people to hide risk, manifest themselves only when external and internal environments motivate them to distort information about real conditions in an organization.

3.5.1 Problem of "Looking Good in the Eyes of Superiors" and Reluctance to Admit Personal Mistakes Because of Fear of Being Seen as Incompetent

(*This occurred in the majority of cases elaborated*)

This problem is closely connected with the "*success at any price*" and "*no bad news*" organizational culture. In order to constantly prove their competence to their superiors, employees prefer to send them soothing reports: they want to look good, to get ahead in the organization, or are simply unwilling to make the possible existence of problems the subject of their communication with senior management.

Closely linked to this desire of subordinates to be seen in a positive light and being rewarded by managers is the reluctance to admit mistakes, even when they are obvious. If the internal and external environments only value success and achievement and do not reward employees for the recognition of their errors, this will obviously not lead to early disclosure of the shortcomings of a system from its creators. The fear of appearing incompetent, of being fired, and of being humiliated in the eyes of colleagues and the public, also pushes people to conceal information.

3.5.2 Unrealistic Projections of Personal Performance

(*This occurred in the majority of cases elaborated*)

The next problem that stems from the desire of employees to look good in the eyes of superiors is that employees make unrealistic claims about, or set unrealistic targets for, their progress in the hope of distinguishing themselves. Experience shows that the actual progress of a project rarely coincides with these initially declared plans. Nevertheless, subordinates fear that superiors will consider such delays as a demonstration of their incompetence and lack of integrity. As a result, they begin to perform their duties poorly, to ignore flaws, in the attempt to fit in with the limited budget or the optimistic and ambitious schedule, desperate to prove by any means that they are able to achieve the goals they declared.

3.5.3 Fear of Criminal Prosecution After Serious Fault

(*Barings bank collapse|Enron case|Subprime mortgage crisis|Sayano-Shushenskaya hydropower station accident|Krymsk flooding*)

The fear of personal criminal prosecution after the occurrence of some disgraceful actions leads to further following distortion of information about the real picture of risks and additional deterioration of a situation.

3.6 Results and Synthesis

If we want everything to stay as it is, everything will have to change

Giuseppe Tomasi di Lamedusa from The Leopard

If we want to have an educated citizenship in a modern technological society, we need to teach them three things: reading, writing, and statistical thinking

Herbert George Wells

We started our exposition with one of the axioms of management theory, which states that managers oversee other people by means of information. The quality of the information conditions the quality of decisions, and later on the adequacy of an organization's response. We documented that this quite simple theoretical explanation of why managers need precise information actually collides with a tough reality: one of the main causes of high-magnitude disasters is the slow response or even absence of reaction of an organization to the evidence concerning mounting risks of a serious incident; a second leading cause of disasters lies in the slow build-up of measures trying to deal with an incident as it unfolds and in its immediate aftermath, which occurred because risks were ignored in the first place. Indeed, our detailed investigations of many accidents show that the initial cause of an inadequate and sluggish response to many major disasters is the communication of distorted or misleading information about risks within an organization before and after an accident.

The fact that an inadequate level of information over the development of a project or the lifetime of the system is often the cause of a disaster does not constitute in itself a surprise or an insight worth mentioning, so obvious is the relationship between the loss that comes at a surprise to many and the absence of awareness of its likelihood. What this book has unearthed is something much more startling and also frightening in its implications, namely that there is in general a widespread voluntary practice of risk information concealment performed consciously by some key actors of the tragedy towards other actors and/or the outside word. By risk information concealment, we have referred to the practice in which decision makers or important personnel in an organization do not transmit or suppress important information about the real condition of a system: it can be the bad design of a nuclear reactor, the risky financial position on certain derivative products, the static fatigue in the turbines at a hydropower station, the defective cement mixture for a deep-water well, and so on. Such deficient information transmission leads to inadequate decisions about, or operation of, the systems involved.

We have documented that, in some of the largest civil nuclear accidents (at the Three Mile Island NPP in the USA, the Chernobyl NPP in the USSR and the Fukushima Daiichi NPP in Japan), risk information concealment was a widespread practice. Similarly, in the three largest bankruptcies in recent US history (Enron, WorldCom and Lehman Brothers), risk concealment took place on a very large scale. And both before and after the second largest industrial accident in

World history in terms of casualty numbers (at a chemical plant in Bhopal, India where up to 15,000 people died), there was serious obfuscation about the level of risk. Likewise, the interruption in the transmission of risk information was one of the leading indirect causes of the largest off-shore oil spill in World history, the Deepwater Horizon platform blowout. We have extended our analyses to these and many other disasters. We have elaborated in details on more than 25 disasters and cover more than 20 additional instances of infamous disasters in the industrial, financial and military sectors where risk concealment led to a catastrophe.

In one group of accidents investigated, we have found a situation where the executives of an organization receive reassuring reports from subordinates with a distorted picture of risks, or of the real size of the actual disaster. They are also given misleading information about possible threats to the surrounding population, and assurances that local staff can manage the mitigation of risks, or the response to an accident, with their own resources and without any help from headquarters or other organizations. Such incomplete and distorted access to risk data leads to a situation where senior management supposes that the scale of a disaster is limited and subordinates do not need any help. While in fact, the organization is losing precious time, when fast, timely and full-scale intervention could reduce the probability of an accident occurring, or could decrease the magnitude of a disaster when it happens. In the majority of infamous accidents, executives and victims soon faced an uncontrollable disaster because risk levels were initially misrepresented within an organization.

In a second group of cases, long before a disaster, there are warnings about risks coming from dedicated and proactive workers, researchers and social activists, but for various reasons managers prefer to ignore them. Most of such alarm bells remain the purview of the management team and are not transmitted within or outside of the organization.

In a third group of cases, executives know about the inherent defects or shortcomings in the design of a technical system or business model, but conceal them from their subordinates who continue to operate the system unaware of the hidden risks. Ultimately, this obfuscation leads to a situation where the workers and the on-site decision makers within the organization make inadequate decisions with respect to the real conditions of the external and internal environment, which catalyze and amplify the size of the disasters.

Our examination of more than 25 cases shows that risk information concealment played a dominant role in creating or aggravating a catastrophe. While the presented cases speak volume by themselves, do they really represent a general syndrome afflicting human management of complex systems? A legitimate concern is that we may be representing a highly biased view of human management, resulting from our ex-post selection of well-known disasters, omitting in this endeavor to discuss the many more examples of systems that are well functioning and in which risk information is communicated seamlessly. Our message could thus be judged as unduly alarmist and misrepresenting the real world of management. In order to address this concern and provide the full picture, we need to consider all together what in statistical jargon are called the false positives (or errors of type I), the false negatives (or errors of type II) as well as the true positives and true negatives.

- **True positives (risk concealment leading to a catastrophe)**: in the second chapter of this book, as already mentioned, we elaborated on some 25 of the most prominent and striking cases of the last 100 years, from the Spanish Flu outbreak to the Fukushima Daiichi nuclear disaster, for which obfuscation of risks occurring over many preceding years ultimately led to a catastrophe.
- **False positives (obfuscation and no catastrophe)**: it is a routine fact that billions of workers in non-critical industries conceal at least some information about the real situation regarding their level of responsibility and negligence toward their duties, with no obvious and direct consequences beyond minor complaints by their customers about the quality of provided services or goods. Many risky technologies or installations may have been functioning in a danger zone for a long time but were decommissioned before anything serious happened and we shall never know whether this was a lucky happening or not. For critical infrastructures, conscious risk concealment would rarely be revealed if not for some external event or some aggravating factor pushing the function beyond the design criteria of the system. There have been several instances of this scenario, just in the civil nuclear industry. If the operators of the Chernobyl NPP had not made an unprecedented experiment with the emergency power supply system, nobody (outside a narrow circle of insiders) would have been aware of shortcomings in the design of the RBMK reactors that only came to light under very unusual circumstances. The world nuclear community would perhaps never have known about the massive falsification of records within the Japanese nuclear industry over the previous decades, or the reluctance of commercial operators to implement international cutting-edge nuclear safety standards on Japanese nuclear plants, if the Tohoku earthquake—the largest event ever recorded in Japan—had not occurred. The full extent of falsification and non-compliance in the Japanese industry could similarly have gone unnoticed to the world community if the Japanese had put the 40-year-old Fukushima Daiichi NPP into the decommission stage by February 2011—one month before the earthquake—but the operators chose to request a permit extension to the authorities to pursue the plant operation for another decade. In every industry and every society, such "sleeping dragons" exist[19,20]: due to risks being hidden, early warnings being ignored, errors accumulating, static fatigue of the system elements and the influence of exogenous events, some of them could be awaken. In some cases, organizations may have hidden risks running over decades

[19]D. Sornette, Dragon-Kings, Black Swans and the Prediction of Crises, International Journal of Terraspace Science and Engineering 2(1), 1–18 (2009) (e-print at http://arXiv.org/abs/0907.4290).

[20]D. Sornette and G. Ouillon, Dragon-kings: mechanisms, statistical methods and empirical evidence, Eur. Phys. J. Special Topics 205, 1–26 (2012) (special issue on power laws and dragon-kings) (e-print at http://arxiv.org/abs/1205.1002).

without any severe consequences. But the accumulating stress, the general erosion and damage of the system together with the ignorance or unawareness of the risks can still ultimately lead to a catastrophe.[21] We will inspect below some ongoing cases where there may be evidence of risk obfuscation. The examples we will cover include the risk concealment dynamics during the development of shale gas and oil formations, as well as genetically modified organisms (GMO), the manipulation and falsification of the statistics about the US debt and the Chinese GDP, and the concealment of vulnerabilities and bugs in the software industry. Such instances of risks being concealed and early warnings being ignored constitute potential large threats to the environmental, financial and political stability of the whole world.

Among false negatives should also be included the class of near misses or narrow escapes, those unplanned events that did not lead to damage or loss.[22] In principle, near misses can provide unique low-cost opportunities for management to receive risk information in timely fashion and react adequately to the revealed risks in order to avoid severe future consequences. Unfortunately, the lack of reporting and concealment of information seems to be also prevalent in near misses, as little incentive in general exists for the involved parties to reveal what is often considered as embarrassing circumstances. While some work has been performed on near misses,[23,24] much more efforts should be dedicated to this extremely important but difficult subject, because most crises and accidents are generally preceded by some kind of warnings or near accidents. (see Footnote 19)

- **True negatives (no obfuscation and no catastrophe)**: this is the normal regime of the typical well-functioning organization and is hopefully the most prevalent.
- **False negatives (no obfuscation and catastrophe)**: there are instances of sudden events that nobody could anticipate, when resources needed to assess the risks promptly and carefully were not available, or when a system was operated without sufficient attention to important features, for which a catastrophe occurred without any conscious risk concealment. This does not mean that there were no tale signs and early warnings, however. This seems to be the case in the Mumbai High North Platform fire, the Concorde supersonic plane crash, the sinking of the Ocean Ranger oil platform and many other disasters.

We have pointed out how and where risk information was concealed in both the build-up and the response to the disasters. Our investigation unveiled the nature of

[21] D. Sornette and P. Cauwels, A Creepy World: How can managers spot and manage systemic crises, Journal of Risk Management in Financial Institutions (JRMFI) 8 (1), (2015) http://ssrn.com/abstract=2388739).

[22] http://en.wikipedia.org/wiki/Near_miss_(safety).

[23] T.W. van der Schaaf (Editor) Near Miss Reporting as a Safety Tool. Butterworth-Heinemann (Jan. 1, 1991).

[24] W.G. Bridges, Gains from Getting Near Misses Reported, Presentation at 8th Global Congress on Process Safety, Houston, TX April 1–4, 2012 (Process Improvement Institute, Inc., 2012).

the real problems within each organization before and after a disaster, the motivation of managers and personnel and the action they took as a result. We saw that, in the majority of the elaborated cases, the causes of accidents are similar. In other words, accidents repeat in different countries and in different contexts according to the same scenario. The problem is not just proximal human mistakes, but result from a failure of the whole system, from the existence of bad design, improper training of personal, unrealistic rosters, and poor quality maintenance, and essentially poor information communication.[25] As a response, risks management should aim at reducing the conditions for errors to occur, in other words, diminish the susceptibility of tasks and processes to errors and miscommunication that are inevitable with complex technology and human fallibility. This can be done by a thorough determination of the factors and organizational issues producing errors to make the organization more resistant to human fallibility.[26,27] This could be helped for instance by initiatives such as that developed by RepRisk (www.reprisk.com), a company that supplies business intelligence on environmental, social, and governance (ESG) risks. RepRisk has developed one of the most comprehensive databases on ESG risks[28] related to companies, projects, sectors, and countries, using open access to thousands of public sources, international and local media, government agencies, nongovernmental organizations, newsletters, blogs, social networks and so on. This is made possible by the increased transparency and inter-connectedness that the Internet and modern electronic era provide. Stockholders can thus break at least in part the veil of concealments and integrate ESG risk analysis to push for more responsible and sustainable governance in the goal of developing resilient organizations[29] that are harmoniously embedded in the environmental and social networks.

The sum of the evidence that we present strongly supports the hypothesis that there is value in comparative history, in the sense that "history" taken in a broad sense repeats itself. In this respect, Diamond and Robinson advocate comparative history as a conduct of 'natural experiments of history' in which the 'perturbations' and their causes (exogenous or endogenous) in the involved cases can be qualitatively (and in some cases quantitatively) analyzed.[30] In this vein, learning

[25]Bennett, S.A., *Human Error—by Design?* Basingstoke: Palgrave-Macmillan (2001).

[26]Reason, J., Managing the Risks of Organizational Accidents. Ashgate Publishing Company, Abingdon, Oxfordshire, 1 edition (Dec. 1, 1997).

[27]Reason, J. and A. Hobbs, Managing Maintenance Error: A Practical Guide, Ashgate Publishing Limited, Abingdon, Oxfordshire (May 1, 2003).

[28]With information concerning criticism, controversies and negative incidents in 14 languages related to 44,000 companies, 10,000 projects, 7000 NGOs and 6000 governmental bodies at the time of writing, according RepRisk sources.

[29]Tatyana Kovalenko and Didier Sornette, Dynamical Diagnosis and Solutions for Resilient Natural and Social Systems. Planet@ Risk 1(1), 7–33 (2013), Global Risk Forum (GRF) Davos (http://arxiv.org/abs/1211.1949).

[30]Jared Diamond and James A. Robinson (eds.) Natural Experiments of History, Belknap Press (2011).

from the cases documented here and realizing/internalizing the ubiquitous processes at work is likely to be one of the most effective ways to prepare decision makers for possible impending disasters. Based on the understanding—elucidated from the studied cases—of what motivates people to conceal risks, we identified 30 common causes of risk information concealment, and proposed a range of recommendations to improve the transmission of risk information in order to avoid the repetition of similar disasters.

We thus depart from Charles Perrow's claim that unexpected failures are intrinsic to society's complex and tightly coupled systems and that such accidents are unavoidable.[31] While we agree that the solution cannot come from better technological design only, recognizing the core issue of information concealment provides a path towards improvement, as suggested by the few positive developments presented in Chap. 5. Yes, people make mistake but a well-functioning process of information sharing can protect against blunders. What is intrinsic is not the unexpected large-scale failures but the existence of errors and dysfunctions that can be integrated into a never ending process of improvement and learning, as exemplified by the Toyota Production System. And the cascades underlying large catastrophes that start from minor origins can be traced if there is a culture of monitoring small trends and weak signals in a comprehensive risk management approach with efficient communication and sharing of information. This provides a process for building strong organizations that self-repair against the unavoidable human faults, similarly to the many flaws continuously occurring in a biological organism that are endlessly repaired.

In the end, there is an inherent conflict between what companies say they want (strong risk management) and how companies act (strong risk management gets in the way). Following John Sneddon,[32,33] we should measure and manage the capability of work processes rather than judge solely on results, because the processes embody the sum of all factors and decisions. Performance measures should emphasize much more the quality and robustness, even resilience, of the path taken by the actors, verifying at each step that none of the 30 main causes of information concealment are seriously present. In contrast, many private and public organizations focus solely on targets, but as Goodhart's Law reminds us, "*When a measure becomes a target, it ceases to be a measure.*" Focusing on targets leads to fail thinking of the whole system or pathway, and instead pushes to focusing on parts of the system, with the result that although each part may be doing its bit, the overall result is terrible, fostering conflicts between parts and opening vulnerabilities. Risk management (integrity, ethics, good governance) usually works within certain environmental parameters, whereas there is an expectation that it works

---

[31]C. Perrow, Normal Accidents: Living with High-Risk Technologies, 2nd edn. Princeton University Press, Princeton, NJ (1999).

[32]John Seddon, Freedom from Command and Control: A Better Way to Make the Work Work. Vanguard Consulting (2003).

[33]John Seddon, Systems Thinking in the Public Sector: The Failure of the Reform Regime… and a Manifesto for a Better Way. Triarchy Press (2008).

uniformly at all times. What do we mean by this? In boom times, risk management gets in the way of making super profits. In times of crisis, well, we have no time to think about risk management. Therefore, risk management works best when companies are on a steady measured course and have the time and inclination to hear a risk manager say "*no*". A prerequisite is a culture of openness and active fight against the strong forces described above pushing towards risk information concealment.

Chapter 4
Major On-going Cases with Information Concealment Practice

History does not repeat itself, but it can rhyme
Mark Twain

Those who cannot remember the past are condemned to repeat it
George Santayana

The synthesis presented in the previous chapter of the main factors facilitating or creating concealment of information, which led to or amplified the previously reviewed disasters, provides the basis for watching, analyzing and understanding some on-going cases for which large-scale information concealment associated with critical technology and pressing social issues are at work. We discuss below four important topics: (i) the shale energy development in the USA, (ii) the business based on genetically modified organisms in the USA, (iii) the real debts and financial liabilities in the US as well as China and (iv) the global cyber arms race and concealment of vulnerabilities in the software industry.

Our sources are varied but necessarily limited by the aura of secrecy and business confidentiality surrounding these sensitive issues. In this respect, the modern electronic era offers useful ways to garner information from otherwise confidential databases. One example concerns nuclear incidents, which are not in general communicated and documented to the public but remain inside confidential databases run by nuclear operators and regulators. Given that only severe accidents, such as Three Miles Island, Chernobyl, and Fukushima, reach the public awareness, one would conclude that the risks associated with the nuclear industry are quite low. The picture is completely different however when using a monetary value of damage severity to make events comparable.[1] This extends the definition usually taken by the nuclear industry and regulators to include incidents that either resulted in the loss of human life or in property damage above some threshold (of, for

---

[1]K. Hsü, Nuclear risk evaluation. Nature 328, 22 (1987); A. Sengör, Evaluating nuclear accidents. Nature 335, 391 (1987).

D. Chernov and D. Sornette, *Man-made Catastrophes and Risk Information Concealment*, DOI 10.1007/978-3-319-24301-6_4

instance, US$ 20 million). By searching historical archives, newspaper and magazine articles, and press wire reports, Prof. Sovacool constructed a database of 99 nuclear incidents worldwide from 1952 to 2009 that occurred in different kinds of nuclear facilities, with an estimation of damage they generated, including loss of production and property damage.[2] Quantifying this database, one of us and collaborators found[3] that this empirical distribution of losses exhibits what statisticians and scientists call a "heavy tail", i.e., a large relative probability for losses of all sizes, including very large losses, which decays much slower than the plant-specific distribution of losses predicted by fault/event trees analysis used in the nuclear industry framework of probabilistic safety assessment (PSA). And the results are robust when breaking the data in distinct epochs, such as separating the pre- from the post-Chernobyl eras. Other detailed statistical studies have almost universally confirmed that the PSA dramatically underestimates the risk of accidents.[4] This discrepancy between the prediction of the nuclear industry probabilistic safety assessment and reality is another case of (unintended) concealment of information (performed often with good intentions), which has several costs: (i) insufficient awareness of the operators concerning the real risks and complacency with respect to the accepted PSA methodology; (ii) often undue fears and exaggerated alarm by the public resulting from this lack of transparency and the apparent discrepancy between predicted and realized accidents. This calls for additional continuous development and validation, making the best use of the experienced incidents, near misses and accidents, which is urgently needed to address the existing known limitations of PSA when aiming at the estimation of total risks in the nuclear industry.[5]

[2]Benjamin K. Sovacool, The costs of failure: A preliminary assessment of major energy accidents, 1907–2007, Energy Policy 36, 1802–1820 (2008).

[3]Didier Sornette, Thomas Maillart, and Wolfgang Kröger, Exploring the limits of safety analysis in complex technological systems, International Journal of Disaster Risk Reduction 6, 59–66 (2013).

[4]D. Smythe. An objective nuclear accident magnitude scale for quantification of severe and catastrophic events. Physics Today: Points of View, 2011; M. Hofert and M. V. Wu ̈thrich. Statistical Review of Nuclear Power Accidents. Asia-Pacific Journal of Risk and Insurance, 7(1):1–18, 2013; L. Escobar Rangel and F. Leveque. How Fukushima Dai-ichi core meltdown changed the probability of nuclear accidents? Safety Science 64 90–98, 2014; M. Ha-Duong and V. Journe. Calculating nuclear accident probabilities from empirical frequencies. Environment Systems and Decisions 34.2, 2014.

[5]Wolfgang Kröger and Didier Sornette, Reflections on Limitations of Current PSA Methodology, ANS PSA 2013 International Topical Meeting on Probabilistic Safety Assessment and Analysis, Columbia, South Carolina, USA, Sep. 22–26, 2013, American Nuclear Society, LaGrange Park, IL (2013),
invited article for the Probabilistic Safety Analysis 2013 (PSA2013) (accepted 5 July 2013) (www.psa2013.org).

4.1 Shale Energy Development in the USA[6]

Shale gas and shale oil production is based on the technology of horizontal drilling and hydraulic fracturing (fracking) of shale formations. Natural gas and oil produced from shale strata in the United States has dramatically changed the structure of the US energy balance in the last decade. According to the U.S. Energy Information Administration (EIA), over the last decade, the proportion of total American gas production provided by shale gas has increased from 1 % to over 40 %—this massive shale gas production has allowed the US to become the largest producer of natural gas in the world. According to EIA projections, the United States could become a net exporter of natural gas before 2020[7] and net exporter of oil and gas by 2040, if the oil prices would be high and the existence of abundant gas resources are confirmed.[8]

However, the technology of horizontal drilling and hydraulic fracturing has several inherent financial and environmental shortcomings. The existence of these shortcomings has forced US government agencies and American energy companies to adopt a sophisticated risk concealment strategy, based on understating political, military, financial and environmental risks of the shale business and deceitfully promoting economic, geopolitical and national security benefits from the development of shale formations. In the following, we explore the different dimensions of these risks and expose the incentives and mechanisms by which they have developed. Since no major crisis has yet developed, this case might be argued not to have its place in this book. But our dissection of the previous crises is useful only if it provides a basis for learning and recognizing future potential crises. We believe that the shale energy development promoted mainly by the USA is one of the present outstanding candidates for possible environmental, financial and/or geopolitical crises to come.

[6]This Sect. 4.1 summarizes the major findings presented in our forthcoming book "The A*merican Shale Bubble, High Energy Prices and Global Political Instability*".

[7]The Annual Energy Outlook 2014, U.S. Energy Information Administration, May 2014, p. MT-22.

[8]Ibid, p. ES-2.

4.1.1 Economics of Exploration of Unconventional Oil and Gas Resources in the US, Geopolitical Challenges and Oil Prices

Research and development of unconventional oil and gas resources were launched after the 1973 oil crisis, when the United States initiated several programs to find alternative ways of extracting energy from American soil in order to be energy independent from world oil shocks and from changes in the political situation in the Middle East.[9] The United States government funded research into new technologies for drilling and exploration, gave preferential tax treatment to alternative energy companies, eased environmental regulation standards and covered the expenses for drilling thousands of exploratory wells.

In spite of the fact that hydraulic fracturing has been used during the last 60 years by the American oil industry (since the end of the 1940s), the combination of horizontal drilling and hydraulic fracturing was a rather new technology, which was tested in 1980s–1990s by Mitchell Energy on Barnett shale play in Texas. Results of test drillings were positive regarding release of hydrocarbons from shale formations, which have extremely low permeability and porosity—thousand times lower in comparison with conventional gas deposits. This was encouraging because shale gas is a highly dispersed natural resource.[10] Nevertheless, further development of the technology was suspended due to lawsuits from landowners of Wise County (Texas) for pollution of residential water by drilling fluids[11] and high production cost from shale formations. In the 1980s–1990s, it was economically impossible to start large-scale alternative energy projects within the United States because of the collapse of energy prices in the mid-1980s: the average nominal U.S. natural gas wellhead price during the 1980s was US $72 [US $152 in 2010 prices]/1000 m^3 and during the 1990s, it was US $67 [US $96 in 2010 prices]/1000 m^3;[12] between 1990 and 1998, the annual average oil price dropped from US $23/bbl [US $38/bbl] to US $12/bbl [US $17/bbl].[13,14] Therefore, in 1990, 90 % of the gas produced in the United States was from low-cost conventional gas fields but, subsequently, the output of

---

[9]Gal Luft, Anne Korin, Energy Security Challenges for the 21st Century: A Reference Handbook, ABC-CLIO, 2009, p. 145.

[10]Andrei Korzhubaev and Alexander Khurshudov, Shale Gas: Great Expectations, Modest Plans, Oil & Gas Eurasia, Dec. 2011, p. 24.

[11]Zhongmin Wang and Alan Krupnick, A Retrospective Review of Shale Gas Development in the United States What Led to the Boom?, Resources for the Future, Apr. 2013, p. 24.

[12]U.S. Natural Gas Wellhead Price (1922–2013), U.S. Energy Information Administration, Apr. 2014.

[13]Oil: Crude oil prices 1861–2009, BP Statistical Review of World Energy 2010, June 2010.

[14]Gal Luft, Anne Korin, Energy Security Challenges for the 21st Century: A Reference Handbook, ABC-CLIO, 2009, p. 146.

low-cost traditional deposits gradually decreased due to the depletion of conventional fields.

These difficulties met by unconventional energy production in the 1980s–1990s supported a thesis that, in a situation of low energy prices, only conventional producers could be profitable. As a matter of comparison, the average production costs of the extraction of natural gas from shale formations within the United States are US $140–230/1000 m^3 on well's pad,[15,16,17,18,19] while the Russian conventional natural gas giant Gazprom's average production costs were US $37–38/1000 m^3.[20] Average costs of shale oil production in the US ranges between US $55/bbl and US $85/bbl.[21,22,23] Canadian oil sands projects present a break-even price of more than $90/bbl,[24] while oil production costs in Saudi Arabia are just US $2/bbl,[25] in Iraq US $1–5/bbl[26,27,28] and at least US $4/bbl in Russia.[29]

The conditions for the full-scale development of unconventional energy sources within the United States was met only after the price hikes caused by 9/11 and the

---

[15]During our calculations we use following conversion rates of natural gas from American metrics into European one: 1 million British Thermal Units (MMbtu) = 1000 cubic foots (Mcf), 35,000 cubic foots (*Mcf*) = 1000 cubic meters (m^3) of natural gas. We use factor 35 for conversation price of 1 MMbtu into 1000 m^3.

[16]The Future of Natural Gas, Appendix 2D: Shale Gas Economic Sensitivities, MIT Energy Initiative, 2011, p. 2.

[17]First 5 years of "shale gas revolution". What we now know for sure? Centre for Global Energy Markets of Energy Research Institute of the Russian Academy of Sciences, Nov. 2012, pp. 25–26, 32.

[18]Euan Mearns, What is the real cost of shale gas? Energy Matters, Nov. 28, 2013.

[19]Ivan Sandrea, US shale gas and tight oil industry performance: challenges and opportunities, The Oxford Institute for Energy Studies, Mar. 21, 2014, p. 4.

[20]Mikhail Korchemkin, East European Gas Analysis, September 11, 2014, http://www.eegas.com/rep2014q1-cost_e.htm.

[21]Selam Gebrekidan, Insight: Peak, pause or plummet? Shale oil costs at crossroads, Reuters, May 17, 2012.

[22]Leonardo Maugeri, The shale oil boom: a U.S. Phenomenon, Belfer Center for Science and International Affairs, Harvard Kennedy School, June 2013, p. 13.

[23]Ivan Sandrea, US shale gas and tight oil industry performance: challenges and opportunities, The Oxford Institute for Energy Studies, Mar. 21, 2014, p. 3.

[24]Leonardo Maugeri, The shale oil boom: a U.S. Phenomenon, Belfer Center for Science and International Affairs, Harvard Kennedy School, June 2013, p. 25.

[25]Interview with Ali Al-Naimi, Saudi Arabian Minister of Petroleum and Mineral Resources and former CEO of Saudi Aramco, from documentary "60 Minutes: Season 41, Episode 11: The Oil Kingdom", CBS television network, 7 Dec. 2008.

[26]Donald L. Barlett and James B. Steele, Iraq's Crude Awakening, The Time, May 10, 2003.

[27]Galina Starynska, Lukoil produced first oil from the West Qurna-2 in Iraq, Vedomosti newspaper, Moscow, Oct. 28, 2013.

[28]Eldar Kasai, Iraq: oil security, Expert, July 7, 2014.

[29]Katya Golubkova, Alexander Winning, David Evans, Russia's Rosneft says OPEC decision won't affect its work, Reuters, Nov. 27, 2014.

following war in Iraq. The start of the "Global War on Terror" as a response to the 9/11 event and the invasion of Iraq were the tipping points for a change of oil price trends, with oil appreciating from the annual average price of US $24/bbl in 2001 to US $97/bbl in 2008. The military and political crises in the Middle East, initiated by the United States, made profitable the American domestic energy production from unconventional energy sources of shale formations and oil sands. Without such a geopolitical context creating oil scarcity and uncertainty, it would be impossible to launch unconventional energy production and contemplate a possible energy independence of the United States based on unconventional gas and oil resources.

4.1.2 Environment Aspects of Hydrologic Fracturing Technology

In the late 1990s and early 2000s, Halliburton, one of the leading service companies in the petroleum industry, was developing improved methods for the extraction of gas and oil from shale formation based on hydraulic fracturing (or "fracking"). Before becoming the CEO of Halliburton from 1995 to 2000, Dick Cheney was the US Secretary of Defense during the term of George H. W. Bush (1989–1993) and the first Iraqi campaign (1990–1991). He became Vice-President of the US during the two terms of George W. Bush (2001–2009). During his time as CEO of Halliburton, there was a strong momentum to improve the technology of hydraulic fracturing. Then, during the eight years of his US Vice-Presidency, Cheney directed the Energy Task Force [officially named as the National Energy Policy Development Group (NEPDG)]. One of actions of the NEPDG was to promote the development of domestic unconventional gas and oil resources, based on the technology of hydraulic fracturing.

In 2001, a NEPDG report described the technology in the following way: "*This is a common procedure used by producers to complete gas wells by stimulating the well's ability to flow increased volumes of gas from the reservoir rock into the wellbore. During a fracture procedure, fluid and a propping agent (usually sand) are pumped into the reservoir rock, widening natural fractures to provide paths for the gas to migrate to the wellbore. In certain formations, it has been demonstrated that the gas flow rate may be increased as much as 20-fold by hydraulic fracturing*".[30] However, the authors of the report did not mention that the most important element of the technology is a mixture of 500 chemicals, which are added to the water and allow it to permeate more effectively through the shale formations during fracturing. This mixture of water and chemicals is pumped at very high pressure (13,500 psi or 920 atm) into a shale well in order to fracture shale rock at

[30]National Energy Policy, Report of the National Energy Policy Development Group, May 2001, p. 5–6.

depth from 400 m to 5200 m, depending on the shale type, to facilitate the release of hydrocarbons from the shale formation to the surface. The fractures are kept open with proppants (generally sand) to ensure the continued flow of resources. Sand and water compose 99.5 % of the total volumes used.[31] These chemicals, invented by Halliburton and kept secret, are extremely harmful to the environment.[32,33,34,35,36,37] Some of these substances are recognized as extremely hazardous for human health and the environment—for example, ethylene glycol, which can damage the kidneys; formaldehyde, which is known to cause cancer; and naphthalene, another possible carcinogen.[38]

The authors of the NEPDG report also did not mention the very large fresh water consumption during hydraulic fracturing. According to Dr. Anthony Ingraffea, a hydraulic fracturing expert from Cornell University, the amount of water required to operate an average shale gas well exceeds the required level for a conventional gas well by a factor 50–100.[39] The amount of water consumed depends on shale formations: for instance, multi-stage fracking operations (a single shale well can be hydraulically fractured up to 20 times) on a typical unconventional well in Texas requires up to 23,000 tons of water, while a California well requires only around 1000 tons[40]; these statistics do not include the necessary subsequent re-fracking procedures needed to stimulate additional hydrocarbons release due to weak intraformational pressure within the shale well several years after the initial fracking.

[31]Modern Shale Gas Development in the United States: A Primer, U.S. Department of Energy, Apr. 2009, p. 61.

[32]Glenn Miller, Review of the Revised Draft Supplemental Generic Environmental Impact Statement on the Oil, Gas and Solution Mining Regulatory Program, Toxicity and Exposure to Substances in Fracturing Fluids and in the Waste Water Associated with the Hydrocarbon Bearing Shale, Consulting Environmental Toxicologist to the Natural Resources Defense Council Dec. 29, 2009.

[33]Glenn Miller, Review of the Revised Draft Supplemental Generic Environmental Impact Statement on the Oil, Gas and Solution Mining Regulatory Program Well Permit Issuance for Horizontal Drilling and High-Volume Hydraulic Fracturing to Develop the Marcellus Shale and Other Low-Permeability Gas Reservoirs, Prepared for Natural Resources Defense Council, Jan. 6, 2012.

[34]Madelon Finkel, Jake Hays, and Adam Law. The Shale Gas Boom and the Need for Rational Policy. American Journal of Public Health 103(7), 1161–1163, July 2013.

[35]Theo Colborna, Carol Kwiatkowskia, Kim Schultza, Mary Bachran, Natural gas operations from a public health perspective. Human & Ecological Risk Assessment, 2011, 17, pp. 1039–1056.

[36]Roxana Witter, Lisa McKenzie, Meredith Towle, Kaylan Stinson, Kenneth Scott, Lee Newman, John Adgate, Health Impact Assessment for Battlement Mesa, Garfield County Colorado, Colorado School of Public Health, University of Colorado Denver, Sep. 2010.

[37]Carol Linnitt, Report "Fracking the Future—How Unconventional Gas Threatens our Water, Health and Climate", DeSmogBlog Project, 2010.

[38]Ohio: Shale drillers must report chemicals locally, Associated Press, Oct. 1, 2013.

[39]Standing Committee on Natural Resources, number 040, 3rd session, 40th parliament, Parliament of Canada, Feb. 1, 2011.

[40]Garance Burke, Colorado's Fracking Woes Show Fight Brewing In Oklahoma, Texas And Other Drought-Ridden Areas, Huffington Post, June 16, 2013.

Not surprisingly, this technology was in clear contravention to the strict American standards on environmental pollution. In 2005, Dick Cheney lobbied to obtain amendments of several acts to exclude fracking fluids and shale wells air pollution from the federal government regulation under the Energy Policy Act.[41] It is noteworthy that the US government removed environmental restrictions to the development of shale formations within the United States at the time when energy prices had already reached a sufficient level to make shale exploration profitable. Indeed, in 2005, the annual average oil price was US $54/bbl[42] and the annual average natural gas price on wellhead in the US exceeded US $255/1000 m^3.[43] Consequentially, the combination of eased environment legislation and high energy prices allowed to drill and frack more than 82,000 unconventional wells on 31 shale plays within the United States from 2005 to 2013. At that time, around 1 billion tons of water were contaminated and more than 1400 km^2 of land were damaged.[44]

The main environmental problem of hydraulic fracturing is (i) the very large quantity of water withdrawn from national consumption for hundreds of years due to its contamination by chemicals and (ii) the possibility that areas surrounding the production sites become later contaminated as a result of defective cementing of shale well walls, corrosion of steel elements and deterioration of the well's integrity over the following decades. In 1992, the U.S. Environmental Protection Agency (EPA) reported that an estimated 1.2 million conventional oil and gas wells were abandoned in the U.S., of which 200,000 were leaking (around 16 % of the total well population)[45] (these conventional wells used much less fluids than unconventional wells used in fracking). In 2003, Schlumberger's Oilfield Review stated that "*Since the earliest gas wells, uncontrolled migration of hydrocarbons to the surface has challenged the oil and gas industry... [It is a] significant problem affecting wells in many hydrocarbon-regions of the world*".[46] Operator-wide statistics in Pennsylvania show that about 6–7 % of new shale wells drilled between 2009 and 2011 had compromised structural integrity.[47] U.S. Government

[41]Renee Lewis Kosnik The Oil and Gas Industry's Exclusions and Exemptions to Major Environmental Statutes, Oil & Gas Accountability Project, a Project of Earthworks, Oct. 2007, p. 2.

[42]Oil: Crude oil prices 1861–2009, BP Statistical Review of World Energy 2010, June 2010.

[43]U.S. Natural Gas Wellhead Price (1922–2013), U.S. Energy Information Administration, Apr. 2014.

[44]Fracking by the Numbers, Key Impacts of Dirty Drilling at the State and National Level, Elizabeth Ridlington, John Rumpler, Environment America Research & Policy Center, Oct. 2013, p. 4.

[45]Roberto Suro, Abandoned Oil and Gas Wells Become Pollution Portals, The New York Times, May 3, 1992.

[46]Claudio Brufatto, Jamie Cochran, Lee Conn, David Power, Said Zaki Abd Alla El-Zeghaty, Bernard Fraboulet, Tom Griffin, Simon James, Trevor Munk, Frederico Justus, Joseph R. Levine, Dominic Murphy, Jochen Pfeiffer, Tiraputra Pornpoch, Lara Rishmani, From Mud to Cement—Building Gas Wells, Schlumberger's Oil Field Review, Aug. 2003, p. 63.

[47]Anthony R. Ingraffea, Fluid Migration Mechanisms Due To Faulty Well Design And/Or Construction: An Overview And Recent Experiences In The Pennsylvania Marcellus Play, October 2012, Physicians, Scientists and Engineers for Healthy Energy, p. 8.

Accountability Office revealed the following: "*Fracturing process itself is unlikely to directly affect freshwater aquifers because fracturing typically takes place at a depth of 6000 to 10,000 feet, while drinking water tables are typically less than 1000 feet deep... The fractures are most commonly vertical and may extend laterally several hundred feet away from the well, usually growing upward until they intersect with a rock of different structure, texture, or strength... For example, for over 200 fractures in the Woodford Shale, the typical distance between the drinking water aquifer and the top of the fracture was 7500 feet, with the highest fracture recorded at 4000 feet from the aquifer. In another example, for the 3000 fractures performed in the Barnett Shale, the typical distance from the drinking water aquifer and the top of the fracture was 4800 feet, and the fracture with the closest distance to the aquifer was still separated by 2800 feet of rock... [Nevertheless,] underground migration of gases and chemicals poses a risk of contamination to water quality. Underground migration can occur as a result of improper casing and cementing of the wellbore as well as the intersection of induced fractures with natural fractures, faults, or improperly plugged dry or abandoned wells. Moreover, there are concerns that induced fractures can grow over time and intersect with drinking water aquifers... [I]nadequate cement in the annular space around the surface casing, and ineffective cement may crack or break down under the stress of high pressures. Casing and cementing practices also apply to conventional oil and gas development. However, wells that are hydraulically fractured have some unique aspects. For example, hydraulically fractured wells are commonly exposed to higher pressures than wells that are not hydraulically fractured. In addition, hydraulically fractured wells are exposed to high pressures over a longer period of time as fracturing is conducted in multiple stages, and wells may be refractured multiple times—primarily to extend the economic life of the well when production declines significantly or falls below the estimated reservoir potential*".[48]

Usually, there is 25-years-guarantee on the integrity of an average fracking well: after the guarantee term, nobody can say what will happen with previously pumped toxic water and possible backwater and methane penetration through old pipes on depleted shale energy fields. In fact, scientific research has demonstrated the existence of very broad distributions of time scales for dispersion of contaminants due to the co-existence of cracks and faults of many lengths scales. The complex distributions of sediments and high contrasts in hydraulic properties present in heterogeneous rock formations naturally give rise to a mixture of preferential pathways and stagnant regions, which generally lead to anomalous transport behavior. For instance, the controlled real-life macrodispersion experiments at the Columbus Air Force in Mississipi showed the existence of a long tail of arrival times of products, much after the main dispersion transit, suggesting together with

[48]Oil and gas: Information on Shale Resources, Development, and Environmental and Public Health Risks, U.S. Government Accountability Office, Sep. 2012, pp. 45–47.

many other experiments as well as theoretical models that the notion of a characteristic time scale for dispersion of contaminant is ill-founded[49]: pollutants are likely to leak for decades and centuries.

A large number of independent researches, of reports from residents of areas where shale plays are located and testimonies from landowners of the plots for shale wells show that hydraulic fracturing technology contaminates water sources not only during deep underground injections, but also during the backflow extraction process, when methane and chemical elements are leaking into surface drinking water sources.[50] In February 2012, the US National Oceanic and Atmospheric Administration and the University of Colorado in Boulder found out that up to 4 % of the methane produced in shale gas fields near Denver was escaping into the atmosphere. The American Geophysical Union reported even higher rates of methane leakage in the Uinta Basin of Utah—up to 9 % of the total production. By comparison, EPA suggested that 2.4 % of the total natural-gas production was lost to leakage in 2009.[51] Some researchers argue that the comparative impact of methane emission on global climate change is up to 72 times greater than carbon dioxide emission[52,53] and "*during the life cycle of an average shale-gas well, 3.6 to 7.9 % of the total production of the well is emitted to the atmosphere as methane. This is at least 30 % more and perhaps more than twice as great as the life-cycle methane emissions we estimate for conventional gas [well], 1.7 % to 6 %*".[54] According to Eli Kintisch, reporter for Science magazine, "*Methane plays an outsized role in climate. Although it is naturally 200 times less abundant in the atmosphere than CO_2, the way its four carbon-hydrogen bonds jiggle when struck by infrared radiation makes it a highly effective warmer... Some of that atmospheric methane comes from natural sources, such as gas seeps or wetlands. But an estimated one-fifth of the global total – and about 30 % of U.S. methane emissions – comes from the natural gas infrastructure, from wells to end users, and the fracking boom is adding thousands of potential new sources of emissions...*

[49]Eggleston, J. and S. Rojstaczer, Identification of large-scale hydraulic conductivity trends and the influence of trends on contaminant transport, Water Resour. Res., 34(9), 2155–2168 (1998); Seong, K. and Y. Rubin, Field investigation of the Waste Isolation Pilot Plant (WIPP) site (New Mexico) using a nonstationary stochastic model with a trending hydraulic conductivity field, Water Resour. Res., 35(4), 1011–1018 (1999); Metzler, R. and J. Klafter, The random walk's guide to anomalous diffusion: a fractional dynamics approach, Physics Reports 339, 1–77 (2000); Berkowitz, B. and H. Scher, The role of probabilistic approaches to transport theory in heterogeneous media. Transport in Porous Media 42, 241–263 (2001).

[50]Documentary "Gasland", 2010, Director Josh Fox.

[51]Jeff Tollefson, Methane leaks erode green credentials of natural gas. Nature 493, 12, Jan. 2, 2013.

[52]Overview of Greenhouse Gases, US Environmental Protection Agency, http://epa.gov/climatechange/ghgemissions/gases/ch4.html.

[53]Eli Kintisch, Plugging methane leaks in the urban maze could be key to making shale gas climate-friendly. Science 344 (6191) 1472, June 27, 2014.

[54]Robert W. Howarth, Renee Santoro, Anthony Ingraffea, Methane and the greenhouse-gas footprint of natural gas from shale formations, Climatic Change 106 (4), 679–690, June 2011.

Overall, some researchers estimate that just 20 % of production leaks could account for some 80 % of emissions...".[55] U.S. Government Accountability Office stated that "*According to EPA analysis, natural gas well completions involving hydraulic fracturing vent approximately 230 times more natural gas and volatile organic compounds than natural gas well completions that do not involve hydraulic fracturing*".[56]

Furthermore, there are risks of fracking-induced earthquakes, and billions of gallons of backflow waste-water injected into disposal wells could also cause earthquakes, with significant losses to other industries. The largest human-caused earthquake in the continental US was associated with wastewater disposal by injection into deep wells, which poses a higher risk than hydraulic fracturing per se, because this practice can induce larger earthquakes.[57] It measured 5.7 on the Richter scale and was registered in November 2011 in Prague, Oklahoma[58]: it destroyed 14 homes and injured two people. The second largest injection-induced earthquake had the magnitude 5.5 on the Richter scale and occurred in Rocky Mountain Arsenal, Colorado in the early 1960s.[59] However, the US Geological Survey (USGS) has not had the budget for a comprehensive research of induced seismicity of wastewater disposal wells until recently.[60] From 1962 to 2012, after decades of a steady earthquake rate (average of 21 events of magnitude ≥ 3 in the U.S. midcontinent per year), the activity increased starting in 2001 and peaked at 188 earthquakes in 2011.[61] Shale energy operators intend to inject annually billions of cubic meters of wastewater by 2030, when the US expects to get 50 % of all its natural gas from shale plays. If they do, nobody can guarantee that a more powerful injection-caused earthquake will not happen, damaging or destroying vulnerable state infrastructure—nuclear plants, chemical plants, oil and gas pipelines, etc.—in areas of massive shale energy exploitation.

The US government uncritical attitude towards shale exploration has made the collection of official information and scientifically confirmed evidence for the risks of shale exploration a very difficult process. The US government has

[55]Eli Kintisch, Plugging methane leaks in the urban maze could be key to making shale gas climate-friendly. Science 344 (6191), 1472, June 27, 2014.

[56]Oil and gas: Information on Shale Resources, Development, and Environmental and Public Health Risks, U.S. Government Accountability Office, September 2012, p. 35.

[57]J.R. Grasso and D. Sornette, Testing self-organized criticality by induced seismicity, J. Geophys.Res. 103 (B12), 29965–29987, 1998 and references therein.

[58]2011 Oklahoma Induced Earthquake May Have Triggered Larger Quake, U.S. Department of the Interior, U.S. Geological Survey, March 6, 2014.

[59]Anne F. Sheehan, Microearthquake study of the Colorado front range: Combining research and teaching in seismology, Department of Geological Sciences and CIRES University of Colorado, March 24, 2000.

[60]Budget Justifications and Performance Information Fiscal Year 2014, U.S. Geological Survey, The United States Department of the Interior, pp. B-34, H-11.

[61]William L. Ellsworth, Injection-Induced Earthquakes, Science 341, 1225942, 1–7, 2013.

eased environmental legislation on the exploration of shale formations; and it has allowed operators to avoid mandatory disclosure of the chemicals used in hydraulic fracturing to extract oil and gas on the argument that the fracking mixture is a "trade secret". In January 2011, energy companies, forced by public pressure, launched FracFocus, an online voluntary chemical disclosure registry, which contains information about pumped chemical elements at 56,000 oil and gas shale wells across the United States. Nevertheless, there is no intention for setting up a nationwide regulation of the industry and for the development of an official national database of all fracked wells. The absence of a united government managed database on the dynamic of well productivity, gas content, geophysical features, injected fracking fluids and development activity on every unconventional well in the US including capital and operating costs (such as leasehold, drilling and completion, maintenance expenses, refracturing economics) contrast with the worldwide standards of conventional energy production that request such information.

Because of national energy security interests, the two government organizations with a duty to monitor the environmental aspects of shale energy production—the EPA and USGS—manifest a rather passive attitude with respect to the risks of hydraulic fracturing. An EPA whistleblower stated that, during the Bush administration, top EPA officials had conflicting interests with shale gas companies, and helped them to neglect possible risks of water contamination from fracking.[62] Under the Obama administration, the EPA has continued to underplay the environmental damage of the shale industry in order to help the country obtain energy independence by any means. For instance, massive shale gas drilling has been carried out since 2005, when Bush canceled the restriction of the Clean Water Act for the shale gas industry; but the EPA had plans to issue a complete report evaluating the "potential" impact of hydraulic fracturing on drinking water resources only by late 2014[63]—after a decade (!) of intensive shale gas exploration in the US, and halfway through the second term of Obama's presidency. The EPA also assesses the volume of every planned wastewater disposal well in the country, gives permits to the shale industry and controls wells, but the regulator's control over wells and the activity of well users seems insufficient.

Moreover, due to the fact that each of the 30-odd shale deposits being "played" is geologically unique, operators persuaded the US government to implement state level regulation of shale exploration activity, instead of a comprehensive integrated nationwide oversight. In this respect, Halliburton stated: "*Every shale is different ... after two decades of development and several iterations of the learning curve, best practices are application-dependent and must evolve locally... Due to the unique nature of shale, every basin, play, well and pay zone may require a unique*

[62]Documentary "Gasland", 2010, Director Josh Fox (evidence that EPA's top-level executives had close relation with shale lobby on 30th minute of the film).

[63]Questions and Answers about EPA's Hydraulic Fracturing Study, US Environmental Protection Agency, http://www2.epa.gov/hfstudy/questions-and-answers-about-epas-hydraulic-fracturing-study.

treatment".[64] The Interstate Oil and Gas Compact Commission, representing the governors of the 37 states that produce oil and natural gas in the US, had views similar to those of Halliburton and offered to leave regulation of shale business at the state level: "*Hydraulic fracturing has been used safely to stimulate oil and gas production in the United States for more than 60 years… Additional study is unnecessary, and in fact, would be a wasteful use of taxpayers' dollars. However, all future studies involving the regulation of oil and natural gas exploration and production must involve leadership by those officials who know it best – state regulators… Further regulatory burdens are unnecessary, and in fact, would delay the development of vital domestic natural gas resources and increase energy costs to the consumer with no resulting environmental benefit... The states feel strongly that additional studies and certainly additional regulatory oversight are unnecessary… Conducting further studies or enacting legislation to address a "problem" that has never been documented – contamination of drinking water as a result of hydraulic fracturing – is simply a waste of taxpayers' money*".[65] This has resulted in uncoordinated regulation with little or no interaction between different regulators, which helps energy companies to hide the whole picture of the risks and prevent authorities from grasping the possible problems nation-wide.

For instance, in 2011, EPA announced for the first time that fracking may be to blame for causing groundwater pollution based on evidences of the existence of fracking chemicals in the groundwater beneath Pavillion, Wyoming. EPA emphasized that the findings are specific to the Pavillion area,[66] while industrial experience confirms that the exploration technology, used by shale operators, are generally similar on different shale plays. Other states, where shale energy production occurs, have their own specific state legislation. Consequently, this leads to situation when regulators have a fragmented picture of risks, which does not allow them to understand all the risks associated with the scale and complexity of exploiting shale formations using innovative, but unproved technology. A similar approach prevailed during the regulation of subprime mortgage deals, when players of the "securitization pipeline" did not understand the whole picture of risks and—because there was no mega-regulator to oversee the whole picture—"*nobody had a 360-degree view*".[67] Moreover, due to massive reduction of government spending on federal and state levels, budgets of many departments of EPA were reduced and many representatives of the agency were fired. This parallels also the

[64]Halliburton Shale Solutions, July 2008, http://www.halliburton.com/public/solutions/contents/Shale/related_docs/H06377.pdf.

[65]Testimony Submitted to the House Committee on Natural Resources Subcommittee on Energy and Mineral Resources, Washington, D.C., The Interstate Oil and Gas Compact Commission on behalf of the Nation's Oil And Gas Producing States, June 18, 2009, pp. 1, 5.

[66]Mead Gruver, EPA: Fracking may cause groundwater pollution, Associated Press, Dec. 8, 2011.

[67]The Financial Crisis Inquiry Report, The Financial Crisis Inquiry Commission, Washington, D.C., Jan. 2011, p. 172.

decrease of funding of the Security Exchange Commission in the decade preceding the financial crisis and great recession, in a global climate of deregulation in the US.[68]

A serious problem has recently been revealed that affects the credibility of previous scientific evidences concerning the real environmental impact of fracking on air and groundwater: it was found that much of the research on these topics has been funded—through speaking fees, grants, and joint research programs with energy companies—by groups with either pro- or anti-energy development agendas. However, it is claimed that "*potential conflicts of interest or sources of bias have not influenced the research*".[69] Reliance on industry–funded reports allowed Shell CEO to say the following: "*The US Environment Protection Agency, together with independent environmental experts, including the Groundwater Protection Council, has classified hydraulic fracturing as a proven and safe technique*".[70] The mentioned Ground Water Protection Council is a research group which, according to tax records, is partly financed by industry[71]: it received from the American Petroleum Institute (the main lobbing organization of American energy industry) US $57,500 and $47,500 for "*Energy Policy Research*".[72] The Ground Water Protection Council was also the author of "*Primer*" (the study was ordered by U.S. Department of Energy and the National Energy Technology Laboratory and published in 2009), which promoted shale gas development and gave an excellent overview of the perspective of fracking and the absence of serious threat of water contamination. In 2011, the U.S. Department of Energy confessed that the "*Primer*" was never intended as a comprehensive review and that further study would continue.[73] In April 2011, the Ground Water Protection Council launched the previously mentioned FracFocus site, which was funded by oil and gas trade groups and the US Department of Energy. As another example, the Massachusetts Institute of Technology issued a report in 2011 on "*The Future of Natural Gas*", which stated that, with around 20,000 shale gas wells in operation during the preceding decade, only 43 "*widely reported*" water contamination incidents had been registered.[74,75] Such statement is surprising given that nobody has reliable statis-

[68]Nolan McCarty, Keith T. Poole, Howard Rosenthal, Political Bubbles: Financial Crises and the Failure of American Democracy, Princeton University Press, May 26, 2013, pp. 123,146.

[69]Russell Gold, Gas Boom Projected to Grow for Decades, Russell Gold, The Wall Street Journal, Feb. 27, 2013.

[70]Peter Voser, End of the oil boom? Notenshtein Dialogue, December 2013, p. 4.

[71]Ian Urbina, Behind Veneer, Doubt on Future of Natural Gas, The New York Times, June 26, 2011.

[72]American Petroleum Institute, Sourcewatch's profile, http://www.sourcewatch.org/index.php/American_Petroleum_Institute.

[73]Ian Urbina, Behind Veneer, Doubt on Future of Natural Gas, The New York Times, June 26, 2011.

[74]Water and Shale Gas Development. Leveraging the US experience in new shale developments, Accenture, 2012, p. 6.

[75]The Future of Natural Gas, MIT Energy Initiative, 2011, pp. 39–40.

tics about the real scale of water contamination by fracking: this would require federal government regulation of the industry, which could provide a unified system of reporting on the chemical composition of water in sources around every shale well in the country during its lifetime (before and after fracking), and disclosing the amount and content of chemical compounds in the fracking mixture. The US Government Accountability Office reacted to this credibility problem of scientific research in the following way: "*[The] extent of risks associated with shale oil and gas development is unknown, in part, because the studies we reviewed do not generally take into account potential long-term, cumulative effects... [T]hese studies are generally anecdotal, short-term, and focused on a particular site or geographic location... Without data to compare predrilling conditions to postdrilling conditions, it is difficult to determine if adverse effects were the result of oil and gas development, natural occurrences, or other activities*".[76] The previously mentioned MIT report was issued under the supervision of Professor Ernest Moniz and funded by oil and gas companies.[77,78] Prof Moniz declared that natural gas from shale formations is the "*cost-effective bridge to such a low-carbon future*" and that "*the large amounts of natural gas available in the U.S. at moderate cost... natural gas can indeed play an important role over the next couple of decades in economically advancing a clean energy system... the development of low-cost and abundant unconventional natural gas resources, particularly shale gas, has a material impact on future availability and price*".[79,80] In May 2013, President Obama offered Ernest Moniz the influential position of US Secretary of Energy, an appointment, which is always associated with large lobbying support of a candidate by American energy companies.

4.1.3 Collapse of Natural Gas Prices in the United States (2008–2009)

From 2005 to 2008, the annual domestic price of natural gas was favorable to shale operators as it ranged from US \$220 to US \$310/1000 m^3 on the main energy hub—the Henry distribution hub in Louisiana, where the price of natural gas is 10–15 % higher than the wellhead price due to additional transportation

[76]Oil and gas: Information on Shale Resources, Development, and Environmental and Public Health Risks, U.S. Government Accountability Office, September 2012, pp. 32–33, 49.

[77]Lynne Peeples, Fracking Industry Conflicts Of Interest With Regulators? Huffington Post, Feb. 27, 2013.

[78]Lynne Peeples, Ernest Moniz: Considered Fracking Shill By Some Environmentalists, Sparks Concern Amid Energy Department Nomination, Huffington Post, June 10, 2013.

[79]Brad Plumer, Is fracking a 'bridge' to a clean-energy future? Ernest Moniz thinks so, The Washington Post, March 4, 2013.

[80]The Future of Natural Gas, MIT Energy Initiative, 2011, p. 2.

costs of the gas from a field to the hub.[81] Based on these prices exceeding the costs of shale gas extraction that promised lucrative return on investment, operators together with Wall Street investment banks invited billions of dollar from local and foreign investors. As a result, both the debts and the market capitalizations of shale gas operators increased significantly.[82] But, during the world financial crisis in 2008, the flee of speculative money from the oil market led to a fourfold oil price drop from US $147/bbl to $37/bbl within half a year. Over the following year, from July 2008 to July 2009, the weekly prices of natural gas in the USA fell by a factor 3.5 (from US $460/1000 to $125/1000 m^3[83]), but did not return to the pre-crisis level as happened for the oil market. During the period from 2009 to 2013, the annual average price of natural gas on the Henry hub wobbled between US $95 and $135/1000 m^3,[84] while the production costs of operators were in the range US $140–$230/1000 m^3.[85,86,87] In 2011, the supply of natural gas within the US exceeded the demand by a factor four.[88]

The main reasons for such low prices included: (i) the long-term economic slowdown of the American economy; (ii) the huge amount of extracted shale gas, which exceeded the demand from corporate consumers inside the USA such as fertilizer and chemical plants or from the national car industry involved in a massive conversion from benzene and diesel to natural gas; (iii) the liability imposed on shale gas operators by the burgeoning futures market to produce the expected amount of gas; (iv) the pressure on operators to produce gas fast enough to pay regular royalties to the owners of leased land plots and avoid losing the plots before their hydrocarbon resources were exhausted; and (v) the large debts of operators, which required them to receive cash flow from any source.

To understand shale business, it is important to take into consideration certain geological properties of shale formations, namely that hydrocarbons have to be extracted immediately after fracking without the possibility of suspending shale

---

[81] Henry Hub Natural Gas Spot Price (1922–2013), U.S. Energy Information Administration, May 2014.

[82] Deborah Rogers, Financial Co-Dependency: How Wall Street Has Kept Shale Alive, Energy Policy Forum, Oct. 23, 2012.

[83] Henry Hub Natural Gas Spot Price (1922–2013), U.S. Energy Information Administration, May 2014.

[84] Ibid.

[85] For our calculations, we use the following conversion rates of natural gas from American units into European metric: 1 million British Thermal Units (MMbtu) = 1000 cubic foots (Mcf), 35,000 cubic foots (*Mcf*) = 1000 cubic meters (m^3) of natural gas. We use the factor 35 for converting the price of 1 MMbtu into 1000 m^3.

[86] The Future of Natural Gas, Appendix 2D: Shale Gas Economic Sensitivities, MIT Energy Initiative, 2011, p. 2.

[87] First 5 years of "shale gas revolution". What we now know for sure? Centre for Global Energy Markets of Energy Research Institute of the Russian Academy of Sciences, Nov. 2012, pp. 25–26, 32.

[88] Deborah Rogers, Shale and Wall Street: was the decline in natural gas prices orchestrated?, Energy Policy Forum, Feb. 2013, p. 1.

well production because of the on-going triggered gas diffusion in the stimulated rock. Active fracking can thus lead to natural gas glut and falling natural gas prices. Since 2008, when the sale price of shale gas crashed, the operators had been working at a loss.[89] According to calculations of Arthur Berman, a veteran in petroleum geology specialized in well assessment, after more than 7 years of horizontal drilling (2005–2011), less than 6 % of Barnett Shale wells have reached or exceeded break-even production volumes (1.5 Bcf [90] at US \$210/1000 m^3 netback gas price); Haynesville play will not be profitable as long as gas prices remain less than US \$245/1000 m^3.[91,92] The absence of gas pipeline operators and low natural gas prices led operators of the Bakken oil fields in North Dakota (they use hydraulic fracturing for production of shale oil from depleted conventional fields) to flare around 30 % of North Dakota's associated natural gas production according to 2011 data.[93] Because the sale of dry shale gas (methane) had proved so disappointing, operators began to extract natural gas liquids (propane, butane, ethane, and so on) and shale oil, which are far more lucrative than dry shale gas.[94] The cost of producing a barrel of shale oil varies between US \$55 and \$85 per barrel; and the prices of natural gas liquids and shale oil are connected to the price of oil. This helped operators to subsidize unprofitable shale gas extraction until the collapse of oil prices that lost 50 % from the last quarter of 2014 to the first quarter of 2015. The fall of oil prices was magnified by the intentional actions of Saudi Arabia and other OPEC members, which did not reduce their productions when evidence of over-supply transpired, taking this as an opportunity to destroy American shale industry and regain the largest share of the oil market.[95,96,97,98,99]

[89]Michael Liebreich, Keynote—Day 2, Bloomberg New Energy Finance Summit, 20 March, 2012, slide 35.

[90]1 Bcf is a unit of gas energy, which means one billion cubic feet equivalent. In terms of energy content, it is approximately equal to 1.028 trillion BTU. The British thermal unit (BTU or Btu) is a traditional unit of energy equal to about 1055 joules.

[91]Arthur E. Berman, Presentation "Shale Gas—The Eye of the Storm", Calgary, Alberta, July 14, 2011, slides 5–8, http://www.artberman.com/presentations/Berman_Shale%20Gas--The%20Eye%20of%20the%20Storm%2020%20July%202011_OPT.pdf.

[92]Arthur E. Berman and Lynn F. Pittinger, U.S. Shale Gas: Less Abundance Higher Cost, The Oil Drum, August 5, 2011, http://www.theoildrum.com/node/8212.

[93]Over one-third of natural gas produced in North Dakota is flared or otherwise not marketed, U.S. Environmental Protection Agency, Nov. 23, 2011.

[94]Chesapeake Energy Corporation Annual Report 2012, April 2013, p. 3.

[95]Will Kennedy and Jillian Ward, OPEC Policy Ensures U.S. Shale Crash, Russian Tycoon Says, Bloomberg, Nov. 27, 2014.

[96]U.S. Shale Oil Output Growth May Stall at \$60 a Barrel, Oil may fall to \$50, EIA Says, Bloomberg, Nov. 18, 2014.

[97]Saudi Arabia to keep politics out of OPEC, will let market stabilize price, RT, Nov. 26, 2014.

[98]Anjli Raval, Prices, not Opec, to balance oil supply, FT, Nov. 28, 2014.

[99]Factbox—OPEC oil ministers positions ahead of Thursday meeting, Reuters, Nov. 26, 2014.

Due to fact that more than a quarter of the costs of shale well working-out are fluid-related expenses,[100] water was selected by the government and the industry to continue to be the fracking fluid for its low cost (a barrel of water costs just US $0.8). The focus on low costs augmented by the pressure from low gas prices prevents mitigation measures using water-free fracking that could reduce environmental damage. For example GasFrac—a small energy company based in Alberta, Canada—offers liquefied propane gas (LPG) fracturing service. This uses a thick propane gel in place of water. They pump a mix of gel and sand into shale formations. By 2014, GasFrac had performed more than 2500 propane fracks in 700 wells.[101] LPG fracking is at least 20–40 % more expensive than hydraulic fracking,[102,103] but uses no water at all. However, because the production costs of shale gas in the United States during the last few years have exceeded the sale price of natural gas, the transition to more environmentally friendly water-free technology could deteriorate even more the economics of the operators, provoke a collapse of the capitalization of shale energy producers, and even halt shale energy production in US due to financial troubles. The survival of the shale energy industry can be secured only by maintaining high energy prices and by keeping production costs no higher than at their present level (i.e., avoiding costly regulations), which is supported by patriotism and, perhaps a misplaced, blind attitude of the US government toward the environmental risks of hydraulic fracturing.

4.1.4 Manipulation of the Estimation of Unconventional Oil and Gas Resources

The value of any oil and gas company is based on the evaluation of its reserves—the amount of technically and economically recoverable oil or gas. In 2004, a scandal erupted concerning the overestimation of the reserves of Royal Dutch/Shell, when the company overbooked its proven reserves by 4.5 billion barrels, or 23 % of Shell's total reserves.[104] Shell was fined US $120 million by the Securities and Exchange Commission (SEC) and £17 million by the Financial Services Authority. Some analysts suggested that this is a common problem for the

---

[100] Ross Tomson, Water use in fracking needs to be refined, Houston Business Journal, March 4, 2013.

[101] Patrick J. Kiger, Green Fracking? 5 Technologies for Cleaner Shale Energy, National Geographic, March 19, 2014.

[102] Matt Goodman, Waterless Fracking Method Targets Natural Gas Industry's Gaze, CBS News, Jan. 12, 2012.

[103] Sean Milmo, Fracking with propane gel, Royal Society of Chemistry, Nov. 15, 2011.

[104] Mark Tran, Shell fined over reserves scandal, The Guardian, July 29, 2004.

oil and gas industry[105] that "*no one wants to talk about*".[106] For decades, overestimation of reserves was recognized as fraud and misrepresentation to investors.

However, everything changed in 2008 during the global financial and economic crisis and liquidity shortage, when prices on natural gas and oil dropped by factor of 3–4 in less than half a year. In order to convince investors and other countries that a "shale revolution" could offer resources exploitable for centuries, and that investing in stocks of shale energy companies during the economic meltdown could be lucrative, or using them as a secured pledge in exchange for credits from commercial banks, the US government changed the rules for the evaluation of unconventional reserves and resources. In early 2009, Bush's administration—by then in the last weeks of its term—implemented new SEC rules. These allowed domestic energy companies to book unconventional oil and gas reserves more optimistically, based on internal company estimates of the amount of hydrocarbons recoverable from a potential oil or gas field—estimates made without test drilling and without regard for the economic viability or technical feasibility of extracting these reserves. According to a New York Times investigation, after the implementation of the new SEC rules in 2009, at least seven companies among the largest 19 shale operators increased their estimated reserves—some by more than 200 %.[107] On the basis of actual well production data filed in various states, some analysts estimated that operators overbooked their shale reserves by 400–500 %.[108] This change in SEC rules resembles the change in regulation in 1992 during the mandate of George H. W. Bush, when the SEC accepted the mark-to-market accounting method for the energy contracts of Enron Gas Services. This later allowed Enron to calculate its own revenue by the market value of derivatives trading—in other words, on the basis of their own estimate of how deals would perform in the future—and to create the illusion of being "larger" than General Electric or IBM.[109]

In parallel with this change of the SEC rules, shale energy operators began to convince investors and the public that the output of shale wells would be similar to conventional gas and oil wells. The life cycle of an average conventional gas well is 20–30 years, when intra-formational pressure helps to push natural gas out of

[105]Laherrere, Jean, Estimates of oil reserves, paper presented at the EMF/IEA/IEW meeting IIASA. Laxenburg, Austria, June 19, 2001 (http://energycrisis.info/laherrere/iiasa_reserves_long.pdf); Forecasting future production from past discovery, International Journal of Global Energy Issues 18 (2–4), 218–238 (2002); Oil and gas: what future? Groningen annual Energy Convention 21 November 2006, ASPO (Association for the Study of Peak Oil & gas) & ASPO France (http://oilcrisis.com/laherrere/groningen.pdf).

[106]Grant T. Olsen, W.John Lee, Thomas A. Blasingame, Reserves Overbooking: The Problem We're Finally Going to Talk About, Society of Petroleum Engineers Economics & Management, Apr. 2011.

[107]Ian Urbina S.E.C. Shift Leads to Worries of Overestimation of Reserves, The New York Times, June 27, 2011.

[108]Nafeez Ahmed, Shale gas won't stop peak oil, but could create an economic crisis, The Guardian, June 21, 2013.

[109]B. G. Dharan, W. R. Bufkins, Red Flags in Enron's Reporting of Revenues & Key Financial Measures, July 23, 2008, p. 3.

the well. But in a shale gas well, the pressure is only high during the fracking process, when fluid is forced into the well by tremendous artificial pressure. The natural intra-formational pressure of typical shale well is not as strong as that in a conventional well over the long term. This is the very reason for the use of fracking and pressurization in the first place. Accordingly, shale wells produce 74–82 % of their lifetime output in the three years of exploitation.[110] Thus, the full life cycle of shale well is only around three years. Nevertheless, shale operators began to claim a very high estimated ultimate recovery (EUR) for their wells by a blatant piece of creative accounting: they simply multiplied the high initial production rate of a shale well during its first weeks after fracking by the three-decade lifetime usually assumed for a conventional well, instead of using the three years typical for the life of a shale well. Moreover, the producers initially prefer to search and drill "sweet spots", where production rates and rates of return are high in comparison with normal or average wells on a shale play, exploration of which are adjourned to a later time. As a result, they started to claim a tremendously high EUR for their wells and overstated reserves of their shale plays. For example, in investor presentations, Chesapeake Energy Corp. declared average EURs of 4.2 billion cubic feet (Bcf)[111] for their wells in the Marcellus region of Pennsylvania, Range Resources claimed 5.7 Bcf, and Cabot Oil & Gas Corporation 15 Bcf on their newest wells. According to the USGS, however, the average EUR for the Marcellus wells is only 1.1 Bcf.[112] US Government Accountability Office notes: "*EIA reports that experience to date shows production rates from neighboring shale gas wells can vary by as much as a factor of 3 and that production rates for different wells in the same formation can vary by as much as a factor of 10*".[113]

These manipulations allowed shale operators to reduce the "official" production costs of exploiting shale formations: they began to divide current expenses for drilling any shale well by their own generously-evaluated EUR for the well. If an operator declared "tremendous" EURs for their wells, then "official" production costs on wells could be calculated as very low. However, in reality this helped the operators to conceal their huge production costs for exploiting shale formations, increase the market value of the companies by unverified resources in order to convince investors to continue investing in the "shale revolution". Resource manipulation also allows the US government to predict a century of natural gas abundance—see for example Barack Obama's declaration that "*We have a supply of natural gas that can last America nearly 100 years*"[114]—though nobody can

[110]J. David Hughes, Drilling Deeper: A Reality Check on U.S. Government Forecasts for a Lasting Shale Boom, PART 1: EXECUTIVE SUMMARY, Post Carbon Institute, Oct. 26, 2014.

[111]1 Bcf is a unit of gas energy, which means one billion cubic feet equivalent. In terms of energy content, it is approximately equal to 1.028 trillion BTU. The British thermal unit (BTU or Btu) is a traditional unit of energy equal to about 1055 joules.

[112]Deborah Rogers, USGS releases troubling EURs for shale, Energy Policy Forum, Sep. 7, 2012.

[113]Oil and gas: Information on Shale Resources, Development, and Environmental and Public Health Risks, U.S. Government Accountability Office, Sep. 2012, p. 24.

[114]Remarks by the President in State of the Union Address, The United States Capitol, Jan. 24, 2012.

actually predict the amount of shale hydrocarbons that could be produced with the current technology of shale energy extraction. This collision with shale gas resources estimations urged US Government Accountability Office to issue a remarkably vague statement for the main governmental comptroller that "*the amount of domestic technically recoverable shale gas could provide enough natural gas to supply the nation for the next 14 to 100 years*".[115]

4.1.5 Stressed Financial Situation of American Shale Operators

These manipulations in the way resources were estimated allowed American shale companies to improve their investment attractiveness within the States. Wall Street investment banks earned billions of dollars in transaction fees from shale mergers and acquisitions—in 2009, the value of deals totaled US $50 billion, in 2010 $38 billion, in 2011 $47 billion.[116] Since the beginning of the shale boom (2006–2013), the total value of deals surpassed US $200 billion.[117] Because the banks' fees are calculated on the value of merger and acquisition (M&A) deals, the banks are incentivized not to reveal the financial problems of shale operators in their reports and presentations. This practice again resembles the behavior of Wall Street investment banks in the Enron case (1996–2001), when the investment banks had considerable revenues from underwriting or merger deals, while broker fees brought insignificant profits. The same situation is being played out in the case of the shale industry: Wall Street investment banks prefer to publish positive or neutral reports about the financial state of shale operators.

According to Deborah Rogers, financial consultant for several major Wall Street firms, "*analysts and investment bankers... emerged as some of the most vocal proponents of shale exploitation*". She also cites the words of Neal Anderson of Wood Mackenzie about the investment community and shale exploration: "*It seems the equity analyst community has played a key role in helping to fuel the shale gas M&A market, acting as chief cheerleaders for shale gas plays*".[118] American and international conventional producers, and also unsophisticated international investors from Asian countries including Japan, China, and South Korea, all bought American shale assets; and from all of these deals, Wall Street investment banks extracted substantial transactional fees.

In 2010, Exxon acquired XTO Energy, one of the leaders of unconventional production in the US, with an estimated value of between US $26 billion and $41

[115]Oil and gas: Information on Shale Resources, Development, and Environmental and Public Health Risks, U.S. Government Accountability Office, Sep. 2012, p. 19.

[116]Deborah Rogers, Shale and Wall Street: was the decline in natural gas prices orchestrated?, Energy Policy Forum, Feb. 2013, p. 13.

[117]Ivan Sandrea, US shale gas and tight oil industry performance: challenges and opportunities, The Oxford Institute for Energy Studies, Mar. 21, 2014, p. 2.

[118]Deborah Rogers, Shale and Wall Street: was the decline in natural gas prices orchestrated?, Energy Policy Forum, Feb. 2013, p. 7.

billion. This deal made Exxon the largest producer of natural gas in the US. Nevertheless, in 2012, Exxon Chief Executive Rex Tillerson complained to investors: "*We are all losing our shirts today... We're making no money [on shale gas]. It's all in the red*".[119] Moreover, after shale "assets" were integrated into Exxon, the company's profit per barrel of oil and gas in 2012 fell by 23 % compared with a year earlier.[120] And Exxon were not alone: in the middle of 2012, BP announced write-downs of US $4.8 billion, the British BG Group debited $1.3 billion on its shale investments, and the Canadian EnCana lost $1.7 billion and informed its shareholders that this amount would increase if gas prices did not return to an "acceptable" level.[121] Shell has spent about $30 billion on US and Canadian shale plays,[122] but in September 2013 they announced the sale of land in Texas, Kansas and Colorado, including the largest field Eagle Ford. The company admitted that 192 wells "*are not in a position to reach the planned volume production*," announced a debit of US $2.1 billion and began a strategic reassessment of investment in oil shale deposits in the United States. Thus, in 2014, Shell implemented a new strategy—"fix or divest"—for its Marcellus shale' assets (Warrendale-based East Resources with investment of US $4.7 billion) after underwhelming results over the past several years.[123] In March 2014, Ben van Beurden, one of the Shell executives, was urged to confess that "*Financial performance there is frankly not acceptable [for US onshore assets] ... some of our exploration bets have simply not worked out*".[124] The Australian BHP Billiton joined the shale race only in 2011, acquiring the $15.1 billion Texan company Petrohawk Energy, but a year later was already reporting devaluation of US shale assets.[125] According to calculations of Arthur Berman, companies with strong shale emphasis have poor return on equity; 16 shale companies wrote down $35 billion in 2009; $65 billion in write-downs and goodwill over 2008–2010.[126]

Nowadays, the main challenge for the US shale industry is the necessity for continuously refracking or drilling new wells in order to keep up the current level of production, because of the geological nature of shale formations: as we have noted, up to 60–80 % of the total volume of a shale well is extracted within the first one or two years. This means that just to maintain the current production level

---

[119] Jerry A. DiColo, Tom Fowler, Exxon: 'Losing Our Shirts' on Natural Gas, The Wall Street Journal, June 27, 2012.

[120] Daniel Gilbert, Justin Scheck, Tom Fowle, Shale-Boom Profits Bypass Big Oil, The Wall Street Journal, Aug. 2, 2013.

[121] F. William Engdahl, The Fracked-Up USA Shale Gas Bubble, Global Research, March 14, 2013.

[122] Deon Daugherty, As Shell sells off even more major shale assets, what could be next? Houston Business Journal, Sep. 30, 2013.

[123] Anya Litvak, Shell to restructure shale assets in U.S., Pittsburgh Post-Gazette, March 13, 2014.

[124] Karolin Schaps, Dmitry Zhdannikov, Shell cuts spending in U.S. to lower shale exposure, Reuters, March 13, 2014.

[125] Daniel Gilbert, Justin Scheck, Tom Fowle, Shale-Boom Profits Bypass Big Oil, The Wall Street Journal, Aug. 2, 2013.

[126] Arthur E. Berman, Presentation "Shale Gas—The Eye of the Storm", Calgary, Alberta, July 14, 2011, slides 5–8, http://www.artberman.com/presentations/Berman_Shale%20Gas--The%20Eye%20of%20the%20Storm%2020%20July%202011_OPT.pdf.

of shale gas and shale oil, according to the Canadian geologist David Hughes, operators need to invest up to US $42 billion per year in 7000 wells to maintain production of natural gas production and US $35 billion per year in 6000 wells in shale oil production.[127] In 2012, the total value of extracted shale gas within the USA was only around $32–33 billion. After 2014, the drop of oil prices reduced considerably the total value of extracted shale oil and gas, which became insufficient to cover the high production costs within shale industry.

The deliberate overestimation of resources has only postponed the bankruptcy of shale operators: it helped them to hide the real production costs of shale exploration by legal creative accounting, but it could not help them to improve financial results, which for some shale operators has reached pre-default state (tremendous liabilities versus modest or negative net income).[128] The dire financial condition of shale operators demonstrates that the impressive growth of unconventional gas and oil production within USA does not have a sound economic basis with current price level and was only financed by a tremendous growth of borrowing on the part of the operators or sale of previously accumulated leasing land plots with rights on drilling of shale wells.

Aubrey McClendon, former CEO of Chesapeake Energy, claimed during an investors' call in late 2008: "*I can assure you that buying leases for x and selling them for 5x or 10x is a lot more profitable than trying to produce gas at $5 or $6 mcf [US $175 or $210/1000 m³]*".[129] According to 2012 article of Jeff Goodell, "*For Chesapeake, the primary profit in fracking comes not from selling the gas itself, but from buying and flipping the land that contains the gas. The company is now the largest leaseholder in the United States, owning the drilling rights to some 15 million acres – an area more than twice the size of Maryland... At Chesapeake, McClendon operated more like a land speculator than an oilman... [B]uying up such huge swaths of land requires huge chunks of cash – and the money often comes not from gas production, but from selling off land or going into debt... According to Arthur Berman, ... Chesapeake and its lesser competitors resemble a Ponzi scheme, overhyping the promise of shale gas in an effort to recoup their huge investments in leases and drilling. When the wells don't pay off, the firms wind up scrambling to mask their financial troubles with convoluted off-book accounting methods. 'This is an industry that is caught in the grip of magical thinking,' Berman says. 'In fact, when you look at the level of debt some of these companies are carrying, and the questionable value of their gas reserves, there is a lot in common with the subprime mortgage market just before it melted down*'".[130] Reuter investigation in 2012 unveiled the following: "*[Chesapeake Energy] is taking in more money from bankers, other investors and its own financial bets than it is from its oil and gas. Most big energy companies, such as*

---

[127]David Hughes, Drill, Baby, Drill: Can Unconventional Fuels Usher in a New Era of Energy Abundance? Post Carbon Institute, Feb. 2013,p. ii.

[128]Financial results of Chesapeake Energy Corp., Devon Energy Corp., Encana Corp., Range Resources Corp. Cabot Oil & Gas Corp., EOG Resources Inc. from Google Finance Database. Oct. 2014. http://www.google.com/finance.

[129]Deborah Rogers, Shale and Wall Street: was the decline in natural gas prices orchestrated? Energy Policy Forum, Feb. 2013, p. 11.

[130]Jeff Goodell, The Big Fracking Bubble, Rolling Stone, March 1, 2012.

Exxon Mobil Corp, typically earn more selling oil and gas than they spend on investments, financing and other costs, making them cash rich. Chesapeake is expanding so fast that it takes in much less revenue from its oil and gas than it spends, leaving it stretched. Hence, its business depends on deal-making: raise money from investors to acquire land and drill wells; sell the rights to the gas and oil in those wells; plow that money into new land and wells; repeat the cycle all over again".[131] According to Bloomberg: "*Shale debt has almost doubled over the last four years [2010–2013] while revenue has gained just 5.6 percent, according to ... analysis of 61 shale drillers. A dozen of those wildcatters are spending at least 10 percent of their sales on interest compared with Exxon Mobil Corp.'s 0.1 percent... In a measure of the shale industry's financial burden, debt hit $163.6 billion in the first quarter [2014], according to company records compiled by Bloomberg on 61 exploration and production companies that target oil and natural gas trapped in deep underground layers of rock*".[132] Bloomberg also discovered the following: "*It's an expensive boom. About $156 billion will be spent on exploration and production in the U.S. this year [2014]... The spending never stops, said Virendra Chauhan, an oil analyst with Energy Aspects in London. Since output from shale wells drops sharply in the first year, producers have to keep drilling more and more wells to maintain production. That means selling off assets and borrowing more money*". At a time when the oil price is languishing at its lowest level in six years [August 2015], [U.S. onshore oil producers] need to find half a trillion dollars to repay debt. Some might not make it.... As companies run low on cash, they may be forced to idle drilling rigs, confront bankruptcy or seek more expensive financing and sell assets. In the past year [2014], U.S. oil producers used 83 % of their operating cash flow to pay for debt service, according to the U.S. Energy Information Administration. A year ago [2013], it was less than 60 %. Producers who hoped for a price rebound later [2015] have so far been disappointed. U.S. oil futures fell to 45 [by September 2015], and are down by half in the past year. A glut of crude may keep oil prices low for the next 15 years, according to Goldman Sachs Group Inc.... Most shale companies cannot make money at Goldman's 50 price forecast.[133,134,135]

Because the United States is regaining its status of a major hydrocarbon producer—and extraction of hydrocarbons from shale formations is at least 5–10 times more expensive than production of conventional gas and oil—it is logical to expect that the US government will support the shale operators by attempting to maintain an environment of high oil prices for decades and keep a flexible position with respect to the environmental risks of hydraulic fracturing. New increase of energy

[131]Carrick Mollenkamp, Special Report: Chesapeake's deepest well: Wall Street, Reuters, May 10, 2012.

[132]Asjylyn Loder, Shakeout Threatens Shale Patch as Frackers Go for Broke, Bloomberg News, May 27, 2014.

[133]Asjylyn Loder, Shale Drillers Feast on Junk Debt to Stay on Treadmill, Bloomberg News, Apr. 30, 2014.

[134]Rakteem Katakey, Luca Casiraghi, Oil Industry Needs Half a Trillion Dollars to Endure Price Slump, Bloomberg, August 27, 2015

[135]Bradley Olson, U.S. Shale Drillers Are Drowning in Debt, Bloomberg, September 18, 2015

prices would be achieved by resumption of monetary policy's easing; by further deregulation of energy futures and conscious stimulation of speculation with oil and gas derivatives; by reducing natural gas supply on domestic market and building up American exports of liquefied natural gas (LNG) to Asia, Latin America and Europe (which are used to much larger energy prices than in the US); and by managing favorably the continuing geopolitical instabilities[136] in the Middle East, North Africa, and Ukraine (main transit hub for supply of Europe by Russian natural gas) in order to increase energy prices and take market share of lucrative energy markets.

Shale Energy Production: Why Risks Are Being Concealed

- The U.S. government wants to reach energy independence, allowing serious changes in American geopolitics, to launch a reindustrialization of the American economy, and to enhance economic growth through new jobs, reducing the trade balance deficit, bringing lower gas costs for American industries, and so on. All these **benefits** have motivated the government to largely ignore environmental damage from shale energy extraction and tacitly support shale energy producers in their intensive drilling.
- Official confirmation, from the EPA or from the US Department of Energy researchers, of environmental damage during shale energy extraction could **increase the production costs for shale energy producers**. Such confirmation would force the implementation of costly environment-friendly solutions and could **bring a wave of lawsuits from the inhabitants of contaminated land**, who were misled by shale companies. It could **provoke a collapse of the capitalization for shale energy producers, and even halt shale energy production in the USA**, causing the country to forgo its aim towards energy independence and pushing it back into dependence on Persian Gulf producers.
- In order to **attract investors and postpone the potential bankruptcy of shale operators** due to the chronic unprofitability of shale exploitation, the US government implemented legal methods allowing the manipulation of the accounting rules of oil and gas reserves. This also enabled the government to declare that a "shale revolution" could form the basis for decades of economically viable production from the world's shale formations under the leadership of the American energy industry.

4.2 Genetically Modified Organisms

The curious task of economics is to demonstrate to men
how little they really know about what they imagine they can design

Friedrich Hayek

[136] Friedman, George, The next 100 years, a forecast for the 21st century. Doubleday, New York, 2009.

The case of genetically modified organisms (GMOs) used in advanced agriculture presents another flagrant risk concealment example with potentially dire consequences that are worth examining in some details.

In the twentieth century, in spite of the tremendous growth of the world population soaring from 1.6 to 6.7 billion people (and passing 7 billion in 2011), Malthusian theory, that global famine would result from the inability for agriculture to grow at the same exponential pace, did not materialize. This resulted from the Green revolution, which includes the scientific approach of optimizing seeds selection, and the use of mineral fertilizers, pesticides and irrigation. This allowed to maximize yield productivity and to release millions of people from agriculture-related business. For instance, corn yield increased from 20 bushels/acre in 1920 to 200 bushels/acre nowadays. A hundred years ago, a US farmer produced food to feed 8–15 people compared to 140 people nowadays.[137] The gene revolution that occurred several decades ago allowed to increase harvests and to reduce production costs. The associated biotechnologies involve genetic manipulation in order to create organisms in which the genetic material (DNA) has been altered in a way that does not occur naturally.[138] Genetically modified organisms (GMOs) are usually selected for (i) their ability to withstand certain herbicides (during dusting of crops by herbicides, all plants die except for the GMOs, which survive) or (ii) their resistance to insects (as a result of genetic manipulations, some plants become insecticides, which are toxic for plant pests) or (iii) their special characteristics (drought tolerance, prolonged shelf life, attractive appearance, oversized products, and so on). In spite of their economic advantage, long-term consequences for human health that could result from the consumptions of GMOs remain a disputed topic among scientists. John Fagan and his colleagues, authors

[137] Michael Pollan, Farmer in Chief, The New York Times, Oct. 9, 2008.

[138] 20 questions on genetically modified foods, World Health Organization, 2002.

of the report "GMO Myths and Truths", write: "*The genetic engineering process is not precise or predictable. Genes do not function as isolated units but interact with each other and their environment in complex ways that are not well understood or predictable. The genetic engineering process can disrupt the host organism's genome or genetic functioning in unexpected ways, resulting in unpredictable and unintended changes in the function and structure of the genetically modified organism... This is illustrated by the fact that altering a single letter of the genetic code of a single gene can be a significant step leading to cancer, a disease that involves alterations in the function of multiple genes, proteins and cellular systems*".[139]

Given the perspective of the risk concealment practice documented in the cases elaborated earlier, in particular associated with the development of innovative solutions and products (Enron case, securitization pipeline in the mortgage market, shale gas, and so on), one can observe similar methods in the current expansion of GMO business, which trigger red flag warnings of possible cover-up practices in this business.

4.2.1 Short-Term Profitability Versus Long-Term Sustainability

Firstly, short-term profitability is clearly taking over long-term sustainability. As an illustration, a representative executive of Monsanto, the leading American multinational agrochemical and agricultural biotechnology corporation, pronounced that "*Monsanto should not have to vouch for the safety of biotech food. Our interest is in selling as much of it as possible. Assuring its safety is the FDA's [US Food and Drug Administration – the American federal regulator of food-related business] job*".[140] Claire Robinson, co-author of the report "GMO Myths and Truths", asserted that "*Claims for the safety and efficacy of GM crops are often based on dubious evidence or no evidence at all. The GMO industry is built on myths. What is the motivation behind the deception? Money. GM crops and foods are easy to patent and are an important tool in the global consolidation of the seed and food industry into the hands of a few big companies. We all have to eat, so selling patented GM seed and the chemicals they are grown with is a lucrative business model*".[141]

[139]John Fagan, Michael Antoniou, Claire Robinson, GMO Myths and Truths, An evidence-based examination of the claims made for the safety and efficacy of genetically modified crops and foods, Earth Open Source, 2014, 2nd edn., pp. 22–23.

[140]Michael Pollan, Farmer in Chief, The New York Times, Oct. 9, 2008.

[141]GMO Myths and Truths, Earth Open Source, May 19, 2014, http://earthopensource.org/index.php/reports/gmo-myths-and-truths.

4.2.2 Cozy Relationship Between Government Regulators and the Industry

Secondly, the cozy relationship between government regulators and the industry, which we documented in many previous accidents, is also very present. Industry representatives have been appointed administrators in public regulatory bodies and have been involved in the regulation of GMO products. Executives of Monsanto and other biotechnical companies have been senior government representatives during the Clinton, Bush and Obama administrations and were responsible for the regulation of the GM food-supply chain.[142,143] Similar conflicts of interests occurred in Europe, where some industry representatives became executive officials of regulatory bodies.[144,145,146] Regarding regulation in developing countries, John Fagan el al state the following: "*[American] agricultural biotechnology corporations have lobbied long and hard on every continent to ensure that the weak safety assessment models developed in the US are the norm globally. Working through the US government or groups that appear to be independent of the GMO industry, they have provided biosafety workshops and training courses to smaller countries that are attempting to grapple with regulatory issues surrounding GMOs. The result has been models for safety assessment that favor easy approval of GMOs without rigorous assessment of health or environmental risks. For example, a report by the African Centre for Biosafety described how the Syngenta Foundation, a nonprofit organization set up by the agricultural biotechnology corporation Syngenta, worked on 'a three-year project for capacity building in biosafety in sub-Saharan Africa'... In India, the US Department of Agriculture led a 'capacity building project on biosafety' to train state officials in the 'efficient management of field trials of GM crops' – the first step towards full-scale commercialization. And in 2010, a scandal erupted when a report from India's supposedly independent national science academies recommending release of GM Bt brinjal (eggplant/aubergine) for cultivation was found to contain 60 lines of text*

[142]Documentary "Food Inc.", Director: Robert Kenner, 2008 (1:16:10–1:17:40) and Documentary "Seeds of Death: Unveiling The Lies of GMO's", Directors: Gary Null, Richard Polonetsky, 2012 (0:05:15–0:08:30).

[143]John Fagan, Michael Antoniou, Claire Robinson, GMO Myths and Truths, An evidence-based examination of the claims made for the safety and efficacy of genetically modified crops and foods, Earth Open Source, 2014, 2nd edn., pp. 58–59.

[144]EU Commission shortlists ex-Monsanto employee for EFSA Management Board, Munich Corporate Europe Observatory and Testbiotech, Mar. 8, 2012.

[145]Frederick William Engdahl, The Toxic Impacts of GMO Maize: Scientific Journal Bows to Monsanto, Retracts anti-Monsanto Study, Global Research, Dec. 6, 2013.

[146]Christophe Noisette, Roumanie—OGM: un ex de Monsanto, ministre de l'Agriculture, Inf'OGM, Feb. 2012.

copy-pasted almost word for word from a biotechnology advocacy newsletter – which itself contained lines extracted from a GMO industry-supported publication".[147]

4.2.3 Lack of Independent Risk Assessment

Thirdly, instead of obtaining independent risk assessments by publicly funded studies, authorities rely on the internal industry's research or on industry-funded research. In a logic of deregulation and of promotion of the biotech industry, the FDA and European Food Safety Authority (EFSA) choose not to test GM products and approve them based on biotechnical companies' internal test results. The assumption is that the industry will protect its own business by implementing safe solutions. This attitude is strongly reminiscent of the thinking and actions of officials that contributed to the financial meldown in 2008 (see Sect. 2.2.3 on the subprime mortgage crisis). With the admission in autumn 2008, by one of the most vocal and powerful proponent of deregulation and of free markets, previous Fed Chairman Greenspan, that "*Those of us who have looked to the self-interest of lending institutions to protect shareholders' equity (myself especially) are in a state of shocked disbelief*",[148] one is led to raise the question: are we going to witness in the GMO industry a repetition of the errors perpetrated in the finance industry, with no lesson learned?

These regulators claim that "*Ultimately, it is the food producer who is responsible for assuring safety*" and "*It is not foreseen that EFSA carry out such [safety] studies as the onus is on the [GM industry] applicant to demonstrate the safety of the GM product in question*".[149] According to Professor David Schubert, "*One thing that surprised us is that US regulators rely almost exclusively on information provided by the biotech crop developer, and those data are not published in journals or subjected to peer review... The picture that emerges from our study of US regulation of GM foods is a rubber-stamp 'approval process' designed to increase public confidence in, but not ensure the safety of, genetically engineered foods*".[150] Consequently, based on voluntary industry's tests, FDA declared that GMOs are "*generally recognized as safe*".[151] Nevertheless, 26 American corn-insect special-

---

[147]John Fagan, Michael Antoniou, Claire Robinson, GMO Myths and Truths, An evidence-based examination of the claims made for the safety and efficacy of genetically modified crops and foods, Earth Open Source, 2014, 2nd edn., p. 78.

[148]Kara Scannell, Sudeep Reddy, Greenspan Admits Errors to Hostile House Panel, The Wall Street Journal, Oct. 24, 2008.

[149]John Fagan, Michael Antoniou, Claire Robinson, GMO Myths and Truths, An evidence-based examination of the claims made for the safety and efficacy of genetically modified crops and foods, Earth Open Source, 2014, 2nd edn., p. 56.

[150]Ibid, p. 59.

[151]Ibid, p. 61.

ists complained to EPA that "*No truly independent research can be legally conducted on many critical questions*", because regarding to investigation of New Times, "*farmers and other buyers of genetically engineered seeds have to sign an agreement meant to ensure that growers honor company patent rights and environmental regulations. But the agreements also prohibit growing the crops for research purposes. So while university scientists can freely buy pesticides or conventional seeds for their research, they cannot do that with genetically engineered seeds. Instead, they must seek permission from the seed companies. And sometimes that permission is denied or the company insists on reviewing any findings before they can be published*".[152]

Lack of comprehensive, independent and long-term studies published in peer-reviewed journals about the influence of GMO on human health also contributes to the fragmented picture of risks among decision makers. In spite of the industry claims about extensive tests showing that GM foods are safe, many of these studies show evidence of risk.[153] According to Arpad Pusztai, who in 1998 found that GM potatoes are harmful to the health of laboratory rats,[154] "*Even a cursory look at the list of references of a recent major review on food safety issues … showed that most of the publications referred to were non-peer-reviewed institutional opinions or envisaged future scientific and methodological developments for safety assessments, but were short on actual published scientific papers on which a reliable database of safety could be founded. Judging by the absence of published data in peer-reviewed scientific literature, apparently no human clinical trials with GM food have ever been conducted. Most attempts to establish the safety of GM food have been indirect*".[155] "*A review of scientific studies on the health risks of GM crops and foods that did investigate funding sources found that either financial or professional conflict of interest (author affiliation to industry) was strongly associated with study outcomes that cast GM products in a favorable light. Conclusions of safety were also found to be associated with studies in which source of funding was not declared. Furthermore, there was a strong connection between undeclared funding and author affiliation to industry. Genuinely independent studies on GM foods and crops are rare, for two reasons: because independent research on GM crop risks is not supported financially; and because industry uses its patent-based control of GM crops to restrict independent research… Even if permission to carry out research is given, GM companies typically retain the right to block publication. An editorial in Scientific American reported, 'Only studies that the seed companies have approved ever see the light of a peer-reviewed journal. In a number of*

---

[152]Andrew Pollack, Crop Scientists Say Biotechnology Seed Companies Are Thwarting Research, The New York Times, Feb. 19, 2009.

[153]John Fagan, Michael Antoniou, Claire Robinson, GMO Myths and Truths, An evidence-based examination of the claims made for the safety and efficacy of genetically modified crops and foods, Earth Open Source, 2014, 2nd edn., pp. 102–103.

[154]Ibid, p. 97.

[155]A. Pusztai, S.Bardoczi, S.W.B. Ewen, Genetically Modified Foods: Potential Human Health Effects, OCAB International 2003. Food Safety: Contaminants and Toxins, p. 347.

cases, experiments that had the implicit go-ahead from the seed company were later blocked from publication because the results were not flattering"'.[156]

The recent attempt to study long-term effects of GMOs on animal by Gilles-Eric Seralini, professor of molecular biology at the University of Caen in France, and his group, is a case in point. In 2009, GMO maize Monsanto NK603, which is resistant to Monsanto's Roundup (a glyphosate-based herbicide) was approved by EFSA without any independent phase of testing because "*data provided [by Monsanto] are sufficient and do not raise a safety concern*". Moreover, "*The EFSA GMO Panel is of the opinion that maize NK603 is as safe as conventional maize. Maize NK603 and derived products are unlikely to have any adverse effect on human and animal health in the context of the intended uses*".[157] After this approval, professor Seralini independently and secretly began to test on 200 rats the impact of Monsanto's maize, which was previously cultivated in Canada with different portions of Roundup's dusting. The test took two years and cost US $3 million. Finally, he concluded that long-term feeding by Monsanto GMO maize leads to a high frequency of cancer in the rat population. Such effects did not occur when feeding rats over short periods (up to 90 days). Before this study, Monsanto had conducted only short-term tests and their results were the basis for the statement that "*NK603 is as safe and nutritious as conventional corn currently being marketed... Roundup Ready corn plants containing corn event NK603 were shown to be as safe and nutritious as conventional corn varieties and to pose no greater environmental impact than conventional corn varieties*".[158] Seralini's article was published in "*The Journal of Food and Chemical Toxicology*" with the title "*Long term toxicity of a Roundup herbicide and a Roundup-tolerant genetically modified maize*" in November 2012.[159] After its publication, it was heavily criticized by the industry, authorities, and several scientists, but found substantial support among journalists and customers worldwide. Subsequently, due to tremendous pressure from the regulators, the industry and the scientific community with accusations to professor Seralini of unacceptable scientific standards (inadequate design, reporting and analysis of the study), the journal retracted the publication in November 2013. Nevertheless, the article was republished by the journal "*Environmental Sciences Europe*" in June 2014.[160] The remarkable fact is that neither the regulators nor

[156]John Fagan, Michael Antoniou, Claire Robinson, GMO Myths and Truths, An evidence-based examination of the claims made for the safety and efficacy of genetically modified crops and foods, Earth Open Source, 2014, 2nd edn., pp. 89–90.

[157]Frederick William Engdahl, The Toxic Impacts of GMO Maize: Scientific Journal Bows to Monsanto, Retracts anti-Monsanto Study, Global Research, Dec. 6, 2013.

[158]Safety Assessment of Roundup Ready Corn Event NK603, Monsanto, Sep. 2002, pp. 3–17.

[159]Gilles-Eric Séralini, Emilie Clair, Robin Mesnage, Steeve Gress, Nicolas Defarge, Manuela Malatesta, Didier Hennequin, Joël Spiroux de Vendômois, RETRACTED: Long term toxicity of a Roundup herbicide and a Roundup-tolerant genetically modified maize. Food and Chemical Toxicology 50(11), 4221–4231 Nov. 2012.

[160]Gilles-Eric Séralini, Emilie Clair, Robin Mesnage, Steeve Gress, Nicolas Defarge, Manuela Malatesta, Didier Hennequin, Joël Spiroux de Vendômois, Republished study: long-term toxicity of a Roundup herbicide and a Roundup-tolerant genetically modified maize, Environmental Sciences Europe 2014, 26:14.

other serious universities took the trouble of initiating new tests of Seralini's observations and claims in order to prove or disprove the danger of GM maize and Roundup for animals and humans during long-term consumptions. This remains in flagrant contradiction with the scientific standards of replication and reproducibility. In the introduction to a special issue of the journal Science, Jasny et al. state "*the confirmation of results and conclusions from one study obtained independently in another—is considered the scientific gold standard. New tools and technologies, massive amounts of data, long-term studies, interdisciplinary approaches, and the complexity of the questions being asked are complicating replication efforts, as are increased pressures on scientists to advance their research*".[161]

4.2.4 Optimistic Statements of Economic Benefits

Fourthly, GMO business is characterized by significant overstatements of economic benefits and understatement of threats. This is very similar to the promotion of shale energy that we have documented in the previous section. The following statement is representative: "*A July [2013], Gallup poll found that 48 percent of respondents believed that GM foods 'pose a serious health hazard', compared to 36 percent who didn't... Within the scientific community, the debate over the safety of GM foods is over. The overwhelming conclusion is, in the words of the American Association for the Advancement of Science, that 'consuming foods containing ingredients derived from GM crops is no riskier than consuming the same foods containing ingredients from crop plants modified by conventional plant improvement techniques'. Major scientific and governmental organizations agree... A focus on the risks and benefits of all new crops could move the debate in a direction that would prompt scientists, companies, and regulators to more clearly justify the role GMOs play in our food supply. To date, consumers nervous about GMOs have been given little reason to think that companies like Monsanto are designing GM crops to solve any problem other than the one of patents and profits. As journalist Mark Lynas put it in his rousing defense of GM foods, for most people GMOs are about a 'big American corporation with a nasty track record, putting something new and experimental into our food without telling us'. But* many *researchers working on GM crops are in fact trying to solve important problems, such as feeding a growing population, keeping food prices affordable worldwide, making healthier fruits and vegetables widely available, confronting the challenging growing conditions of a changing climate, saving Florida's oranges or Hawaii's papaya from pests, and fighting malnourishment in the*

[161] Barbara R. Jasny, Gilbert Chin, Lisa Chong, Sacha Vignieri, Again, and Again, and Again... Science 334(6060), 1225 (Dec. 2, 2011): introduction to a special issue on scientific replication.

developing world. For many of these problems, genetic engineering is faster, more cost-effective, and more reliable than conventional breeding methods".[162]

Genetically Modified Organisms: Why Risks Are Being Concealed

- Governments of many countries try to **increase agriculture output and reduce food price** for continuous growing population by **implementing innovative, but unproved technologies**.
- **Priority of short-term profitability** of multinational agrochemical and agricultural biotechnology corporations versus long-term sustainability of human beings.
- Testing the real consequences of the distribution of GMO products is difficult because of the cozy **relationship between government regulators** and industry and the **absence of independent risk assessment**.

4.3 Real Debt and Liabilities of US Government and Real GDP of China

4.3.1 Challenges in Assessing the Real National Debt of the United States

Estimating the real national debt of the United States involves the complex accounting question of how to recognize all the explicit as well as the less obvious liabilities of the US government. In general, there are several descriptions of the national/governmental debt: "*the total financial obligations of a national government*" or "*the total outstanding borrowings of a nation's central government*" or "*the total financial obligations incurred by all governmental bodies of a nation*".

The US Treasury uses the following description of the gross government debt: "*all of the federal government's outstanding debt, measured by outstanding bills, notes, bonds, and other debt instruments of the U.S. government*".[163] This excludes counting the sum of all social unfunded liabilities of entitlement programs like Social Security, Medicare and Medicaid program. By Q2 2014, the US government debt counted on the basis of issues of US Treasury exceeded US $17 trillion (or 101.7 % of American GPD).[164] But, this does not take into account the potential governmental liabilities to millions of pensioners, ill and disable persons in future decades.

---

[162]Michael White, The Scientific Debate About GM Foods Is Over: They're Safe, Pacific Standard: The Science of Society, Sep. 24, 2013.

[163]Federal Debt. Answers to Frequently Asked Questions, U.S. Government Accountability Office, GAO-04-485SP, Aug. 2004, p. 5.

[164]Federal Debt: Total Public Debt as Percent of Gross Domestic Product, Federal Reserve Bank of St. Louis, Sep. 29, 2014.

Researchers have calculated such unfunded liabilities and found that the real debt of the US government exceeds official measures by a factor from 6 to 15, when taking into account the sum of all social liabilities. For example, James D. Hamilton estimates that US federal liabilities not included in the officially reported figures come to a total of about US $70 trillion[165]; in March 2015, Michael D. Tanner from the Cato Institute provided his estimations of US federal liabilities—US $90.6 trillion[166]; Laurence J. Kotlikoff assesses the US fiscal gap at around $222 trillion during the next 75 years and concluded: "*Bankruptcy is a strong term. But it is apt... Anyone who thinks the U.S. is immune from fiscal meltdown and high inflation, if not hyperinflation, should think again. Too many countries, big and small, rich and poor, have learned that, sooner or later, fiscal profligacy comes at a very high price... In the last century, 20 countries ended up with hyperinflations by using the printing press to avoid taking fiscal responsibility. Now we have virtually the entire developed world printing money out the wazoo. Krugman's response is, 'Where is the inflation? Where are the high interest rates?' My answer is, give it time. The long history of fiscal and financial crises is that they happen abruptly, at a moment that seems not to matter, but in retrospect it was easy to see that the crisis would eventually materialize*".[167,168] As some point in the future, such conclusions could lead to a change of the standing and rating of US government debt in the eyes of sovereign funds, corporate and private investors, likely to question the ability of the US government in the middle and long-term to honor its debt and liabilities while avoiding large inflation. According

---

[165]James D. Hamilton, Off-Balance-Sheet Federal Liabilities, National Bureau of Economic Research, July 2013.

[166]Michael D. Tanner, Medicare and Social Security Tabs Coming Due, Cato Institute, March 2015, http://www.cato.org/publications/commentary/medicare-social-security-tabs-coming-due.

[167]Kotlikoff, Laurence J., The US Fiscal cliff—When economists recklessly endanger the economy, CESifo Forum 2/2013 (June), pp. 3–8.

[168]Kotlikoff, Laurence J., The Emperor's Dangerous Clothes, Economists, Voice 5, 2, April 2008.

to the Economist's world debt comparison rating, the current US public debt (without counting the unfunded liabilities mentioned above) is the largest in the World, at more than 26 % of the global public debt,[169] which of course is not surprising given that its GDP is also the largest, representing 22.3 % of the world in nominal terms (and 19.3 % in Purchasing Power Parity). The time may come when serious doubts about the solvency of the US government could provoke a global reserve crisis, when the quality of the international monetary reserves of hundred countries, which are denominated in large part in US treasury debt and in US$, will be found wanting. Moreover, the widespread recognition of the scale of unfunded liabilities of the US government could shake the position of the American dollar as the world reserve currency.

4.3.2 Challenges in Assessing the Real Chinese GDP

Another illustration of the manipulation with governmental statistics is the Chinese GDP data. Provincial governments in China try to convince local businessmen to increase their companies' revenue by the process of "creative accounting" of their books, in order to demonstrate to the Chinese federal government that provincial governments can support two-digit annual growth of the GDP.[170,171,172] Analysts of Stratfor concluded: "*[A]ccounting changes can easily add a few percentage points to a country's GDP. For example, changes in U.S. accounting [in 2013] added 3 percent to the country's GDP. China's 2004 economic census led to an upward revision of GDP of 16.8 percent, and the 2009 census led to a 4.4 percent upward revision, mostly on the basis of incorporating China's larger-than-expected services sector... [China] still shows a GDP growth rate above 7 percent, which is considerably higher than that of other developing nations. In addition, there are questions about discrepancies between central and provincial statistics as well as doubts about industrial output, nonblank lending and extremely low unemployment and inflation rates. The belief that China's accounting is inaccurate, often due to manipulation and falsification, has led many outsiders to question the real size and growth of the Chinese economy*".[173] A tip of the iceberg may be revealed by the investigations of accounting irregularities at Chinese companies listed on U.S. stock exchanges, by the US department of

---

[169]The global debt clock, The Economist, http://www.economist.com/content/global_debt_clock.

[170]Keith Bradsher, Chinese Data Mask Depth of Slowdown, Executives Say, The New York Times, June 22, 2012.

[171]Chinese Companies Forced to Falsify Data, Government Says, Bloomberg News, March 16, 2012.

[172]Keith Bradsher, Chinese Data Mask Depth of Slowdown, Executives Say, The New York Times, June 22, 2012.

[173]China's Plans to Revise Its National Accounting System, Stratfor, Nov. 20, 2013.

Justice: "*Not having proper accounting and reliable audit review for publicly traded companies with operations in China is just not acceptable. We have to find a path to resolution of this issue*," said Robert Khuzami, director of enforcement at the SEC.[174]

The practice of hiding debts revealed by the cases of Enron, Parmalat, Lehman Brothers and others should also keep alert to the possible concealment of the real debts of the Chinese banking system and companies.[175,176,177] The liquidity shortage in June 2013 in China was one of the first red flags for a possible future instability in the Chinese banking system after decades of immense lending activity.[178]

This makes one think of a similar developing question facing global investors, including pension funds, mutual funds and private households, namely what can sustain the continuation of the US equity performance relative to international markets since March 2009? Perhaps, a significant part of the answer lies in the divulgation of non-GAAP accounting practices[179] by one of the largest asset manager in the world[180]: "*It becomes tempting to take on too much leverage, use financial wizardry to reward shareholders or even stretch accounting principles. S&P 500 profits are 86 % higher than they would be if accounting standards of the national accounts were used*". To fully appreciate the full implication of this diagnostic, one should remember that a firm valuation is essentially dependent on its earnings. It was the ENRON and Worldcom scandals of creating accounting of their earnings that led to the hurried passage of the Sarbanes–Oxley Act of 2002[181] to calm panicked investors who had the right to doubt in the sound functioning of the capitalistic system. It seems that, after slashing costs since the financial crises, firms have difficulties in finding ways to generate the analysts' expected 10 % earning-per-share growth when nominal GDP growth is 4 % or less…

---

[174]Andrea Shalal-Esa and Sarah N. Lynch, Exclusive: Justice Department probing Chinese accounting Reuters, Washingtion, Fri Sep 30, 2011, http://www.reuters.com/article/2011/09/30/us-china-usa-accounting-idUSTRE78S3QM20110930.

[175]Henny Sender, Finance: Money for nothing, The Financial Times, July 11, 2013.

[176]Matthew Forney, Laila F. Khawaja, Due Diligence: Don't Get Fooled Again, Fathom China, July 2014.

[177]Carl E. Walter, Fraser J. T. Howie, Red Capitalism: The Fragile Financial Foundation of China's Extraordinary Rise, John Wiley & Sons, 2012, pp. 27–95.

[178]In China's Banking Problems, a Challenge for Reform, Stratfor, June 25, 2013.

[179]Generally accepted accounting principles (GAAP) refer to the standard framework of guidelines for financial accounting used in any given jurisdiction. These include the standards, conventions, and rules that accountants follow in recording and summarizing and in the preparation of financial statements (http://en.wikipedia.org/wiki/Generally_accepted_accounting_principles).

[180]BlackRock Investment Institute, Dealing with divergence, 2015 investment outlook (Dec. 2014) (https://www.blackrock.com/corporate/en-us/literature/whitepaper/bii-2015-investment-outlook-us.pdf).

[181]called the "Public Company Accounting Reform and Investor Protection Act" (in the Senate) and the "Corporate and Auditing Accountability and Responsibility Act" (in the House).

Real Debt of US Government: Why Risks Are Being Concealed

- The concealment of the real magnitude of the total US government liabilities by legal manipulation in accounting to make the official debt as small as possible **allows the US government to borrow money at one of the lowest interest rates and to enjoy one of the highest credit ratings in the World**. In addition, it also allows the US to continue to pay in fiat dollar for imported commodities and products (raw materials and imported goods).

Real GDP of China: Why Risks Are Being Concealed

- Adherence of the Chinese government to overstated country's GDP data allows **maintaining lucrative indicators of a large growth of the Chinese economy** in comparison with other countries in order to increase China's investment attractiveness (direct investment, stocks and renminbi as possible future new currency for world trade) as well as contributing to a general positive atmosphere catalyzing the support of the Chinese governing bodies by the public.

4.4 The Global Cyber Arms Race and Concealment of Vulnerabilities in the Software Industry

4.4.1 The Tumultuous Affair Between Iran and the USA

The relatively new cyber risk landscape is, like many other risks, rooted in history. One crucial thread goes back to the 1970s, when the United States recognized Iran—under the leadership of Mohammad Reza Pahlavi, Shah of Iran—as a strategic ally of America and Israel in the Middle East. Political matters remained under the total control of the Shah and his Organization of Intelligence and National Security (SAVAK), hated by many Iranians for torturing and executing opponents of the Pahlavi regime. On one occasion, US President Nixon emphasized that the "*[Iranian political] system has worked for them. It is time for us to recognize that much as we like our own political system, American style democracy is not necessarily the best form of government for people in Asia ... with entirely different backgrounds*".[182] The US provided Iran with a range of modern and sophisticated weapons in exchange for billions of petrodollars. James R. Schlesinger, an executive member of Nixon's team, commented on US military support of the Shah's regime: "*We were going to make the Shah the Guardian of the Gulf. Well, if we*

[182] Andrew Scott Cooper, The Oil Kings: How the U.S., Iran, and Saudi Arabia Changed the Balance of Power in the Middle East. Simon and Schuster, New York, 2011, p. 30.

were going to make the Shah the Guardian of the Gulf, we've got to give him what he needs—which comes down to giving him what he wants".[183] So Nixon was ready to provide both nuclear weapons and nuclear power stations to the loyal regime.[184] But when the anti-Shah Islamic revolution took place in February 1979, the Iranian Republic—and its government, with a democratic mandate to pursue a strict anti-American agenda—had the same nuclear ambitions as the Shah's regime but became one of the most hostile countries in the world towards America.

In June 2012, David E. Sanger, chief Washington correspondent of The New York Times, published a book entitled "*Confront and Conceal: Obama's Secret Wars and Surprising Use of American Power*". He revealed detailed information about a secret cyber military program called "Olympic Games" under the supervision of the National Security Agency (NSA) of the United States, which was launched by George W. Bush and continued by Barack Obama. The main goal of this program was to set up a remote cyber-attack on Iranian nuclear facilities—especially the centrifuges for uranium enrichment located at the Natanz underground nuclear site—as a substitute of air attacks of this site by the Israeli air force. To carry out this cyber-attack, American and Israeli military programmers allegedly wrote a sophisticated virus, later known to the world public as Stuxnet. The virus was introduced into the software of Siemens's programmable logic controllers for the centrifuges, changing their rpm-mode regime so that they reached damaging regimes of function. Ultimately, 984 out of around 5000 uranium enrichment centrifuges on the site were destroyed.[185,186] Hillary Clinton, the US Secretary of State, declared that the Stuxnet virus had set Iran's nuclear programme back by several years.[187]

4.4.2 History-Making Cyber Warfare Between Nation States

In fact, Stuxnet was not the first instance of government sponsored cyber attack. Such a distinction is reportedly attributed to the introduction of malware by a U.S. agency into the software controlling oil pipeline pumps in Siberia, as one of the many development of the cold war era. The resulting explosion in 1982 is considered the largest non-nuclear explosion recorded.[188] There are numbers of alleged

[183]Ibid, p. 20.

[184]Ibid, pp. 66, 210.

[185]David E. Sanger, Confront and Conceal: Obama's Secret Wars and Surprising Use of American Power, Crown Publ. Group, 2012, pp. 188–209.

[186]David E. Sanger, Obama Order Sped Up Wave of Cyberattacks Against Iran, The New York Times, June 1, 2012.

[187]Con Coughlin, Stuxnet virus attack: Russia warns of 'Iranian Chernobyl', The Telegraph, Jan 16, 2011.

[188]http://en.wikipedia.org/wiki/Siberian_pipeline_sabotage and Cyber War: The Next Threat to National Security and What to Do About It at http://www.amazon.com/Cyber-War-Threat-National-Security/dp/0061962244.

state-sponsored attacks with wide media coverage since then,[189] which show the versatility of cyber weapons, from foreign intelligence, to industrial intelligence, to hacktivism, and so on.

However, the Stuxnet case stands out. It is perhaps too early to fully realize the long term implications but, when looking back in twenty or fifty years from now, it might not be then exaggerated to state that the significance of this event was comparable to that of the nuclear bombardment of Japan in 1945 by the USAF, which ignited the nuclear arms race between the United States and the Soviet Union. Former CIA chief Michael Hayden put it this way: "*Somebody has crossed the Rubicon*".[190] Stuxnet used four zero-day vulnerabilities in Microsoft's Windows platform to infect the computer system of Iran's nuclear scientists with malicious code. But due to a programming error, the virus revealed itself when the infected computers were connected to the Internet: the virus recognized the Internet as a local network, escaped to the web and infected ordinary computer users, where in due course it was detected by antivirus specialists from Belarus.

In May 2012, the Russian anti-virus company Kaspersky Lab discovered a new virus called Flame—which was far more sophisticated than Stuxnet, but contained code segments similar to the earlier virus, allowing experts to conclude that both viruses had been developed by the same programmers.[191,192] The Flame virus had been distributed in Middle Eastern countries but the majority of computers infected were within the Iranian Oil Ministry. The virus also used vulnerabilities in Microsoft Windows, which allowed it to masquerade within the local network as part of the Windows Update service during regular updates implemented in the Microsoft software. In October 2012, Kaspersky Lab announced that they had detected the "Red October" virus, which used vulnerabilities in Microsoft Office and Microsoft Excel to allow cyber-espionage within government embassies, research firms, military installations, energy providers, and nuclear and other critical infrastructure, mainly in Russia and the former Soviet Republics. Kaspersky Lab's Global Research & Analysis Team concluded that "*[Red October's] configuration rivals in complexity the infrastructure of the Flame malware*".[193]

Edward Snowden, former NSA agent and now famous whistleblower, provided the German magazine Spiegel with NSA documents about plans by the United

---

[189]https://en.wikipedia.org/wiki/Titan_Rain
https://en.wikipedia.org/wiki/2007_cyberattacks_on_Estonia.

[190]David E. Sanger, Confront and Conceal: Obama's Secret Wars and Surprising Use of American Power. Crown Publ. Group, 2012, p. 200.

[191]Ellen Nakashima, Greg Miller, Julie Tate, U.S., Israel developed Flame computer virus to slow Iranian nuclear efforts, officials say, The Washington Post, June 19, 2012.

[192]Flame virus hit Iran's oil industry but officials say antidote found. Israeli vice-prime minister suggests Israel might have been behind cyberattack, The Associated Press, May 30, 2012.

[193]The "Red October" Campaign—An Advanced Cyber Espionage Network Targeting Diplomatic and Government Agencies. Kaspersky Labs' Global Research & Analysis Team, Jan. 14, 2013.

States to use cyber weapons in future wars. They expected to be able to "*remotely degrade or destroy opponent computers, routers, servers and network enabled devices by attacking the hardware*", "*erase the BIOS on a brand of servers that act as a backbone to many rival governments*" and "*paralyze computer networks and, by doing so, potentially all the infrastructure they control, including power and water supplies, factories, airports or the flow of money*".[194] By 2013, US Cyber Command comprised more than 40,000 employees, responsible for both digital spying and destructive network attacks because "*Part of [US] defense has to consider offensive measures*".[195,196] In February 2015, Barack Obama remarked that "*We [Americans] have owned the Internet. Our companies have created it, expanded it, perfected it in ways that they [non-American vendors] can't compete*".[197] The United States produce more than 90 % of all commercial software in the world; such military plans from a nation with a near-monopoly in software and Internet solutions make a range of countries, which may have very different views on domestic and foreign policy to Washington, vulnerable to a potential American hybrid attack (i.e. a combination of cyber attacks with conventional military measures).

However, the U.S. itself is extremely exposed to cyber attacks as their civilian and military infrastructure is heavily dependent on networks and software, more than that of any other country.

World history shows that, if one country develops an advanced and effective weapon, other countries will eventually develop such weapons too. While it is not clear who started first, it seems that the American attacks on the military infrastructure of Iran catalyzed more visible retaliatory action from China, Russia, Iran and other countries, igniting a digital arms race and the active development of their cyber forces: in 2008, China's cyber forces consisted of 10,000 "cyber-troops" while Russia had 7300,[198] but in recent years the quality and quantity of these forces have been strengthened. In August 2012, Saudi Aramco—the largest oil company in the world and the top Saudi taxpayer—was attacked by the Shamoon virus. Data from more than 2000 servers and 30,000 desktop computers

[194]Jacob Appelbaum, Aaron Gibson, Claudio Guarnieri, Andy Müller-Maguhn, Laura Poitras, Marcel Rosenbach, Leif Ryge, Hilmar Schmundt and Michael Sontheimer, The Digital Arms Race: NSA Preps America for Future Battle, Spiegel, Jan. 17, 2015.

[195]Jacob Appelbaum, Aaron Gibson, Claudio Guarnieri, Andy Müller-Maguhn, Laura Poitras, Marcel Rosenbach, Leif Ryge, Hilmar Schmundt and Michael Sontheimer, The Digital Arms Race: NSA Preps America for Future Battle, Spiegel, Jan. 17, 2015.

[196]Tom Simonite, Welcome to the Malware-Industrial Complex, The U.S. government is developing new computer weapons and driving a black market in "zero-day" bugs. The result could be a more dangerous Web for everyone, MIT Technology Review, Feb. 13, 2013, http://www.technologyreview.com/news/507971/welcome-to-the-malware-industrial-complex.

[197]Kara Swisher, White House. Red Chair. Obama Meets Swisher, February 2015, ReCode, http://recode.net/2015/02/15/white-house-red-chair-obama-meets-swisher.

[198]Ward Carroll, Russia's Cyber Forces, May 27, 2008, Defense Tech, http://defensetech.org/2008/05/27/russias-cyber-forces/ and Ward Carroll China's Cyber Forces, May 8, 2008, Defense Tech, http://defensetech.org/2008/05/08/chinas-cyber-forces/.

was wiped. The virus was far simpler than Stuxnet, and looked as if it had been written by private "hacktivists" rather than nation-state developers with unlimited resources.[199,200,201] Nobody took responsibility for the attack, but the US pointed to Iran as the obvious beneficiary of an attack on its main rival in the Middle East, although there was no strong evidence.[202,203] The goal of the attack was to disrupt oil production in Saudi Arabia[204] but, like Stuxnet and its subsequent developments, the Shamoon virus relied on bugs in Microsoft software and damaged only computers with Windows operating systems, while computers operating under the supervisory control and data acquisition (SCADA) system were not damaged.[205,206] Several weeks later, Shamoon was used in an attack on the network of RasGas, a joint venture between Qatar Petroleum and ExxonMobil.[207]

In 2013, more than 740 million data files were potentially viewed or stolen worldwide[208] but, according to a Kaspersky Lab expert, the main focus of cybercriminals is financial profit; the usage of destructive malware (malicious software) like Stuxnet or Shamoon is a very rare case.[209] Nevertheless, the active development of malware by US Cyber Command helped develop further, by being strongly on the demand side of the supply-demand market, a world black-market for the research and discovery of "zero-day" vulnerabilities in widely used software. "Zero-day" vulnerabilities are software bugs that are unknown to the software vendor (and a fortiori to the users and to the public), so that the vendor and

---

[199]Kim Zetter, The NSA Acknowledges What We All Feared: Iran Learns From US Cyberattacks, WIRED, Feb. 10, 2015, http://www.wired.com/2015/02/nsa-acknowledges-feared-iran-learns-us-cyberattacks/.

[200]Dmitry Tarakanov, Shamoon The Wiper: Further Details (Part II), Kaspersky Labs' Global Research & Analysis Team, September 11, 2012.

[201]Norman Johnson, Shamoon Wiper: The most damaging corporate attack—Did you miss it?, http://www.cyber50.org/blog/shamoon-wiper.

[202]Kim Zetter, The NSA Acknowledges What We All Feared: Iran Learns from US Cyberattacks, WIRED, February 10, 2015, http://www.wired.com/2015/02/nsa-acknowledges-feared-iran-learns-us-cyberattacks/.

[203]Sean Lawson, Anonymous Sources Provide No Evidence of Iran Cyber Attacks, Forbes, Oct. 31, 2012.

[204]Saudi Arabia says cyber attack aimed to disrupt oil, gas flow, Reuters, Dec. 9, 2012.

[205]Thomas Rid, Cyber War Will Not Take Place. Oxford University Press, 2013, p. 66.

[206]Heather MacKenzie, Shamoon Malware and SCADA Security—What are the Impacts, Tofino Security, Oct. 25, 2012, https://www.tofinosecurity.com/blog/shamoon-malware-and-scada-security-%E2%80%93-what-are-impacts.

[207]Chris Bronk, Eneken Tikk-Ringas, Hack or Attack? Shamoon and the Evolution of Cyber Conflict, Baker Institute for Public Policy, February 1, 2013.

[208]Eric Chabrow, Records Exposed Hit New High in 2013, Data Breach Today, http://www.databreachtoday.com/interviews/records-exposed-hit-new-high-in-2013-i-2166.

[209]Shamoon the Wiper - Copycats at Work, Kaspersky Labs' Global Research & Analysis Team, Aug. 16, 2012, http://securelist.com/blog/incidents/57854/shamoon-the-wiper-copycats-at-work/.

any potential users will have "zero days" to prepare for a security breach.[210] Zero-day vulnerabilities will also be unknown to conventional computer security programs, allowing the vulnerabilities to be used for national defense purposes: as long as a zero-day vulnerability is kept secret, a weapon exploiting it remains usable.[211] According to a MIT Technology Review article, "*information about such flaws can command prices in the hundreds of thousands of dollars from defense contractors, security agencies and governments ... An intelligence agency or military force might steal diplomatic communications or even shut down a power plant ... Governments and companies in the United States and around the world have begun paying more and more for the exploits needed to make such weapons work ... On the one hand, the government is freaking out about cyber-security, and on the other the U.S. is participating in a global market in vulnerabilities and pushing up the prices ... Around 100 countries already have cyber-war units of some kind, and around 20 have formidable capabilities*".[212] Companies such as Microsoft already tell the government about gaps in their product security before issuing software updates, reportedly to give the NSA a chance to exploit those bugs first. "*But the NSA is also reaching into the Web's shadier crevices to procure bugs the big software vendors don't even know about — vulnerabilities that are known as "zero-days*".[213] Among active countries, China is believed to be unique in its large-scale theft of foreign technology.[214] And the US has blamed directly the Chinese Army for deploying cyber resources for espionage against defense and industrial targets in the United States and elsewhere.[215] There are often political motives for a country to publicly identify the origin of an attack, in particular, a rival country. This is called the "naming and shaming" policy. The alledged goal of this policy is to provide leverage for maintaining or increasing military spendings in cyber related aspects of warfare and homeland security.

There are problematic dilemmas associated with the various actors' Incentives: if a U.S. agency finds a zero-day vulnerability in a prevalent software product, it can coordinate the information with the software industry to protect 320 million

---

[210]Stefan Frei, Security Econometrics. The Dynamics of (In)Security, A dissertation submitted to the ETH ZURICH for the degree of Doctor of Science, 2009, p. 94.

[211]Hardin Tibbs, The Global Cyber Game: Achieving strategic resilience in the global knowledge society, The Defence Academy Cyber Inquiry Report, 2013, p. 53, http://www.futurelens.com/wp-content/uploads/2014/04/The-Global-Cyber-Game.pdf.

[212]Tom Simonite, Welcome to the Malware-Industrial Complex, The U.S. government is developing new computer weapons and driving a black market in "zero-day" bugs. The result could be a more dangerous Web for everyone, MIT Technology Review, Feb. 13, 2013, http://www.technologyreview.com/news/507971/welcome-to-the-malware-industrial-complex.

[213]Brian Fung, The NSA hacks other countries by buying millions of dollars' worth of computer vulnerabilities, Washington Post Aug. 31, 2013.

[214]Eckert P. and A. Yukhananov, U.S. opens China talks with cyber complaints, vow to boost trust. Reuters (July 10, 2013).

[215]Sanger, D.E., U.S. blames China's military directly for cyber attacks. NY Times, Section A, p 1 (May 6, 2013).

Americans. Alternatively, it can go to the president and brag that the U.S. can now compromise 2 billion Chinese machines. Such opportunity would be available for almost a year, as a compromise by zero-day exploit stays undetected on average for 312 days.[216] It is clear that intelligence agencies' reason for existence is to gather information and develop methods and infrastructure to support intelligence gathering. Nowadays, much of this is via information communication technology (ICT). It is also clear that they will do anything technologically possible, and quite a lot is technologically possible. Intelligence agencies are supposed to conceal information, thus they are not the targets of our book. But the interesting point is that their mission is greatly facilitated, even made possible in the explosively developing ICT universe in which the concealment of vulnerabilities by software vendors occurs.

4.4.3 Explosive Growth of Complexity in Software

All this is possible because prominent software applications contain many defects, such as in the applications made by the US-based software giants Microsoft, Oracle, Cisco, Apple, and Adobe, which produce a significant proportion of the global digital software infrastructure. All the previously mentioned malwares developed to attack critical infrastructure used vulnerabilities in Microsoft systems. Windows dominates the world market in desktop operating systems, with more than 91 % of desktop and laptop computers using it according to statistics from Net Applications in January 2015.[217] Yet Microsoft produces software in which hundreds of bugs are revealed every year. The same is true for other prevalent products such as Oracle Java, Adobe Reader and Flash player, with billions of installations worldwide[218] and hundreds of new vulnerabilities discovered every year. For instance, the attack on RSA, which exposed the widely use SecureID two-factor authentication token, used an Adobe Flash zero-day exploit activated through Microsoft Excel.[219]

The history of Microsoft is connected with the development of Intel. In 1975 Gordon E. Moore, co-founder of the Intel Corporation, made a prediction that came to be known as "Moore's Law". Based on the observation of the development of computing hardware so far, he estimated that the number of transistors

[216] Secunia Yearly Report 2011, The evolution of software security from a global enterprise and end-point perspective. Feb. 14, 2012 (secunia.com/resources); The Known Unknowns—Analysis of zero-days (http://techzoom.net/Publications/Papers/knownunknowns).

[217] Desktop Operating System Market Share, January 2015, http://www.netmarketshare.com/.

[218] The market share of Windows Desktop Programs (2011 Data) are, according to the Secunia Yearly Report 2012: Adobe Flash Player 98 %, Oracle/Sun Java 86 %, Adobe Reader 82 %, Apple Quicktime 52 %.

[219] Elinor Mills, Attack on RSA used zero-day Flash exploit in Excel, CNet, April 5, 2011, http://www.cnet.com/news/attack-on-rsa-used-zero-day-flash-exploit-in-excel/.

that could be fitted into a dense integrated circuit would double every two years.[220] The subsequent development of the hardware industry over the next several decades proved his prediction true. Intel and Microsoft, as monopolists on the market during the closing decades of the 20th century, established a long-term hardware/software technological alliance for the mutual development of new products, especially on the desktop and laptop computer market. This "Wintel" alliance enabled the writing of ever more sophisticated software to match hardware whose capabilities were rising exponentially. Obviously, both corporations are (were?) interested in non-stop progress for financial reasons, despite the fact that the ever-rising capabilities of hardware probably exceed the everyday needs of most users. Along with the rising performance of Intel processors, the complexity of Microsoft's software rose exponentially too. According to Nathan Myhrvold, Director of Microsoft's Advanced Technology Group: "*We have increased the size and complexity of software even faster than Moore's Law. In fact, this is why there is a market for faster processors – software people have always consumed new capability as fast or faster than the chip people could make it available*". He provided statistics: in 1975, Basic had 4000 lines of code, but 20 years later it had around 500,000. In 1982, Microsoft Word contained 27,000 lines of code, and about 2 million lines 20 years later.[221] In 1993, Windows NT consisted of 4.5 million lines of code, but a decade later Windows Server 2003 had 50 million lines.[222]

It should be stressed that the software industry tends toward dominant firms, largely for good reasons, such as the benefits of interoperability, user base, or dominant platform, protocols, and formats.[223] Thus, just a few vendors provide the majority of widely used software products, which in turn are the most interesting targets for criminals. In fact, the top 10 software vendors Microsoft, IBM, Oracle, Cisco, Apple, Adobe, Google, Mozilla, Linux, and RedHat account for 32 % percent of all vulnerabilities published in the last 12 months as of March 2015. These vendors jointly represent more than 80 percent of the market share of operating systems, web browsers, mail clients, and office applications. Compared to the past five years. Only four of these ten vendors managed to reduce the number of vulnerabilities in their products.

The information technology sector suffers from the chronic "featuritis" syndrome, which contributes significantly to the growing complexity of software. You should ask yourself: how many features should your product have? The truth is that the marginal benefit of a new feature is often concentrated in some target

---

[220] Andrey Mezhiba, Eby G. Friedman, Power Distribution Networks in High Speed Integrated Circuits. Berlin/Heidelberg/New York, Springer, 2003, p. 2.

[221] Bob Schaller, The Origin, Nature, and Implications of "MOORE'S LAW", The Benchmark of Progress in Semiconductor Electronics, Sep. 26, 1996, http://research.microsoft.com/en-us/um/people/gray/moore_law.html.

[222] Larry O'Brien, How Many Lines of Code in Windows? Dec. 6, 2005, http://www.knowing.net/index.php/2005/12/06/how-many-lines-of-code-in-windows/.

[223] We are grateful to Dr. Stefan Frei for many constructive comments and supportive evidence on these aspects (http://www.techzoom.net/BugBounty/SecureSoftware).

market or user group, while the marginal cost is spread over all users. This results in more features and larger complexity, less usability and larger insecurity for all.

Inevitably, the continuous rise in the complexity of software leads to an ever-greater number of vulnerabilities,[224] and Microsoft also typically leads the world regarding the number of high severity vulnerabilities in its software. It is known that the Common Vulnerability Scoring System (CVSS) of vulnerabilities[225] is systematically biased towards lower risk by various vendors (e.g. Cisco, Oracle). Microsoft typically leads by high-risk vulnerabilities numbers, but paired with effective exploit mitigation methods introduced much earlier than Apple (or still not present with other vendors), this may distort the picture based solely on vulnerability numbers. Further, there is a rather high fluctuation of vulnerability numbers by vendor and year, so that a single year may be overemphasizing or underestimating the vulnerability shares of the different vendors.[226] From 2005 to 2014, the numbers of Common Vulnerabilities and Exposures (CVE) for Microsoft and Apple fluctuated between 150 and 400 per year, while Google has caught up with less then 10 CVEs per year until 2008 to almost 300 in 2011. Cisco, IBM and Oracle exhibit the largest number of CVEs per year, in the range

[224]National Vulnerability Database (NVD), http://nvd.nist.gov.

[225]Common Vulnerability Scoring System, V3 Development Update, First.Org, June 10th, 2015, https://www.first.org/cvss.

[226]In 2013, around 500 high severity bugs came to light in Microsoft's operating systems and applications, of which 371 were in Microsoft's Windows series; by comparison, Apple's operating systems (installed on only 7 % of desktop computers around the globe) had just 24 high severity vulnerabilities, less than a fifteenth of the figure for Windows, while Linux open code operating systems (with 2 % global penetration) had 34. Adobe had around 230 high severity flaws, mainly in applications like Acrobat Reader and Flash Player. Oracle had more than a hundred bugs, mainly in its Java application [Cristian Florian, Report: Most vulnerable operating systems and applications in 2013, GFI Software, February 3, 2014, http://www.gfi.com/blog/report-most-vulnerable-operating-systems-and-applications-in-2013/].

of 350–450 from 2012 to 2014.[227] The trends are not showing any improvement for most of the vendors, if anything an increase of the number of CVEs.

In reality, nobody knows the real number of zero-day vulnerabilities in the software of the top vendors: some of them are still unknown to everybody while some are known, but stay in the arsenals of state cyber forces as secret cyber weapons or with private hackers as sources of potential enrichment. Nevertheless, it is important to emphasize that Microsoft's almost monopolistic market share, combined with the large number of high severity vulnerabilities in its software, have led to the company's products being the main channels for recent sophisticated cyber attacks—a position that will clearly continue for the foreseeable future. To be fair, we should also mention that Microsoft did a lot to improve security in the last decade, while one cannot say the same for Oracle, Apple, Cisco and others, whose relative lack of investment in better processes for working with the community has favored the explosion of vulnerabilities. The prevalence of Microsoft products, paired with the reluctance of users to frequently apply updates allows for many successful attacks using known (and therefore tested and robust) exploits.[228] Prevalent software such as web browsers (Internet Explorer, Google Chrome, Firefox, Apple Safari), Adobe Flash, and Java are the most prevalent attack vector (direct vs. indirect attack).

Other industries also suffer form the ever-increasing complexity of software. For example, the explosion in the number of lines of code (LOC)[229] in fighter aircrafts is documented by A.T. Kearney (2012),[230] with the revealing quote: "The B-52 lived and died on the quality of its sheet metal. Today our aircraft will live or die on the quality of our software." Private conversations with Swiss fighter pilots[231] reveal that the Swiss F/A 18 fighters get updates form Boeing several times a year. We can assume that, in case of a (extremely unlikely) conflict with the U.S., the Swiss weapon systems would not work as intended against U.S. targets. The growing awareness of this situation motivates other nation states around the world to diversify their weapon systems.

---

[227]Data provided by Dr. Stefan Frei, based on the NVD vulnerability database for the last 10 years for all vulnerabilities, and vulnerabilities with CVSS > 7.

[228]Dr. Stefan Frei's personal communication on the data of unpatched programs (http://www.techzoom.net/BugBounty/SecureSoftware) and Thomas Maillart, Didier Sornette, Stefan Frei, Thomas Duebendorfer and Alexander Saichev, Quantification of deviations from rationality from heavy-tails in human dynamics, Phys. Rev. E 83, 056101 (2011).

[229]While a useful indicative metric, the LOC metric remains rather fuzzy, as for instance discussed at http://security.stackexchange.com/questions/21137/average-number-of-exploitable-bugs-per-thousand-lines-of-code.

[230]Software: the brains behind U.S. defense systems: as the U.S. military shifts its focus from metal and mechanics to unmanned vehicles, drones, and smart bombs, software is becoming a crucial piece of weaponry. by A.T. Kearney (2012) http://www.atkearney.com/documents/10192/247932/Software-The_Brains_Behind_US_Defense_Systems.pdf/69129873-eecc-4ddc-b798-c198a8ff1026.

[231]Stefan Frei, private communication, Mar. 16, 2015.

4.4.4 The Special Business Model of Software

The fundamental causes for this state of affairs are straightforward and result from the special nature of software that can be both delivered and altered remotely as explained below. First, the drive to maintain the relentless miniaturization described by Moore's Law, and the pressure of competition in general, force corporations to launch new solutions in a constant rush, which also implies negligence and concealment about minor flaws. In 1994, Intel met with public outrage for concealing the "floating point" bug in its Pentium processor, which caused errors during mathematical calculations; this was only revealed to the public after 5 million processors had already been installed, in spite of the fact that Intel had known about the bug many months before the story appeared in the professional IT press.[232,233] The recall was disastrous for Intel's reputation, because the company required customers to prove that they needed better accuracy than the erroneous chip provided in order to get a free replacement.[234]

Thus, companies like Intel, GM, Toyota and Apple have to achieve close to faultless design and production quality at their plants. Once a product is launched, any recall to eliminate mistakes in the design of goods or defects in manufacturing will bring millions in losses: retail hardware producers have to announce recalls to customers and the media, and later serve millions of angry clients who have been using defective or even risky products for some time. We have already described the pain of such recalls in the case of the Toyota pedal crisis and the iPhone4 antenna problem. By contrast, software vendors have been able to operate an "invisible recall" option. Because, unlike retail goods, software products can be both delivered and altered remotely, software vendors make constant updates to their products. This gives them the opportunity to launch products that are initially less reliable than they ought to be in order to get them onto the market faster than their competitors, skimming the cream off as pioneers in the field and preserving their market share. After launching new software, vendors correct or compensate for its defects by providing hundreds of patches through the update function, remotely and barely noticed by their customers—effectively performing the function of a product recall but with no damage to their reputation. Of course, this must be nuanced by noting that in many cases, critical updates are not so silent and require rebooting, which is something that consumers dislike if it disrupts too much their workflow: this has an impressive measurable consequence in the form of procrastination of users in their updating habits.[235]

[232]Nick Wingfield, Chip Firms No Longer Ignore Even the Least Offensive Bugs. The Wall Street Journal, Nov. 15, 1997.

[233]Pentium FDIV bug, Wikipedia, http://en.wikipedia.org/wiki/Pentium_FDIV_bug.

[234]Capers Jones, Quantifying Software Failures and Disasters, Part 2: 1990–1999 Version 2.0 Sep. 12, 2012, http://www.itmpi.org/Portals/10/PDF/Failures-Capers-2.pdf.

[235]Thomas Maillart, Didier Sornette, Stefan Frei, Thomas Duebendorfer and Alexander Saichev, Quantification of deviations from rationality from heavy-tails in human dynamics, Phys. Rev. E 83, 056101 (2011).

This leads to a very particular situation in which software producers are in the almost unique position that they do not face liability for their products. As Dan Geer, who is currently the chief information security officer for In-Q-Tel, a not-for-profit venture capital firm that invests in technology to support the Central Intelligence Agency,[236] stated in his keynote at the BlackHat USA conference in 2014: "*There are only two industries without product liability: religion and the software industry*". This lack of liability is arguably as important as the ever-increasing complexity or perhaps even a more serious driver of the cyber-security problems we face today. And there is an intense lobbying by software vendors against software liability insurance.[237]

Microsoft, among others, uses the update function to determine the lifecycle of its products. For instance, when the company canceled its ongoing support of the Windows XP update in 2014, governmental and business organizations stopped using Windows XP almost overnight because it was impossible to download fresh patches for the protection of what was fundamentally a poorly developed operating system. In fact, it is impossible to write software with no possibility of bugs being discovered; but it is in the developers' interest that the majority of customers do not realize that the continuous update function is effectively an invisible recall of software, and that a high frequency of updates shows that the software was poorly developed. Software companies can continue to work in an environment of constant rush, forever launching new products and phasing out the old ones—a business model that suits them very well—without drawing too much public attention. There is also evidence that vendors tend to suppress the release of individual patches in the six month before a major new software release (unless the vulnerabilities become public).[238]

The danger of cyber attacks is further enhanced by the delays with which vulnerabilities are patched. One recent study found that about 30 % of released vulnerabilities are still not patched after one month. The precise figures are: Microsoft, 31 %; Apple, 26 %; and Linux, 35 %.[239] Using anonymized daily log files of Google web servers (constituting more than 70 % of the worldwide daily searches collected over more than three years), one of us and collaborators performed a study of the persistence of the use of outdated (and thus vulnerable) Web browsers (Firefox, Opera, Chrome and Safari) after the release of a new browser

[236]Additional information about Daniel Earl Geer, Jr., http://en.wikipedia.org/wiki/Dan_Geer.

[237]Liabilities and Software Vulnerabilities, personal website of Bruce Schneier, https://www.schneier.com/blog/archives/2005/10/liabilities_and.html.

[238]Analysis of the patch release dynamics of Microsoft and Apple between 2002 and 2008 (http://techzoom.net/Publications/Papers/0daypatch).

[239]Muhammad Shahzad, Muhammad Zubair Shafiq, Alex X Liu, A large scale exploratory analysis of software vulnerability life cycles, Proceedings of the 34th International Conference on Software Engineering, IEEE Press, June 2, 2012, pp. 771–781.

version.[240] We found that most users update or patch quickly upon receiving the alert message notifying the user of the pending update but a surprisingly large fraction of users takes a very long time (characterized technically by a "heavy-tailed power law" decay of the number of unpatched computers). And hundreds of thousands of computers never update their old browser versions. In their paper entitled "An Empirical Study of Zero-day Attack in The Real World", Leyla Bilge and Tudor Dumitra of Symantec Research Labs recently identified 18 vulnerabilities exploited in the real world before full disclosure. They detailed examples of vulnerabilities being exploited in the wild for an average of 312 days before petering out. Their study measured the duration and prevalence of these attacks in the real world before the disclosure of the corresponding vulnerabilities.[241] Analyzing ten years of data from two major vulnerability purchase programs, NSS Labs reported in a 2013 publication that "*the market for vulnerability and exploit information has grown significantly in recent years... privileged groups have had access to at least 58 vulnerabilities targeting Microsoft, Apple, Oracle, or Adobe... these vulnerabilities remained private for an average of 151 days... such access would have allowed these groups to compromise all vulnerable systems without public knowledge... Specialized companies are offering zero-day vulnerabilities for subscription fees that are well within the budget of a determined attacker (for instance, 25 zero-days per year for USD $2.5 million); this has broken the monopoly that nation-states historically have held regarding ownership of the latest cyber weapon technology. Jointly, half a dozen boutique exploit providers have the capacity to offer more than 100 exploits per year*".[242] The exploitations of such zero-day vulnerabilities are however often delayed by the fact that their use today may well prevent them from being available for use later. There is thus a strategic decision to exploit the capacity immediately or wait for a more propitious time.[243]

Regarding the economics of security, the information technology sector is driven by demand for features, for new options, by time to market. It has no standards for security or quality and no liability for insecure software. In the language of economics, insecurity can be seen as a "negative externality" in the sense that the group that profits from a product is not the group that bears the negative impact of it. It is a side effect of the business, like environmental pollution. It is also important to realize that the software industry has zero marginal cost while

---

[240] Thomas Maillart, Didier Sornette, Stefan Frei, Thomas Duebendorfer and Alexander Saichev, Quantification of deviations from rationality from heavy-tails in human dynamics, Phys. Rev. E 83, 056101 (2011).

[241] Dinesh Theerthagiri, Zero-Day World, Symantec Official Blog, Oct. 30, 2012, http://www.symantec.com/connect/blogs/zero-day-world; Bilge, Leyla, and Tudor Dumitras. "Before we knew it: an empirical study of zero-day attacks in the real world." Proceedings of the 2012 ACM conference on Computer and communications security. ACM, 2012.

[242] Stefan Frei, The Known Unknowns in Cyber Security (An empirical analysis of publicly unknown security vulnerabilities), NSS Labs Analyst Brief (2013), http://techzoom.net/Publications/Papers/knownunknowns.

[243] Robert Axelrod and Rumen Iliev, Timing of cyber conflict, Proceedings of the National Academy of Science USA 111 (4), 1298–1303 (2014).

being completely dominated by the "network effect", so that time to market is critical and "the winner takes it all". This is a strong lock-in effect as switching between platform (Windows Office, iTunes) is expensive. Moreover, most of users cannot tell good security from bad security. All this contributes to a situation where users are disempowered with respect to the control of their security in using their software products.

4.4.5 Collaboration Between Software Vendors and National Security Agencies

This business model gives state cyber forces the opportunity to use installed Microsoft software with the update function enabled for one-off attacks on millions of computers: if they can locate a zero-day vulnerability through which to attack, they can disguise their attack as a legitimate Windows update (this was how the Flame virus operated). The same is true for prevalent Cisco networking devices, Adobe Flash Player, Huawai (China) networking devices, Apple Mobiles (zero-days for iOS are the highest priced) and there are also "backdoors" built directly into the hardware design of chips. In other words, governmental and corporate users are held captive by the current business model of the top software developers: they cannot reject the update function because the installed software requires the constant downloading of repair and maintenance updates to save computers from ordinary hackers, but using the update function is also risky because state hackers could use hacked security certificates during regular updates for instance for national security purposes.

Edward Snowden, the whistleblower we have mentioned already, revealed that Microsoft collaborated closely with the NSA's Prism program and gave US intelligence access to encrypted messages within web and communication solutions like Outlook.com, Skype, and Skydrive.[244] If such collaboration took place regarding Microsoft's flagship internet solutions, why should the company not have provided U.S. intelligence with back door entry points in its sophisticated and closed-code software and constant update function—installed in millions of computers worldwide—making possible a full-fledged cyber attack by the US Army on any adversary in parallel with more conventional weapons? In 2013, Bloomberg issued a report where the following was mentioned: "*Microsoft ... provides intelligence agencies with information about bugs in its popular software before it publicly releases a fix, according to two people familiar with the process. That information can be used to protect government computers and to access the computers of terrorists or military foes. [It] allowed the U.S. to exploit vulnerabilities in software sold to foreign governments, according to two U.S. officials. Frank Shaw, a*

[244]Glenn Greenwald, Ewen MacAskill, Laura Poitras, Spencer Ackerman and Dominic Rushe, Microsoft handed the NSA access to encrypted messages, The Guardian, July 12, 2013.

spokesman for Microsoft, said those releases occur in cooperation with multiple agencies and are designed to give government "an early start" on risk assessment and mitigation.[245] *In an e-mailed statement, Shaw said there are "several programs" through which such information is passed to the government, and named two which are public, run by Microsoft and for defensive purposes… Intel Corp.'s McAfee unit, which makes Internet security software, regularly cooperates with the NSA, FBI and the CIA, for example, and is a valuable partner because of its broad view of malicious Internet traffic, including espionage operations by foreign powers*".[246] After the Bloomberg report, Jesse Emspak, a contributor to the online newsletter TechNewsDaily, initiated an important discussion: "*So what if an American company discovered a security flaw on its own? Would it have to ask the government whether it could disclose it publicly? Probably not. The NSA, in particular, wouldn't tell a company like Microsoft or Verizon to hide a security flaw. If it turned out that a software firm deliberately concealed a known flaw, the act of omission might reveal more than the NSA would want other countries to know. On top of that, ignoring flaws, disclosed or not, would mean that a company was not fixing problems. If that policy were exposed, it would give its customers reason to pursue an alternate vendor. There have been questions recently … about long time lags between when major Microsoft software vulnerabilities have been discovered and when they've finally been fixed [once it took 600 days for Microsoft to fix a critical vulnerability in its Microsoft Explorer browser*[247]*]. Was Microsoft keeping the vulnerabilities open for the NSA? … There's little hard evidence for that*".[248] But perhaps there is! In March 2015, it was revealed that Microsoft's latest batch of fixes addressed another longstanding threat to Windows: Stuxnet. It turned out the patch provided by Microsoft back in 2010 to stop the Stuxnet worm from spreading did not quite do the job, according to a new report by HP's TippingPoint security wing. The upshot is that countless Windows machines have been left vulnerable to Stuxnet and other similar attacks for the five years since 2010.[249] Another interesting statistics is that, for 2013 alone, the NSA secretly spent more than US $25 million to procure "software vulnerabilities' from private malware vendors," according to a wide-ranging report on the NSA's offensive work by the Washington Post's Barton Gellman and Ellen Nakashima.[250]

---

[245]Microsoft Active Protections Program (MAPP) is the public face of it: https://technet.microsoft.com/en-us/security/dn467918.

[246]Michael A Riley, U.S. Agencies Said to Swap Data With Thousands of Firms, Bloomberg, June 15, 2013.

[247]Dan Goodin, Explorer info leak festers for 2 years Microsoft's 600-day bug bite, The Register, November 1, 2010, http://www.theregister.co.uk/2010/11/01/internet_explorer_600_day_bug/.

[248]Jesse Emspak, Does Microsoft Help the NSA Hack Your Computer? TechNewsDaily, July 3, 2013, http://mashable.com/2013/07/02/microsoft-nsa-hack/.

[249]Stuxnet Redux: Microsoft patches Windows vulnerability left open for five years http://www.theregister.co.uk/2015/03/10/stuxnet_vulns_finally_patched/.

[250]Brian Fung, The NSA hacks other countries by buying millions of dollars' worth of computer vulnerabilities, Washington Post Aug. 31, 2013.

The possibility that back door entries had been left open to allow the NSA access to computer hard drives was revealed in February 2015, when Kaspersky Lab published an investigation report about the Equation group malware, which had a similar code as Stuxnet and was deeply hidden within hard drives produced by a total of 12 corporations including Western Digital, Seagate, Toshiba and other top manufacturers. Equation infected personal computers in business and administration—including government and military institutions, telecommunication companies, banks, energy companies and nuclear researchers—in 30 countries including Russia, Iran, Pakistan, Afghanistan, China, Mali, Syria, Yemen and Algeria. According to Kaspersky Lab, the malware could give its developer full control over any infected computer to access data, control operations or re-format the infected hard disk. A former NSA employee commented to Reuters that "*Kaspersky's analysis was correct, and that people still in the intelligence agency valued these spying programs as highly as Stuxnet. Another former intelligence operative confirmed that the NSA had developed the prized technique of concealing spyware in hard drives*".[251] In fact, there is an official policy and political position in favor of backdoors in Washington DC, which fuels a long-lasting and still ongoing battle with security experts.[252,253,254,255]

4.4.6 Worst-Case Scenarios and Needed International Collaboration

It seems clear that the problem of the production of software with hundreds of vulnerabilities, in parallel with an intensifying cyber arms race based on the use of these vulnerabilities for state-to-state cyber attacks, is a key threat to our

[251]Joseph Menn, UPDATE 3-Russian researchers expose breakthrough U.S. spying program, Reuters, Feb. 16, 2015 □ and Equation: The Death Star of Malware Galaxy, Kaspersky Labs' Global Research & Analysis Team, Feb. 16, 2015, http://securelist.com/blog/research/68750/equation-the-death-star-of-malware-galaxy and https://securelist.com/files/2015/02/Equation_group_questions_and_answers.pdf.

[252]S. Landau, Making Sense of Snowden: What's Significant in the NSA Surveillance Revelations, IEEE Security and Privacy, Vol. 11, No. 4, July/August 2013, pp. 54–63.

[253]S. Landau, Making Sense of Snowden Part II: What's Significant in the NSA Surveillance Revelations, IEEE Security and Privacy, Vol. 12, No. 1, January/February 2014; a short version, Highlights from Making Sense of Snowden Part II: What's Significant in the NSA Surveillance Revelations, (behind payment wall) appeared in IEEE Security and Privacy, Vol. 12, No. 1, January/February 2014, pp. 62–64.

[254]S. Landau, Under the Radar: NSA's Efforts to Secure Private-Sector Telecommunications Infrastructure, Journal of National Security Law and Policy, Vol. 7, No.3 (2014).

[255]H. Abelson, R. Anderson, S. Bellovin, J. Benaloh, M. Blaze, W. Diffie, J. Gilmore, M. Green, S. Landau, P. Neumann, R. Rivest, J. Schiller, B. Schneier, M. Specter, D. Weitzner, Keys Under Doormats: Mandating insecurity by requiring government access to all data and communications, MIT-CSAIL-TR-2015-026 (tech report), July 6, 2015.

information-dependent world society—constructed as it is on US-based infrastructure with back door entry points for US intelligence, and critical vulnerabilities known to cyber forces from different countries. One of the worst case scenarios of the use of a cyber weapon would be if state hackers took control, through software vulnerabilities, of several nuclear plants and allowed the cores to melt down, potentially leading to several accidents on the scale of Fukushima or Chernobyl. Just a few such cases would already be catastrophic for an entire continent and could determine the outcome of an interstate war in the XXI century…

Unsurprisingly, the United States—which introduced the World with the global cyber arms race with the Stuxnet attack on critical infrastructure in Iran in 2010—admitted several years later that cyber attacks now pose "*one of the gravest national-security dangers*" to America. The Director of National Security, James R. Clapper, listed cyber security first among the threats facing America today[256]: "*We judge that there is a remote chance of a major cyber attack against US critical infrastructure systems during the next two years that would result in long-term, wide-scale disruption of services, such as a regional power outage. The level of technical expertise and operational sophistication required for such an attack—including the ability to create physical damage or overcome mitigation factors like manual overrides—will be out of reach for most actors during this time frame. Advanced cyber actors— such as Russia and China —are unlikely to launch such a devastating attack against the United States outside of a military conflict or crisis that they believe threatens their vital interests. However, isolated state or non-state actors might deploy less sophisticated cyber attacks as a form of retaliation or provocation*".[257] According to a statement by Barack Obama in February 2014, governments, businesses and individuals are increasingly data-dependent; digital infrastructure is becoming ever more complex, making it harder to police; and highly sophisticated hacking groups have suddenly risen in number.[258] In addition, Edward Snowden's documents showed that the NSA is worried that Iran has learned from attacks like Stuxnet, Flame and others to improve its own capabilities for retaliatory cyber attacks against the United States.[259] Doubtless with this in mind, in February 2015, the US President proposed the development of an agreement between countries with advanced cyber weapon capabilities, similar to the restrictive conventions during the American-Soviet nuclear arms race: "*We have*

[256]James R. Clapper, Statement for the Record: Worldwide Threat Assessment of the US Intelligence Community, US Senate Select Committee on Intelligence, Washington, DC, Mar. 12, 2013.

[257]Ibid, p. 1.

[258]Statement by the President on the Cybersecurity Framework, The White House Office of the Press Secretary, Feb. 12, 2014, http://www.whitehouse.gov/the-press-office/2014/02/12/statement-president-cybersecurity-framework.

[259]Kim Zetter, The NSA Acknowledges What We All Feared: Iran Learns From US Cyberattacks, Wired, Feb. 10, 2015, http://www.wired.com/2015/02/nsa-acknowledges-feared-iran-learns-us-cyberattacks/.

great capabilities here. But there are other countries that have great capabilities, as well. Eventually, what we're going to need to do is to find some international protocols that, in the same way we did with nuclear arms, set some clear limits and guidelines, understanding that everybody's vulnerable and everybody's better off if we abide by certain behaviors".[260] Note that, at the time of writing, there is no emergency line between the US and Chinese Presidents, like the Washington-Moscow Direct Communications Link (famously but incorrectly known at the "red phone") that existed between US and Soviet leaders during the cold war. In principle, both sides agree that it is a very important feature, but it seems to be a difficult diplomacy tool to set up, which boils down to the following simple but tricky question: what the called side should answer when the other one calls in the context of a cyber issue, provided that the answer should come quickly?

But it is not clear how this could be implemented if our societies choose to continue functioning with an open Internet. Indeed, state of the art weapon systems such as aircraft carriers, satellites, and fighter airplanes have traditionally been available to nation states only. State of the art cyber weapons, on the contrary, are available to almost any interested group with decent funding. Thus, the historic monopoly of nation states to exclusively access state of the art weapons is now broken.[261] Richard Alan Clarke, the former National Coordinator for Security, Infrastructure Protection, and Counter-terrorism for the United States, and Robert Knake discussed the challenges of international protocols on the limitations of cyber arms in his book "Cyber War".[262] Unlike the protocols limiting traditional weapons (such as tanks, artillery, battleships) or nuclear weapons, cyber has a set of own challenges:

(i) An adversary's cyber arms arsenal cannot be reliably/independently counted;
(ii) There is a blurry or no line between government cyber arms, and the cyber arms of cyber criminals;
(iii) Plausible deniability: while traditional weapons leave a smoking gun when used, no attribution of origin of attack exists if the cyber attack is executed with care.

At the time when we were completing this book and reviewing its final content for accuracy, a crucial decision is being vigorously discussed in the US, putting senior law enforcement officials, including FBI Director James B. Comey, at odds with a coalition of tech firms including Apple, Facebook, Google and Microsoft, security experts and a number of civil society organizations. According to an open letter to President Obama dated May 19, 2015 and signed by these above parties,[263]

[260]Kara Swisher, White House. Red Chair. Obama Meets Swisher, Feb. 2015, ReCode, http://recode.net/2015/02/15/white-house-red-chair-obama-meets-swisher.

[261]We are grateful to Dr. Stefan Frei for useful discussions on this point.

[262]Richard A. Clarke and Robert Knake, Cyber War: The Next Threat to National Security and What to Do About It, Ecco; Reprint (Apr. 10, 2012).

[263]https://static.newamerica.org/attachments/3138--113/Encryption_Letter_to_Obama_final_051915.pdf.

"*Those* [Administration] *officials have suggested that American companies should refrain from providing any products that are secured by encryption, unless those companies also weaken their security in order to maintain the capability to decrypt their customers' data at the government's request.*" Some senior government officials have indeed advocated weakening the use of encrypted technologies to give law enforcement agencies the ability "*to do their jobs*". It is claimed that there is a strong threat to public safety from the loss of access to data and communications associated with strong encryption technologies. For instance, the move by Apple and Google in 2014 to offer forms of smartphone encryption so secure that even law enforcement agencies could not gain access even with a warrant is viewed by enforcement officials as "*allowing people to place themselves beyond the law*".[264]

Opposed to this view are more than 140 tech companies, prominent technologists and civil society groups, which summarize the problem in the letter to President Obama in the following terms[265]: "*Strong encryption is the cornerstone of the modern information economy's security. Encryption protects billions of people every day against countless threats—be they street criminals trying to steal our phones and laptops, computer criminals trying to defraud us, corporate spies trying to obtain our companies' most valuable trade secrets, repressive governments trying to stifle dissent, or foreign intelligence agencies trying to compromise our and our allies' most sensitive national security secrets… Encryption thereby protects us from innumerable criminal and national security threats. This protection would be undermined by the mandatory insertion of any new vulnerabilities into encrypted devices and services. Whether you call them "front doors" or "back doors", introducing intentional vulnerabilities into secure products for the government's use will make those products less secure against other attackers. Every computer security expert that has spoken publicly on this issue agrees on this point, including the government's own experts… If American companies maintain the ability to unlock their customers' data and devices on request, governments other than the United States will demand the same access, and will also be emboldened to demand the same capability from their native companies. The U.S. government, having made the same demands, will have little room to object. The result will be an information environment riddled with vulnerabilities that could be exploited by even the most repressive or dangerous regimes. That's not a future that the American people or the people of the world deserve.*"

The core of the dispute lies in the intrinsic technology incompatibility between, on the one hand, the desire by officials to get lawful access to data and, on the

---

[264]Nakashima, E., Tech giants don't want Obama to give police access to encrypted phone data, The Washington Post, May 19. 2015, http://www.washingtonpost.com/world/national-security/tech-giants-urge-obama-to-resist-backdoors-into-encrypted-communications/2015/05/18/11781b4a-fd69-11e4-833c-a2de05b6b2a4_story.html.

[265]https://static.newamerica.org/att8achments/3138--113/Encryption_Letter_to_Obama_final_051915.pdf.

other hand, the unavoidable cyber vulnerabilities that this would entail and which could be exploited by hackers and unfriendly governments. And history has shown that, without doubts, what can be done, will occur one way or another. The well-intentioned proximate desire to ensure public safety comes unfortunately with the fundamental flaw of weakening the whole security infrastructure of the digital age, with unintended (possibly long-term) future consequences of possible frightful extreme impact. We are at crossroads of what will be the twenty-first century digital age: as recalled in the letter to President Obama,[266] during the so-called "Crypto Wars", U.S. policymakers at the end of the last century, and the President's Review Group on Intelligence and Communications Technologies in their 2013 report, unanimously recommended to promote encryption technology. We expect this fight not to be settled permanently but to become a recurrent theme as a result of the intrinsic conflict between (i) the need to combat terrorism via increased top-down intelligence gathering, (ii) privacy rights concerns, and (iii) cyber security.

[266]Ibid.

Chapter 5
Successful Risk Information Management

5.1 Toyota's Production System

> *The key to the Toyota Way and what makes Toyota stand out is not any of the individual elements.... But what is important is having all the elements together as a system. It must be practiced every day in a very consistent manner— not in spurts*
>
> Fujio Cho

The Japanese carmaker Toyota is one of the most successful firms of all times. Until the recall problems discussed in Sect. 2.4.2, its widely acknowledged unique way, the "Toyota Production System" (TPS), has been viewed as the main source of its success. One of the key principles of the TPS that is particular relevant to the discussion of this book was characterized by Professor J. Liker in his book[1] as "*Build[ing] a culture of stopping to fix problems to get quality right the first time*" in order to enhance productivity in the long run. Fujio Cho, previous President of Toyota Motor Company and Chairman of the Board and Representative Director of Toyota Motor Corporation from June 2006 to June 2013, commented as follows on the differences in culture between Toyota plants in Japan and those in the US, as reported by J. Liker: "*He did not hesitate to note that his number one problem was getting group leaders and team members to stop the assembly line. They assumed that if they stopped the line, they would be blamed for doing a bad job. Cho explained that it took several months to 're-educate' them that it was a necessity to stop the line if they want to continually improve the process. He had to go down to the shop floor every day, meet with his managers, and, when he noticed a reason to stop the line, encourage the team leaders to stop it*".[2] This is in striking contrast with the instructions of the management of competing carmakers, which,

[1]Jeffrey K. Liker, The Toyota Way (14 management principles from the world's greatest manufacturer). New York: McGraw Hill (2004).

[2]Ibid, p. 129.

D. Chernov and D. Sornette, *Man-made Catastrophes and Risk Information Concealment*, DOI 10.1007/978-3-319-24301-6_5

before adopting even in part the TPS, rewarded uninterrupted assembly lines so as to maximize production and minimize (short-term) costs. As Toyota managers put it, it is impossible to have zero defects. The best response is to acknowledge this and develop a countermeasure process as embodied by the TPS. In contrast, an absence of assembly line interruption, previously saluted as the ultimate goal, is a strong signal that defects are either missed, inadequately addressed or outright concealed.

In the light of the many cases of concealment of information about defects that led to disasters reported in this book, this TPS principle of creating systematic operations that bring problems to full light strikes a sensitive chord. But in order to be successful in its implementation, other ingredients need to be considered, as shown by the failures of other U.S. carmakers to implement the TPS until they recognized the following. In order for a problem to be first recognized and then appropriately addressed, the employees at the level of the assembly chain must be empowered and trusted. The temptation to use sophisticated technology to detect and diagnose defects is grossly insufficient if people are not prepared and incentivized to use it. Moreover, technology often introduces complexity, which prevents from a practical and efficient implementation. The Toyota way is about simplicity and can be summarized through the two pillars that support it: "Continuous Improvement" and "Respect for People." This is implemented within the 4P of the Toyota way:

(a) Philosophy: long-term thinking (possibly at the cost of short-term costs);
(b) Process: eliminate waste by process improvement;
(c) People and partners: putting employees at the heart of the process and developing a culture of partnership with providers; and
(d) Problem solving: empowering and motivating employees to continuously improve and learn.

And "*the true value of continuous improvement is in creating an atmosphere of continuous learning and an environment that not only accepts, but actually embraces change… Defining and explaining what the goal is, sharing a path to achieving it, motivating people to take the journey with you, and assisting them by removing obstacles—those are management's reasons for being*".[3] As a corollary, "*a common Toyota quality tactic is to front-load projects of all kinds, to anticipate problems as early as possible and put in place countermeasures before the problems even occur*".[4]

In search of ever greater performance and safety, when Toyota let complexity overwhelm its principles, the ax of retribution fell inexorably with the series of recall problems in 2009–2011, as discusses in Sect. 2.4.2. There is an important lesson here. A company or an institution, or even an individual, may use a sound, reliable and successful process but, as the Toyota example illustrates, the

[3]Ibid, pp. xi–xii.

[4]Ibid, p. 139.

challenge then lies in keeping it going without betraying it. This requires perhaps super-human dedication and continuous questioning of what is supposed to be well known and established. These are undertakings contradicting the drive of most humans to achieve a laudable goal, and then to enjoy the fruit of success. Ironically, this last phase is often the beginning of the end, so to speak, when vulnerabilities emerge, building up towards the next crisis.

5.2 The Sony Battery Recall in 2006

On August 14, 2006, Dell, the world's largest PC maker, recalled 4.1 million notebook computer batteries made by Sony, because it documented six instances since December 2005 in which notebooks overheated or caught fire.[5] These lithium-ion batteries are widely used in electronic devices such as cellphones, portable power tools, camcorders, digital cameras and MP3 players. Experts have known for years that such batteries could catch fire in extremely rare circumstances but the recent cases involving notebook computers have led to the largest safety recall in the history of the consumer electronics industry, according to the Consumer Product Safety Commission.[6] Dell, helped financially by Sony, did not hesitate in launching the recalls. Softpedia reported on October 13, 2006 that the number of Sony-made batteries being replaced is 8 million since August 2006.[7]

Mr. Hideki Takayasu, senior scientist at Sony Computer Science Laboratories and a professor at Meiji University, Japan, shared the following personal experience[8]: "*Several years ago, I had a chance to be partly involved in Sony's risk management about personal computers' batteries. Before the official announcement, Mr. Chubachi, CEO at that time, called me to his office and he explained to me the detailed facts. He asked me, as a specialist of probability analysis, whether Sony should recall all suspicious batteries although the probability of catching fire is much less than the risk of an airplane accident. I mentioned that the customers of Sony believe that the probability of a computer catching fire must be zero, while airplane customers accept (or at least are aware of) a non-zero probability of a crash, and this difference between zero and nearly zero is very important, so Sony have to recall all suspicious objects immediately, although it may cost a lot. I also advised him not to hide any scientific fact about the cause of the fires, the needle-like crystallization that can break some batteries, which was an unexpected*

[5]Damon Darlin, Dell Recalls Batteries Because of Fire Threat, The New York Times, Published: August 14, 2006 (http://www.nytimes.com/2006/08/14/technology/14cnd-battery.html?_r=2&oref=slogin&).

[6]Ibid.

[7]Sony Exploding Batteries—The Chronicles, http://archive.news.softpedia.com/news/Sony-Exploding-Batteries-Chronicles-37848.shtml.

[8]Personal communication with Professor Hideki Takayasu (December 20, 2014).

physical phenomenon. Mr. Chubachi may have asked similar questions to many people, anyway he chose the cost-full way of disclosing all details about the troubles and accepted the largest-scale recall. I believe this decision was right. If he had tried to conceal the trouble, Sony might have appeared in the company of the other top stories in your book. I think you can add Sony's case as a positive outcome".

Mr. Hideki Takayasu also mentioned the following additional insight[9]: "*As Mr.Chubachi is also a scientist (he got his PhD from Tohoku University majoring in the Physics of magnetic fluids, he entered Sony as a high-class engineer), he knows the importance of open information. I am afraid that Sony might have gone into a wrong way if the president at that time was not him or someone like him with a PhD in science*". This remark echoes an analysis that one of us provided in the aftermath of the 2008 financial crisis,[10] namely that "*a little-discussed reason for the present crisis is the lack of adequate education of top managers on risk in all its dimensions and implications… CEOs with MBAs have training that make them cognitively blind to the technical aspects found in the quantitative reports and recommendations of serious risk managers. How then can these decision makers weight in the importance of risk management in the face of the attractive short-term profits of alluring opportunities? Only CEOs and decision makers with a solid training in quantitative engineering sciences and strong notions about quantitative metrics can have the full picture of the risks entailed in their decisions*". This calls for the development of a genuine culture of risks, which should be obligatory training for managers in governments, in regulatory bodies, and financial institutions.

5.3 The Seveso Directive and Beyond

In July 10, 1976, at the chemical plant ICMESA (owned by Swiss multinational Roche Group) in the Milan suburb of Seveso (Northern Italy), a cloud of 2,3,7,8-tetrachlorodibenzodioxin gas (known as the carcinogenic dioxin or TCDD) was released into the atmosphere. Fortunately, nobody among the residents of Seveso and of other neighboring cities lost their life due to dioxin poisoning. However, the land and vegetation were contaminated and more than 2000 people received treatment against dioxin poisoning.

Several risk management challenges characterize this case. First, the plant has been producing chemicals for more than 30 years but local communities were not aware of the hazardous nature of the site. Second, the threat of dioxin to the human health was a much discussed topic of disagreement between scientists and

---

[9]Personal communication with Professor Hideki Takayasu (December 21, 2014).

[10]D. Sornette, Risk Management and Governance Lessons and Prospects From The 2007–20XX Crisis, REVUE für Postheroisches Management 7, 44–53 (2010).

the global chemical industry, right after the Vietnam war during which dioxin has been used as a component of the defoliant "Agent Orange". Third, the managers of the plant were aware of the leak immediately after it occurred, but had difficulties in assessing the amount of toxic substances that were released.[11] The managers in fact concealed the existence of the release of dioxin from the authorities and the people of nearby communities for ten days after its occurrence, until an external independent laboratory established the existence of dioxin contamination near the plant.[12] This case thus constitutes a milder instance of concealment of risk information, compared with the accidents and crises analyzed above.

A positive aspect of this event, which makes quite singular, is that it became a pivotal point in changing the European regulation regarding the prevention of industrial disasters in the chemical industry. The legislation, called Seveso Directive, includes risks ranking and risks assessment, the exchange of information on near-misses and incidents within the industry, a procedure of obligatory risk information disclosure implemented on more than ten thousand facilities in Europe where dangerous substances are located, and more. The Seveso Directive was first passed in 1982, and has been continuously updated in the following three decades. A key principle of the Seveso Directive is the use of the precautionary approach expressed at the "need to know", which involves the preventive transmission of information concerning existing risks of a hazardous object to all associated internal and external audiences.[13]

In a similar spirit, the Three Miles Island nuclear accident had positive consequences in the management of potential crises, in particular in the country, France, whose primary source of energy is nuclear energy. Indeed, the lack of reporting of incidents and the bad procedures towards other utility owners has led to a worldwide shared database of incidents. The problem of the huge delay before identifying properly the causes of the TMI accident has led to the setup of two independent crisis groups that work independently at Electricité de France (EDF), based on the data reported by the nuclear power plant operators: one is on site, one is distant.[14] Finally, the risk communication problems are probably better managed now, with dedicated people who only manage the real-time communication to the public.

In conclusion, informed by the understanding and awareness resulting from our analyses of tens of disastrous cases that took place in different countries and industries, a better management of risk transmission can be built within

---

[11]B. De Marchi, S. Funtowicz, and J. Ravetz, Seveso: A paradoxical classic disaster, Mitchell (note 5), 1996.

[12]Barbara Pozzo, The implementation of the Seveso Directives in an enlarged Europe: a look into the past and a challenge for the future, 2009, p. xx.

[13]B. De Marchi, S. Funtowicz, and J. Ravetz, Seveso: A paradoxical classic disaster, Mitchell (note 5), 1996.

[14]Bruno Sudret, private communication (19 Feb. 2015).

organizations and among external audiences in many industries and regulator's bodies based on the approach of learning from history. This will help avoid repeating many managerial mistakes regarding risk communications, which happened in the past.

Erratum to: Man-made Catastrophes and Risk Information Concealment

Erratum to:
D. Chernov and D. Sornette, *Man-made Catastrophes and Risk Information Concealment*, DOI 10.1007/978-3-319-24301-6

On page xv the text "About the Authors" was not complete. Please find hereafter the complete text.

About the Authors

Dmitry Chernov is a researcher in the Chair of Entrepreneurial Risks at ETH Zurich. He has more than 15 years of experience as a corporate communication consultant in the former Soviet Union Republics, focusing on improving investor and government relations, as well as crisis communication within the following industries: oil and gas, chemical, electric power, metals and mining, telecommunication, transport, etc. He specializes on researching solutions for efficient risk information transmission that enables timely decision-making before and during industrial disasters.

The updated original online version for this book can be found at
DOI 10.1007/978-3-319-24301-6

D. Chernov (✉)
ETH Zurich, Zurich, Switzerland
e-mail: chernovd@student.ethz.ch

D. Sornette
ETH Zurich, Zurich, Switzerland
e-mail: dsornette@ethz.ch

D. Chernov and D. Sornette, *Man-made Catastrophes and Risk Information Concealment*, DOI 10.1007/978-3-319-24301-6_6

Didier Sornette is Professor on the Chair of Entrepreneurial Risks at the ETH Zurich, Director of the Financial Crisis Observatory, Co-founder of the ETH Risk Center and member of the Swiss Finance Institute. He has received many prizes and honors, among them the 2000 Research McDonnell award and the Risques-Les Echos prize 2002. He is a Fellow of the World Innovation Foundation (2004) and, since 2013, of the American Association for the Advancement of Science. Prof. Sornette has edited and authored a number books, many of them published by Springer and is a member of the Editorial and Advisory Board of the Springer Complexity publishing program.

In Chapter 2:
On page 15 in Sect. 2.1.2.1 several corrections have been incorporated in the text:

The first two sentence in lines 1–6 should read:

At 5:00 p.m. on March 7, 1979, during regular servicing of the feedwater system on Unit 2, the polisher machines—which remove dissolved minerals from the system—were being repaired when a leakage of water whose causes are not fully understood occurred into the air-controlled system that opens and closes the polisher valves. Eleven hours later, at 4:00 a.m. on March 28, 1979, this problem triggered a stoppage of the secondary feedwater pumps, which were responsible for sending re-heated water to the steam generators to remove heat from the reactor core of Unit 2.

The sentence starting in line 10 should read:

Immediately after shutdown, the decay heat power generation was about 160 MW of the thermal power of 2575 MW, corresponding to the 850 MW generating capacity of TMI-2.

In line 13 (4 %) has to be deleted and in line 14 (2 %). The sentences should read:

One hour after the reactor shutdown, decay heat power generation was approximately 33 MW. Ten hours after shutdown, it was about 15 MW.

The sentences in lines 1 and 2 of the second paragraph in this section should read:

Because the secondary feedwater pumps had tripped, heat was not anymore being removed from the reactor. An activation of three auxiliary pumps occurred but the valves have been closed earlier for routine maintenance; the system was unable to pump any water, in violation of a key US Nuclear Regulatory Commission rule. This led to rising pressure within the system, so a relief

Footnote [25] should read:

[25] Ibid, p. 90.

Footnote [26] should read:

[26] Three Mile Island: Report to the Commissioners and to the Public, M. Rogovin and G. Frampton, U.S. Nuclear Regulatory Commission, Jan. 1980, Vol. II-Part 2, p. 311.

On page 16 the sentence in lines 2 and 3 should read:

Instruments in the control room of TMI-2, however, indicated to the plant operators that the valve was closed as the light did not indicate the position of the valve, only the status of the solenoid being powered or not.[28]

On page 16 in line 8 it should read:

Operators of the next shift

On page 16 in line 12 it should read:

level due to fears that the system was overfilled.

On page 16 in line 23 it should read:

within the reactor vessel likely produced